Springer Series on
Atoms+Plasmas 6
Editors: Arthur L. Schawlow and Koichi Shimoda

Springer Series on
Atoms+Plasmas

Editors: G. Ecker   P. Lambropoulos   I. Sobel'man   H. Walther
Managing Editor: H. K. V. Lotsch

Volume 1 **Polarized Electrons**   2nd Edition
By J. Kessler

Volume 2 **Multiphoton Processes**
Editors: P. Lambropoulos and S. J. Smith

Volume 3 **Atomic Many-Body Theory**   2nd Edition
By I. Lindgren and J. Morrison

Volume 4 **Elementary Processes in Hydrogen-Helium Plasmas**
By R. K. Janev, W. D. Langer, K. Evans, Jr. and D. E. Post, Jr.

Volume 5 **Pulsed Electrical Discharge in Vacuum**
By G. A. Mesyats and D. I. Proskurovsky

Volume 6 **Atomic and Molecular Spectroscopy**
Basic Aspects and Practical Applications
By S. Svanberg

Volume 7 **Interference of Atomic States**
By E. B. Alexandrov, M. P. Chaika and G. I. Khvostenko

Volume 8 **Plasma Physics**   Basic Theory with Fusion Applications
By K. Nishikawa and M. Wakatani

Sune Svanberg

Atomic and Molecular Spectroscopy

Basic Aspects and Practical Applications

With 331 Figures

Springer-Verlag

Berlin Heidelberg New York London
Paris Tokyo Hong Kong Barcelona

Professor Sune Svanberg
Department of Physics, Lund Institute of Technology, P.O. Box 118, S-22 100 Lund, Sweden

Guest Editors:

Professor Arthur L. Schawlow, Ph. D.
Department of Physics, Stanford University, Stanford, CA 94305-4060, USA

Professor Dr. Koichi Shimoda
Faculty of Science and Technology, Keio University, 3-14-1 Hiyoshi,
Kohoku-ku, Yokohama 223, Japan

Series Editors:

Professor Dr. Günter Ecker
Ruhr-Universität Bochum, Institut für Theoretische Physik, Lehrstuhl I, Universitätsstrasse 150,
D-4630 Bochum-Querenburg, Fed. Rep. of Germany

Professor Peter Lambropoulos, Ph. D.
University of Crete, P.O. Box 470, Iraklion, Crete, Greece, and
Department of Physics, University of Southern California, University Park,
Los Angeles, CA 90089-0484, USA

Professor Igor I. Sobel'man
Lebedev Physical Institute, USSR Academy of Sciences,
Leninsky Prospekt 53, SU-Moscow, USSR

Professor Dr. Herbert Walther
Sektion Physik der Universität München, Am Coulombwall 1,
D-8046 Garching/München, Fed. Rep. of Germany

Managing Editor: Dr. Helmut K. V. Lotsch
Springer-Verlag, Tiergartenstrasse 17, D-6900 Heidelberg, Fed. Rep. of Germany

ISBN 3-540-52594-7 Springer-Verlag Berlin Heidelberg New York
ISBN 0-387-52594-7 Springer-Verlag New York Berlin Heidelberg

Library of Congress Cataloging-in-Publication Data. Svanberg, S. R. (Sune R.), 1943– . Atomic and molecular spectroscopy / Sune Svanberg. p. cm.– (Springer series on atoms + plasmas ; 6) Includes bibliographical references (p.) and index. ISBN 3-540-52594-7 (alk. paper).– ISBN 0-387-52594-7 (alk. paper) 1. Atomic spectroscopy. 2. Molecular spectroscopy. I. Title. II. Series. QC454.A8S85 1990 539'.6–dc20 90-10207

This work is subject to copyright. All rights are reserved, whether the whole or part of the material is concerned, specifically the rights of translation, reprinting, reuse of illustrations, recitation, broadcasting, reproduction on microfilms or in other ways, and storage in data banks. Duplication of this publication or parts thereof is only permitted under the provisions of the German Copyright Law of September 9, 1965, in its current version, and a copyright fee must always be paid. Violations fall under the prosecution act of the German Copyright Law.

© Springer-Verlag Berlin Heidelberg 1991
Printed in Germany

The use of registered names, trademarks etc. in this publication does not imply, even in the absence of a specific statement, that such names are exempt from the relevant protective laws and regulations and therefore free for general use.

The text was prepared using the PS™ Technical Word Processor

2154/3140-543210 – Printed on acid-free paper

Preface

Atomic and molecular spectroscopy has provided basic information leading to the development of quantum mechanics and to the understanding of the building blocks of matter. It continues to provide further insight into the statics and dynamics of the microcosmos, and provides the means for testing new concepts and computational methods. The results of atomic and molecular spectroscopy are of great importance in astrophysics, plasma and laser physics. The rapidly growing field of spectroscopic applications has made considerable impact on many disciplines, including medicine, environmental protection, chemical processing and energy research. In particular, the techniques of electron and laser spectroscopy, the subjects of the 1981 Nobel prize in physics, have contributed much to the analytical potential of spectroscopy.

This textbook on *Atomic and Molecular Spectroscopy* has been prepared to provide an overview of modern spectroscopic methods. It is intended to serve as a text for a course on the subject for final-year undergraduate physics students or graduate students. It should also be useful for students of astrophysics and chemistry. The text has evolved from courses on atomic and molecular spectroscopy given by the author since 1975 at Chalmers University of Technology and at the Lund Institute of Technology. References are given to important books and review articles which allow more detailed studies of different aspects of atomic and molecular spectroscopy. No attempt has been made to cover all important references, nor have priority aspects been systematically considered.

It is assumed that the reader has a basic knowledge of quantum mechanics and atomic physics. However, the completion of a specialized course on atomic and molecular physics is not required. The present treatise (disregarding Chap.4) is not particularly mathematical, but emphasizes the physical understanding of the different techniques of spectroscopy. In the course given by the author, the time for solving calculational problems has been reduced to allow a more complete overview of the field in the time available. Particular emphasis has been given to technical applications. However, by increasing the allotted problem-solving time or by omitting certain areas of spectroscopy, a more problem-oriented course can easily be taught based on this book. In his courses, the author has combined lectures with a number of 5-hour laboratory experiments (performed on research equipment) and a number of 1-2 hour visits to local research groups in physics, chemistry and astronomy.

Part of the material is reworked from the Swedish textbook *Atomfysik* by I. Lindgren and S. Svanberg (Universitetsförlaget, Uppsala 1974). The author is very grateful to his teacher Prof. I. Lindgren for contributions and support through the years. He would also like to thank many colleagues, including Prof. D. Dravins, Dr. Å. Hjalmarsson, Prof. I. Martinson, Prof. J. Nordgren, Prof. C. Nordling, Dr. W. Persson, Prof. A. Rosén, Prof. H. Siegbahn and Dr. C.-G. Wahlström for valuable suggestions and corrections.

Special thanks are due to Mrs. C. Holmqvist for typing numerous versions of the manuscript and Dr. H. Sheppard for correcting the English and assisting with the figures. Mr. Å. Bergqvist and Mr. G. Romerius helped by drawing some of the figures. Finally, the kind help and support of Dr. H. Lotsch of Springer-Verlag is gratefully acknowledged.

Lund, September 1990   Sune Svanberg

Contents

1. Introduction 1
2. Atomic Structure 4
   2.1 One-Electron Systems 4
   2.2 Alkali Atoms 5
   2.3 Magnetic Effects 7
       2.3.1 Precessional Motion 7
       2.3.2 Spin-Orbit Interaction 8
   2.4 General Many-Electron Systems 9
   2.5 The Influence of External Fields 16
       2.5.1 Magnetic Fields 16
       2.5.2 Electric Fields 20
   2.6 Hyperfine Structure 21
       2.6.1 Magnetic Hyperfine Structure 21
       2.6.2 Electric Hyperfine Structure 23
   2.7 The Influence of External Fields (hfs) 25
   2.8 Isotopic Shifts 27
3. Molecular Structure 29
   3.1 Electronic Levels 29
   3.2 Rotational Energy 32
   3.3 Vibrational Energy 34
   3.4 Polyatomic Molecules 35
   3.5 Other Molecular Structures 36
4. Radiation and Scattering Processes 37
   4.1 Resonance Radiation 37
   4.2 Spectra Generated by Dipole Transitions 46
       4.2.1 Atoms 48
       4.2.2 Molecules 50
             a) Rotational Transitions 51
             b) Vibrational Transitions 51
             c) Vibrational-Rotational Spectra 52
             d) Electronic Transitions -
                The Franck-Condon Principle 54
   4.3 Rayleigh and Raman Scattering 56
   4.4 Raman Spectra 58
       4.4.1 Vibrational Raman Spectra 58
       4.4.2 Rotational Raman Spectra 59
       4.4.3 Vibrational-Rotational Raman Spectra 59

	4.5	Mie Scattering	60
	4.6	Atmospheric Scattering Phenomena	61
	4.7	Comparison Between Different Radiation and Scattering Processes	64
	4.8	Collision-Induced Processes	65

5. Spectroscopy of Inner Electrons 66
- 5.1 X-Ray Spectroscopy 66
  - 5.1.1 X-Ray Emission Spectroscopy 68
  - 5.1.2 X-Ray Absorption Spectroscopy 73
- 5.2 Photo-Electron Spectroscopy 75
  - 5.2.1 XPS Techniques and Results 77
  - 5.2.2 Chemical Shifts 80
- 5.3 Auger Electron Spectroscopy 83

6. Optical Spectroscopy 85
- 6.1 Light Sources 85
  - 6.1.1 Line Light Sources 86
  - 6.1.2 Continuum Light Sources 94
  - 6.1.3 Synchrotron Radiation 95
  - 6.1.4 Natural Radiation Sources 99
- 6.2 Spectral Resolution Instruments 101
  - 6.2.1 Prism Spectrometers 101
  - 6.2.2 Grating Spectrometers 104
  - 6.2.3 The Fabry-Pérot Interferometer 107
  - 6.2.4 The Fourier Transform Spectrometer 112
- 6.3 Detectors 115
- 6.4 Optical Components and Materials 119
  - 6.4.1 Interference Filters and Mirrors 119
  - 6.4.2 Absorption Filters 122
  - 6.4.3 Polarizers 125
  - 6.4.4 Optical Materials 127
  - 6.4.5 Influence of the Transmission Medium 128
- 6.5 Optical Methods of Chemical Analysis 131
  - 6.5.1 The Beer-Lambert Law 132
  - 6.5.2 Atomic Absorption/Emission Spectrophotometry . 134
  - 6.5.3 Burners, Flames, Sample Preparation and Measurements 137
  - 6.5.4 Modified Methods of Atomization 138
  - 6.5.5 Multi-Element Analysis 139
  - 6.5.6 Molecular Spectrophotometry 141
  - 6.5.7 Raman Spectroscopy 144
- 6.6 Optical Remote Sensing 145
  - 6.6.1 Atmospheric Monitoring with Passive Techniques ... 146
  - 6.6.2 Land and Water Measurements with Passive Techniques 150
- 6.7 Astrophysical Spectroscopy 152

7. Radio-Frequency Spectroscopy ... 159
7.1 Resonance Methods ... 159
7.1.1 Magnetic Resonance ... 159
7.1.2 Atomic-Beam Magnetic Resonance ... 160
7.1.3 Optical Pumping ... 168
7.1.4 Optical Double Resonance ... 172
7.1.5 Level-Crossing Spectroscopy ... 174
7.1.6 Resonance Methods for Liquids and Solids ... 181
   a) Nuclear Magnetic Resonance ... 182
   b) Electron Spin Resonance ... 186
   c) Electron-Nuclear Double-Resonance ... 187
7.2 Microwave Radiometry ... 188
7.3 Radio Astronomy ... 190

8. Lasers ... 195
8.1 Basic Principles ... 195
8.2 Coherence ... 198
8.3 Resonators and Mode Structure ... 199
8.4 Fixed-Frequency Lasers ... 203
8.4.1 The Ruby Laser ... 203
8.4.2 Four-Level Lasers ... 205
8.4.3 Pulsed Gas Lasers ... 207
8.4.4 The He–Ne Laser ... 209
8.4.5 Gaseous Ion Lasers ... 210
8.5 Tunable Lasers ... 212
8.5.1 Dye Lasers ... 212
8.5.2 Colour-Centre Lasers ... 221
8.5.3 Tunable Solid-State Lasers ... 221
8.5.4 Tunable CO_2 Lasers ... 222
8.5.5 Semiconductor Lasers ... 224
8.6 Nonlinear Optical Phenomena ... 226

9. Laser Spectroscopy ... 235
9.1 Basic Principles ... 235
9.1.1 Comparison Between Conventional Light Sources and Lasers ... 235
9.1.2 Saturation ... 236
9.1.3 Excitation Methods ... 237
   a) Single-step excitation ... 237
   b) Multi-step excitation ... 237
   c) Multi-photon absorption ... 237
9.1.4 Detection Methods ... 238
   a) Fluorescence ... 238
   b) Photoionization ... 239
   c) Collisional ionization ... 239
   d) Field ionization ... 239
9.1.5 Laser Wavelength Setting ... 239

9.2 Doppler-Limited Techniques 241
  9.2.1 Absorption Measurements 242
  9.2.2 Intra-Cavity Absorption Measurements 243
  9.2.3 Absorption Measurements on Excited States 244
  9.2.4 Level Labelling 245
  9.2.5 Two-Photon Absorption Measurements 246
  9.2.6 Opto-Galvanic Spectroscopy 248
  9.2.7 Single-Atom Detection 251
  9.2.8 Opto-Acoustic Spectroscopy 251
9.3 Optical Double-Resonance and Level-Crossing Experiments
    with Laser Excitation 252
9.4 Time-Resolved Spectroscopy 258
  9.4.1 Generation of Short Optical Pulses 258
  9.4.2 Generation of Ultra-Short Optical Pulses 259
  9.4.3 Measurement Techniques for Optical Transients 262
      a) Transient-digitizer technique 262
      b) Boxcar technique 263
      c) Delayed-coincidence techniques 263
      d) Streak-camera techniques 266
      e) Pump-probe techniques 266
  9.4.4 Background to Lifetime Measurements 267
  9.4.5 Survey of Methods of Measurement for Radiative
        Properties 268
      a) Linewidth measurements 268
      b) ODR and LC 268
      c) Beam-foil techniques 269
      d) Beam-laser techniques 269
      e) Time-resolved spectroscopy with pulsed lasers 270
      f) Time-resolved spectroscopy with pulsed electron
         beam excitation 270
      g) Phase-shift method 271
      h) The emission method 272
      i) The hook method 272
  9.4.6 Quantum-Beat Spectroscopy 274
9.5 High-Resolution Laser Spectroscopy 278
  9.5.1 Spectroscopy on Collimated Atomic Beams 279
      a) Detection through fluorescence 280
      b) Detection by photoionization 282
      c) Detection by the recoil effect 282
      d) Detection by magnetic deflection 283
  9.5.2 Saturation Spectroscopy and Related Techniques 285
  9.5.3 Doppler-Free Two-Photon Absorption 293
  9.5.4 Spectroscopy of Trapped Ions and Atoms 298

10. Laser-Spectroscopic Applications 302
10.1 Diagnostics of Combustion Processes 302
  10.1.1 Background 302
  10.1.2 Laser-Induced Fluorescence
         and Related Techniques 305

  10.1.3 Raman Spectroscopy 310
  10.1.4 Coherent Anti-Stokes Raman Scattering 312
  10.1.5 Velocity Measurements 315
 10.2 Laser Remote Sensing of the Atmosphere 318
  10.2.1 Optical Heterodyne Detection 319
  10.2.2 Long-Path Absorption Techniques 320
  10.2.3 Lidar Techniques 322
 10.3 Laser-Induced Fluorescence and Raman Spectroscopy
   in Liquids and Solids 329
  10.3.1 Hydrospheric Remote Sensing 330
  10.3.2 Monitoring of Surface Layers 331
 10.4 Laser-Induced Chemical Processes 335
  10.4.1 Laser-Induced Chemistry 336
  10.4.2 Laser Isotope Separation 337
 10.5 Spectroscopic Aspects of Lasers in Medicine 341

References 351

Subject Index 395

1. Introduction

By *spectroscopy* we usually mean experimental charting of the energy-level structure of physical systems. For that purpose, the transition processes, spontaneous or induced, between different energy states are studied and spectroscopy therefore normally means analysis of various types of radiation - electromagnetic or particle emission. Spectroscopic investigations can be of a fundamental or an applied nature. In fundamental spectroscopy experimentally determined energy levels, transition probabilities, etc. are employed for obtaining an understanding of the studied systems using adequate theories or models. Usually, certain primary quantities (wavelengths, intensities, etc.) are measured in spectroscopic investigations. These quantities are then used to evaluate more fundamental quantities. This process is schematically illustrated in Fig. 1.1.

Fundamental quantities, such as wavelengths and transition probabilities, determined using spectroscopy, for atoms and molecules are of direct importance in several disciplines such as astro-physics, plasma and laser physics. Here, as in many fields of applied spectroscopy, the spectroscopic information can be used in various kinds of analysis. For instance, optical atomic absorption or emission spectroscopy is used for both qualitative and quantitative chemical analysis. Other types of spectroscopy, e.g. electron spectroscopy methods or nuclear magnetic resonance, also provide information on the chemical environment in which a studied atom is situated. Tunable lasers have had a major impact on both fundamental and applied spectroscopy. New fields of applied laser spectroscopy include remote sensing of the environment, medical applications, combustion diagnostics, laser-induced chemistry and isotope separation.

In principle, a set-up for spectral studies consists of three components: a radiation source, an analyser and a detection system. In many modern techniques the system under investigation is subjected to different types of static or oscillatory fields and the influence of these fields on the system is studied in order to obtain a more complete picture of the system. Resonance methods are of special importance since they provide high accuracy

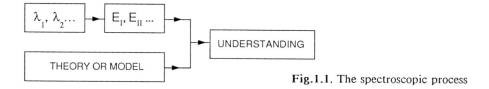

Fig. 1.1. The spectroscopic process

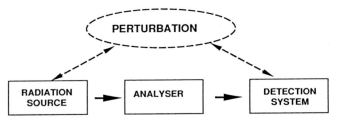

Fig.1.2. Basic arrangement of a spectroscopic set-up

in the determination of small energy splittings. The basic arrangement of a spectroscopic set-up is shown in Fig.1.2.

The choice of spectroscopic method is primarily determined by the energy range of the phenomenon to be studied. In Fig.1.3 the spectral ranges that are of interest in atomic and molecular spectroscopy are shown. The energy ranges for different types of structures and transitions are also indicated. Using the simple relations

$$\Delta E = h\nu, \quad \lambda = c/\nu, \quad 1/\lambda = \nu/c, \quad \nu = c(1/\lambda) \tag{1.1}$$
$$\text{Energy} \quad \text{Wavelength} \quad \text{Wavenumber} \quad \text{Frequency}$$

(h: Planck's constant, c: velocity of light) an energy interval ΔE can be uniquely expressed in eV (energy), nm (=10Å) (wavelength), cm^{-1} (wavenumber) or Hz (frequency). 1 cm^{-1} is sometimes called 1 Kayser. In Table 1.1 conversion factors between different units are given.

The choice of unit depends, to a great extent, on the energy region and on traditional factors:

X-ray region	keV ,
Visible and UV regions	nm, Å (in solid-state physics: eV) ,
Infrared region	μm, cm^{-1} ,
Radio-frequency region	MHz, cm^{-1} .

It is practical to memorize the following approximate relations

$$1 \text{ eV} \leftrightarrow 8000 \text{ cm}^{-1} \leftrightarrow 12000 \text{ Å},$$
$$1 \text{ cm}^{-1} \leftrightarrow 30 \text{ GHz},$$

and kT at T = 300 K (room temperature)

$$kT_{300} \simeq 1/40 \text{ eV}.$$

(k is Boltzmann's constant, and T is the absolute temperature).

Transitions between inner electron orbitals normally occur in the keV range (X rays) while the energies for transitions between outer orbitals are in the eV region (visible or near UV and IR regions). The fine structure of atoms is of the order of 10^{-3} eV (~10cm^{-1}) and hyperfine structures are typically about 10^{-6} eV (~300MHz). molecular vibrational energies split-

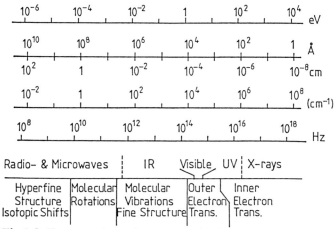

Fig.1.3. Energy scales and spectroscopic phenomena

Table 1.1. Conversion factors between different energy units

Unit	Joule	cm^{-1}	Hz	eV
1 Joule (1 J)	1	$5.03378 \cdot 10^{22}$	$1.50919 \cdot 10^{33}$	$6.24150 \cdot 10^{18}$
1 cm^{-1}	$1.98658 \cdot 10^{-23}$	1	$2.99792 \cdot 10^{10}$	$1.23992 \cdot 10^{-4}$
1 Hz	$6.62608 \cdot 10^{-34}$	$3.33565 \cdot 10^{-11}$	1	$4.13567 \cdot 10^{-15}$
1 eV	$1.60218 \cdot 10^{-19}$	$8.06502 \cdot 10^{3}$	$2.41799 \cdot 10^{14}$	1

tings are of the order of 10^{-1} eV, while rotational splittings are typically 10^{-3} eV. Of course, these energies vary widely and the above values are given only to provide a first estimate of typical orders of magnitude.

As a background to atomic and molecular spectroscopy a survey of atomic and molecular structure is given in Chaps.2 and 3. Chapter 4 deals with the fundamental radiative and scattering processes that are encountered in spectroscopy. Chapter 5 is devoted to the study of inner electrons while the techniques of basic and applied optical spectroscopy are treated in Chap.6. The precision methods of radio-frequency spectroscopy are described in Chap.7. Finally, the last three chapters deal with lasers and their application to fundamental and applied spectroscopy.

2. Atomic Structure

In this chapter a brief description of the energy-level structure of atomic systems will be given. This will not include a rigorous quantum-mechanical treatment, but will deal with more qualitative aspects. The atomic structures that are explored with the spectroscopic techniques discussed in this text will be described. For a more complete treatment the reader is referred to standard textbooks on atomic physics and quantum mechanics [2.1-18].

2.1 One-Electron Systems

In hydrogen and hydrogen-like systems a single electron moves around the nucleus of charge Ze at a distance r in a *central field*, and its potential energy is given by

$$V(r) = -\frac{Ze^2}{4\pi\epsilon_0 r} \, . \tag{2.1}$$

The Hamiltonian of the system is then

$$\mathcal{H}_0 = \frac{1}{2m}\mathbf{p}^2 + V(r) \tag{2.2}$$

($\mathbf{p}$ is the momentum operator, m is the electron mass) and the energy eigenvalues are obtained by solving the Schrödinger equation

$$\left[-\frac{\hbar^2}{2m}\Delta + V(r)\right]\psi = E\psi \tag{2.3}$$

(Δ is the Laplace operator, and $\hbar = h/2\pi$). The eigenvalues depend only on the principal quantum number n and are independent of the azimuthal quantum number ℓ and its projection m_ℓ

$$E_n = -hcR_y \frac{Z^2}{n^2} \quad (n = 1, 2, 3 \ldots) \, , \tag{2.4}$$

where

$$R_y = \frac{me^4}{4\pi(4\pi\epsilon_0)^2 c\hbar^3} \, . \tag{2.5}$$

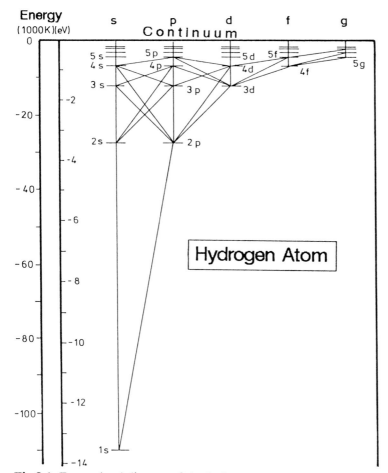

Fig.2.1. Energy level diagram of the hydrogen atom

However, the ℓ value specifies the wavefunctions and the states are characterized according to

$$\ell = \begin{array}{cccccc} 0 & 1 & 2 & 3 & 4 & 5 \\ s & p & d & f & g & h \end{array}.$$

If relativistic and quantum electrodynamic (QED) effects are considered, the far-reaching degeneracy for hydrogen-like systems is lifted. In Fig.2.1 the simple energy-level diagram of hydrogen is shown.

2.2 Alkali Atoms

In alkali atoms there is a single electron outside a spherically symmetric inner electron cloud (core). The field in which the outer electron moves is

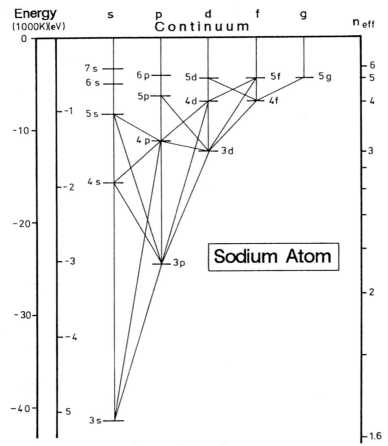

Fig.2.2. Energy level diagram of the sodium atom

not Coulombic, but is still essentially central. The energy for such a system can be written

$$E = -\frac{hcR_y}{n_{eff}^2},\qquad(2.6)$$

where n_{eff} = n-d is called the *effective quantum number*, and d the *quantum defect*. The s electrons may penetrate into the core (they can even be found at the surface or actually inside the nucleus) thus experiencing an increased force from the less shielded nucleus, which results in a stronger binding. This behaviour is reflected in the quantum defect, which for s-electrons is of the order of 2. For high-ℓ orbits (*non-penetrating electrons*) the quantum defect approaches zero. These effects are illustrated for the case of sodium in Fig.2.2.

2.3 Magnetic Effects

2.3.1 Precessional Motion

An angular momentum vector **L**, associated with a magnetic moment μ_L will precess in a magnetic field, as illustrated in Fig.2.3. Frequently we encounter this phenomenon in the description of atomic and molecular processes. To demonstrate the phenomenon, imagine magnetic poles q in analogy with the electrical case, for which we have the same mathematical description. For the mechanical moment **M** we then have

$$|\mathbf{M}| = qB\frac{d}{2}\sin\theta + qB\frac{d}{2}\sin\theta = qBd\sin\theta, \qquad (2.7)$$

that is

$$\mathbf{M} = \mu_L \times \mathbf{B} \quad \text{with} \quad |\mu_L| = qd. \qquad (2.8)$$

But

$$\mu_L = -g\mu_B \mathbf{L} \qquad (2.9)$$

where μ_B is the *Bohr magneton* and g is a constant of proportionality. (Throughout this book we use *dimensionless* angular momentum vectors. μ_B has the dimension of a magnetic moment). The laws of motion then yield

$$\frac{d\mathbf{L}}{dt} = \mathbf{M} = -g\mu_B \mathbf{L} \times \mathbf{B}. \qquad (2.10)$$

Thus

$$d\mathbf{L} = -(g\mu_B \mathbf{L} \times \mathbf{B})dt, \qquad (2.11)$$

showing that the tip of **L** moves perpendicularly to both **L** and **B**, i.e. **L** precesses with an angular frequency $\omega = g\mu_B B$.

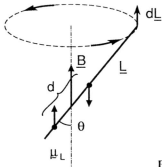

Fig.2.3. Precession of an angular momentum vector and an associated magnetic moment in the magnetic field

2.3.2 Spin–Orbit Interaction

The electron has a spin angular momentum **s** and an associated magnetic moment $\boldsymbol{\mu}_s$, i.e.,

$$\boldsymbol{\mu}_s = -g_s \mu_B \mathbf{s} \ . \tag{2.12}$$

As the electron moves in the electric field of the nucleus it is subject to a magnetic field $\mathbf{B}_\ell$, which is proportional to the angular momentum **l** of the electron (Fig.2.4). The electron then has an orientational energy E_{so} in the field

$$E_{so} = -\boldsymbol{\mu}_s \cdot \mathbf{B}_\ell = \zeta \mathbf{l} \cdot \mathbf{s} \ , \tag{2.13}$$

where ζ is a constant of proportionality. **l** and **s** couple and precess about their mutual resultant **j** with a frequency proportional to the strength of the field. We have

$$\mathbf{j} = \mathbf{l} + \mathbf{s} \ . \tag{2.14}$$

For the corresponding quantum numbers describing the length of the vector (e.g., $|\mathbf{j}| = [j(j+1)]^{1/2}$, dimensionless angular momentum! The normally occurring $\hbar$ in the length expression of an angular momentum vector is included in ζ !) we have

$$s = 1/2 \ ,$$
$$\ell = 0, 1, 2, \ldots \ ,$$
$$j = l \pm 1/2 \ .$$

Using the vector cosine theorem

$$\mathbf{j}^2 = (\mathbf{l} + \mathbf{s})^2 = \mathbf{l}^2 + \mathbf{s}^2 + 2\mathbf{l} \cdot \mathbf{s}$$

we obtain

$$E_{so} = \zeta[j(j+1) - \ell(\ell+1) - s(s+1)]/2 \ . \tag{2.15}$$

Here we have inserted the "quantum-mechanical squares" or, to use a more modern term, used first-order perturbation theory. The so-called *fine-structure splitting* between the two possible j levels is given by

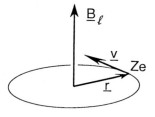

Fig.2.4. Magnetic field associated with orbital motion

Table 2.1. Fine-structure splittings in ^{2}P states of the alkali atoms

n	Na	K	Rb	Cs
3	17			
4	5.6	58		
5	2.5	19	238	
6	1.2	8.4	78	554
7	0.7	4.5	35	181
8	0.4	2.7	19	83
9		1.7	11	45
10		1.2	7	27

$$\Delta E_{so} = \zeta(\ell+1/2) \ . \tag{2.16}$$

In a simple calculation, ζ is found to be a positive quantity, scaling as $1/n_{eff}^3$. The high-j level then has a higher energy than the low-j level. This is found to be true for p-states of alkali atoms, and in Table 2.1 some experimentally determined fine-structure splittings are given. For $\ell > 1$ *level inversion* sometimes occurs, i.e. the high-j level has the lower energy. This is due to *core polarization*. The outer electron polarizes the closed shells so that a dominating, oppositely directed magnetic field contribution is obtained from these shells. In Fig.2.5 the level ordering for the alkali-atoms doublets is indicated.

2.4 General Many-Electron Systems

The structure of the general many-electron system with i electrons is obviously much more difficult to calculate. As a starting point the following approximate Hamiltonian is used

$$\mathcal{H} = \sum_i \left(\frac{1}{2m}p_i^2 - \frac{Ze^2}{4\pi\epsilon_0 r_i} \right) + \sum_{i<j} \frac{e^2}{4\pi\epsilon_0 r_{ij}} + \sum_i \zeta_i(r_i)l_i \cdot s_i \ . \tag{2.17}$$

ALKALI ATOM FINE-STRUCTURE

	Li	Na	K	Rb	Cs
^{2}P	↑	↑	↑	↑	↑
^{2}D	↑	↓	↓	↑*	↑
^{2}F	↑	↑	↓	↓	↓
^{2}G	↑	↑			

NORMAL: ↑
INVERTED: ↓

Fig.2.5. Level ordering in alkali atom doublet series. (The asterisc * indicates that the lowest member of the Rubidium series is inverted)

Table 2.2. Ground configurations of the atoms [2.1]

Atomic number Z	Element		Shells											LS configuration of the ground state	First ionisation potential [eV]
			K n=1 s	L n=2 s	p	M n=3 s	p	d	N n=4 s	p	d	O n=5 s	p		
1	Hydrogen	H	1											$^2S_{1/2}$	13.60
2	Helium	He	2											$^1S_0$	24.58
3	Lithium	Li	2	1										$^2S_{1/2}$	5.39
4	Beryllium	Be	2	2										$^1S_0$	9.32
5	Boron	B	2	2	1									$^2P_{1/2}$	8.30
6	Carbon	C	2	2	2									$^3P_0$	11.26
7	Nitrogen	N	2	2	3									$^4S_{3/2}$	14.54
8	Oxygen	O	2	2	4									$^3P_2$	13.61
9	Fluorine	F	2	2	5									$^2P_{3/2}$	17.42
10	Neon	Ne	2	2	6									$^1S_0$	21.56
11	Sodium	Na	2	2	6	1								$^2S_{1/2}$	5.14
12	Magnesium	Mg	2	2	6	2								$^1S_0$	7.64
13	Aluminium	Al	2	2	6	2	1							$^2P_{1/2}$	5.98
14	Silicon	Si	2	2	6	2	2							$^3P_0$	8.15
15	Phosphorous	P	2	2	6	2	3							$^4S_{3/2}$	10.55
16	Sulphur	S	2	2	6	2	4							$^3P_2$	10.36
17	Chlorine	Cl	2	2	6	2	5							$^2P_{3/2}$	13.01
18	Argon	Ar	2	2	6	2	6							$^1S_0$	15.76
19	Potassium	K	2	2	6	2	6		1					$^2S_{1/2}$	4.34
20	Calcium	Ca	2	2	6	2	6		2					$^1S_0$	6.11
21	Scandium	Sc	2	2	6	2	6	1	2					$^2D_{3/2}$	6.56
22	Titanium	Ti	2	2	6	2	6	2	2					$^3F_2$	6.83
23	Vanadium	V	2	2	6	2	6	3	2					$^4F_{3/2}$	6.74
24	Chromium	Cr	2	2	6	2	6	5	1					$^7S_3$	6.76
25	Manganese	Mn	2	2	6	2	6	5	2					$^6S_{5/2}$	7.43
26	Iron	Fe	2	2	6	2	6	6	2					$^5D_4$	7.90
27	Cobalt	Co	2	2	6	2	6	7	2					$^4F_{9/2}$	7.86
28	Nickel	Ni	2	2	6	2	6	8	2					$^3F_4$	7.63
29	Copper	Cu	2	2	6	2	6	10	1					$^2S_{1/2}$	7.72
30	Zinc	Zn	2	2	6	2	6	10	2					$^1S_0$	9.39
31	Gallium	Ga	2	2	6	2	6	10	2	1				$^2P_{1/2}$	6.00
32	Germanium	Ge	2	2	6	2	6	10	2	2				$^3P_0$	7.88
33	Arsenic	As	2	2	6	2	6	10	2	3				$^4S_{3/2}$	9.81
34	Selenium	Se	2	2	6	2	6	10	2	4				$^3P_2$	9.75
35	Bromine	Br	2	2	6	2	6	10	2	5				$^2P_{3/2}$	11.84
36	Krypton	Kr	2	2	6	2	6	10	2	6				$^1S_0$	14.00
37	Rubidium	Rb	2	2	6	2	6	10	2	6		1		$^2S_{1/2}$	4.18
38	Strontium	Sr	2	2	6	2	6	10	2	6		2		$^1S_0$	5.69
39	Yttrium	Y	2	2	6	2	6	10	2	6	1	2		$^2D_{3/2}$	6.38
40	Zirconium	Zr	2	2	6	2	6	10	2	6	2	2		$^3F_2$	6.84
41	Niobium	Nb	2	2	6	2	6	10	2	6	4	1		$^6D_{1/2}$	6.88
42	Molybdénum	Mo	2	2	6	2	6	10	2	6	5	1		$^7S_3$	7.13
43	Technetium	Tc	2	2	6	2	6	10	2	6	6	1		$^6D_{9/2}$	7.23
44	Ruthenium	Ru	2	2	6	2	6	10	2	6	7	1		$^5F_5$	7.37
45	Rhodium	Rh	2	2	6	2	6	10	2	6	8	1		$^4F_{9/2}$	7.46
46	Palladium	Pd	2	2	6	2	6	10	2	6	10			$^1S_0$	8.33
47	Silver	Ag	2	2	6	2	6	10	2	6	10	1		$^2S_{1/2}$	7.57
48	Cadmium	Cd	2	2	6	2	6	10	2	6	10	2		$^1S_0$	8.99
49	Indium	In	2	2	6	2	6	10	2	6	10	2	1	$^2P_{1/2}$	5.79
50	Tin	Sn	2	2	6	2	6	10	2	6	10	2	2	$^3P_0$	7.33
51	Antimony	Sb	2	2	6	2	6	10	2	6	10	2	3	$^4S_{3/2}$	8.64
52	Tellurium	Te	2	2	6	2	6	10	2	6	10	2	4	$^3P_2$	9.01
53	Iodine	J	2	2	6	2	6	10	2	6	10	2	5	$^2P_{3/2}$	10.44
54	Xenon	Xe	2	2	6	2	6	10	2	6	10	2	6	$^1S_0$	12.13

Transition elements: Z = 21–30, 39–48

Atomic number Z	Element		Shells											LS configuration of the ground state	First ionisation potential [eV]	
			N $n=4$				O $n=5$				P $n=6$			Q $n=7$		
			s	p	d	f	s	p	d	f	s	p	d	s		
55	Cesium	Cs	2	6	10		2	6			1				$^2S_{1/2}$	3.89
56	Barium	Ba	2	6	10		2	6			2				$^1S_0$	5.21
57	Lanthanum	La	2	6	10		2	6	1		2				$^2D_{3/2}$	5.61
58	Cerium	Ce	2	6	10	2	2	6			2				$^3H_4$	5.6
59	Praseodymium	Pr	2	6	10	3	2	6			2				$^4I_{9/2}$	5.46
60	Neodymium	Nd	2	6	10	4	2	6			2				$^5I_4$	5.51
61	Promethium	Pm	2	6	10	5	2	6			2				$^6H_{5/2}$	
62	Samarium	Sm	2	6	10	6	2	6			2				$^7F_0$	5.6
63	Europium	Eu	2	6	10	7	2	6			2				$^8S_{7/2}$	5.67
64	Gadolinium	Gd	2	6	10	7	2	6	1		2				$^9D_2$	6.16
65	Terbium	Tb	2	6	10	9	2	6			2				—	5.98
66	Dysprosium	Dy	2	6	10	10	2	6			2				$^5I_8$	6.8
67	Holmium	Ho	2	6	10	11	2	6			2				$^4I_{15/2}$	
68	Erbium	Er	2	6	10	12	2	6			2				$^3H_6$	6.08
69	Thulium	Tm	2	6	10	13	2	6			2				$^2F_{7/2}$	5.81
70	Ytterbium	Yb	2	6	10	14	2	6			2				$^1S_0$	6.22
71	Lutetium	Lu	2	6	10	14	2	6	1		2				$^2D_{3/2}$	6.15
72	Hafnium	Hf	2	6	10	14	2	6	2		2				$^3F_2$	5.5
73	Tantalum	Ta	2	6	10	14	2	6	3		2				$^4F_{3/2}$	7.7
74	Tungsten	W	2	6	10	14	2	6	4		2				$^5D_0$	7.98
75	Rhenium	Re	2	6	10	14	2	6	5		2				$^6S_{5/2}$	7.87
76	Osmium	Os	2	6	10	14	2	6	6		2				$^5D_4$	8.7
77	Iridium	Ir	2	6	10	14	2	6	9						$^2D_{5/2}$	9.2
78	Platinum	Pt	2	6	10	14	2	6	9		1				$^3D_3$	9.0
79	Gold	Au	2	6	10	14	2	6	10		1				$^2S_{1/2}$	9.22
80	Mercury	Hg	2	6	10	14	2	6	10		2				$^1S_0$	10.43
81	Thallium	Tl	2	6	10	14	2	6	10		2	1			$^2P_{1/2}$	6.11
82	Lead	Pb	2	6	10	14	2	6	10		2	2			$^3P_0$	7.42
83	Bismuth	Bi	2	6	10	14	2	6	10		2	3			$^4S_{3/2}$	7.29
84	Polonium	Po	2	6	10	14	2	6	10		2	4			$^3P_2$	8.43
85	Astatine	At	2	6	10	14	2	6	10		2	5				9.5
86	Radon	Rn	2	6	10	14	2	6	10		2	6			$^1S_0$	10.75
87	Francium	Fr	2	6	10	14	2	6	10		2	6		1		4
88	Radium	Ra	2	6	10	14	2	6	10		2	6		2		5.28
89	Actinium	Ac	2	6	10	14	2	6	10		2	6	1	2		
90	Thorium	Th	2	6	10	14	2	6	10		2	6	2	2		
91	Protactinium	Pa	2	6	10	14	2	6	10	2	2	6	1	2		
92	Uranium	U	2	6	10	14	2	6	10	3	2	6	1	2		
93	Neptunium	Np	2	6	10	14	2	6	10	4	2	6	1	2		
94	Plutonium	Pu	2	6	10	14	2	6	10	6	2	6		2		
95	Americium	Am	2	6	10	14	2	6	10	7	2	6		2		
96	Curium	Cm	2	6	10	14	2	6	10	7	2	6	1	2		
97	Berkelium	Bk	2	6	10	14	2	6	10	8	2	6	1	2		
98	Californium	Cf	2	6	10	14	2	6	10	10	2	6		2		
99	Einsteinium	Es	2	6	10	14	2	6	10	11	2	6		2		
100	Fermium	Fm	2	6	10	14	2	6	10	12	2	6		2		
101	Mendelevium	Md	2	6	10	14	2	6	10	13	2	6		2		
102	Nobelium	No	2	6	10	14	2	6	10	14	2	6		2		
103	Lawrencium	Lw	2	6	10	14	2	6	10	14	2	6	1	2		
104	Kurchatovium		2	6	10	14	2	6	10	14	2	6	2	2		
105	Hahnium		2	6	10	14	2	6	10	14	2	6	3	2		

Rare earths (Z = 57–70)

Transition elements (Z = 71–80)

Actinides (Z = 89–105)

Here r_{ij} is the distance between the electrons i and j. As an approximation to (2.17) we assume that every electron moves independently of the other electrons in an average field, generated by the nucleus and the other electrons (the *independent particle model*). The field is assumed to be *central* (dependent only on r). This is the *central-field approximation*. The assumption of a central field combined with the *Pauli exclusion principle* results in a shell structure for the electrons and successively heavier elements can be constructed using the building-up principle (the total energy is minimized). The atom can be characterized by its *electron configuration*, e.g. for the lowest state of sodium we have

$1s^2 2s^2 2p^6 3s$.

In Table 2.2 the ground configurations of the atoms are listed. Obviously, the field in atoms is only approximately central. Besides a *central* part, the electrostatic repulsion between electrons causes a *non-central* contribution, which can be treated as a perturbation. The spin-orbit interaction must also be taken into account. If the non-central electrostatic part strongly dominates over the spin-orbit interaction, the latter is neglected as a first approximation. A coupling between the individual angular momenta is then obtained

$$\mathbf{L} = \sum_i \mathbf{l}_i \, , \quad \mathbf{S} = \sum_i \mathbf{s}_i \, . \qquad (2.18)$$

Depending on which values are possible for the corresponding quantum numbers L and S for a certain configuration, a number of electrostatically split *terms* are obtained. Such terms are designated

^{2S+1}X .

The quantity 2S+1 is called the *multiplicity*. In analogy with the one electron case we have

$L = 0 \quad 1 \quad 2 \quad 3 \quad 4$
$X = S \quad P \quad D \quad F \quad G$.

This type of coupling is called *LS coupling*.

The spins of two electrons can be arranged such that S = 0 (singlet state) or S = 1 (triplet state). In Fig.2.6 the energy-level diagram for helium is shown (ground configuration $1s^2$). The alkaline-earth atoms Be, Mg, Ca, Sr and Ba in their ground configuartion all have two s electrons in the outer shell and thus have similar energy-level diagrams. As an example, the energy-level diagram for calcium is given in Fig.2.7. Zn, Cd and Hg also have similar energy-level schemes as they also have two electrons in the outer shell. For atoms with more electrons in the outer shell, coupling the individual orbital and spin angular momenta to the resultants **L** and **S** is more complicated. For electrons of the same type (e.g., p electrons) it is

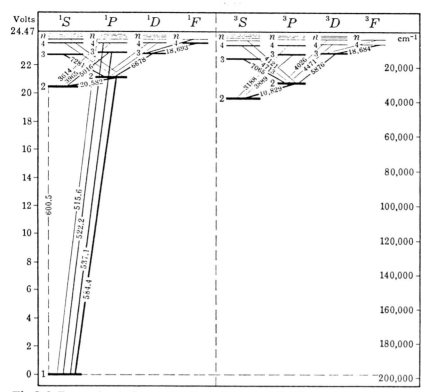

Fig.2.6. Energy level diagram for the helium atom [2.12]

important to distinguish between *equivalent* electrons (n quantum numbers also the same) or *non-equivalent* electrons. In the former case the Pauli principle strongly reduces the number of possible states. In Table 2.3 the LS terms obtained for different electron configurations are listed.

The description LS coupling may be somewhat confusing as there is primarily no spin-orbit interaction. However, we can now introduce this interaction as an additional, small perturbation. Through this perturbation, **L** and **S** couple to produce a resulting **J**. As magnetic fields and magnetic moments can be associated with both **L** and **S**, there should, according to Sect.2.3.1, be a precession of **L** about **S** and of **S** about **L**. As none of the vectors is fixed in space, a precession about the resultant **J** occurs as shown in Fig.2.8. For the corresponding J quantum number we have the relation

$$J = L+S, L+S-1, \ldots\ldots |L-S| \ . \tag{2.19}$$

In analogy with the one-electron case we have

$$E_{so} = A(L,S)[J(J+1) - L(L+1) - S(S+1)]/2 \ . \tag{2.20}$$

For the interval ΔE_{so} we have the *Landé interval rule*

$$\Delta E_{so(J,J-1)} = AJ \ . \tag{2.21}$$

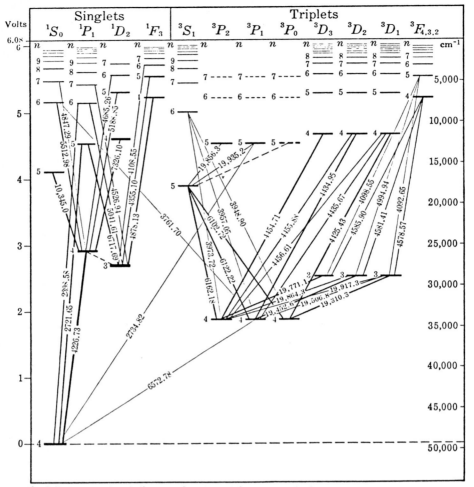

Fig.2.7. Energy level diagram for the calcium atom [2.12]

The resulting energy levels for the case of two equivalent p electrons in LS coupling is shown in the left-hand part of Fig.2.9.

If, contrary to what was assumed above, the spin-orbit interaction dominates over the non-central electrostatic interaction, we obtain primarily a coupling between the orbital and spin angular momenta of the individual electrons

$$\mathbf{j}_i = \mathbf{l}_i + \mathbf{s}_i \ . \tag{2.22}$$

In this case we talk about *jj coupling*. If the weaker electrostatic interaction is then applied, the individual $\mathbf{j}_i$:s couple to a resultant **J**. The resulting energy-level diagram is quite different compared with the case of LS coupling (i.e., there is no interval rule). In the right-hand part of Fig.2.9 the jj structure for two equivalent p electrons is shown. In the outer shells

Table 2.3. Examples of LS terms for some configurations. The designation $^1D(2)$ means that there are two ^1D terms [2.12]

Electron Configuration	Terms
Equivalent	
s^2	^1S
p^2	$^1S, ^1D, ^3P$
p^3	$^2P, ^2D, ^4S$
p^4	$^1S, ^1D, ^3P$
p^5	^2P
p^6	^1S
d^2	$^1S, ^1D, ^1G, ^3P, ^3F$
d^3	$^2P, ^2D(2), ^2F, ^2G, ^2H, ^4P, ^4F$
d^4	$^1S(2), ^1D(2), ^1F, ^1G(2), ^1I, ^3P(2), ^3D, ^3F(2), ^3G, ^3H, ^5D$
d^5	$^2S, ^2P, ^2D(3), ^2F(2), ^2G(2), ^2H, ^2I, ^4P, ^4D, ^4F, ^4G, ^6S$
Non-equivalent	
s s	$^1S, ^3S$
s p	$^1P, ^3P$
s d	$^1D, ^3D$
p p	$^1S, ^1P, ^1D, ^3S, ^3P, ^3D$
p d	$^1P, ^1D, ^1F, ^3P, ^3D, ^3F,$
d d	$^1S, ^1P, ^1D, ^1F, ^1G, ^3S, ^3P, ^3D, ^3F, ^3G$

of light elements there is generally LS coupling, whereas jj coupling is typically exhibited by inner shells of heavy atoms. In general, the situation is an intermediate of the two cases described and *intermediate coupling* arises. By experimentally determining the relative positions of the energy levels of a given configuration using spectroscopy, the coupling conditions can be investigated. The energy-level diagrams and level designations of atoms and singly and multiply ionized atoms have been determined in a process that is still continuing. A lot of this material is collected in [2.19-22]. Theoretical calculations, employing *self-consistent field* methods and other advanced techniques for describing the basic atomic energy-level structure, have been described in [2.13,23-27]. The energy-level structure for atoms with two outer electrons is frequently parametrized using the so-called *multichannel quantum defect theory* (MQDT) [2.28-30].

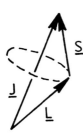

Fig.2.8. Precession of L and S about J

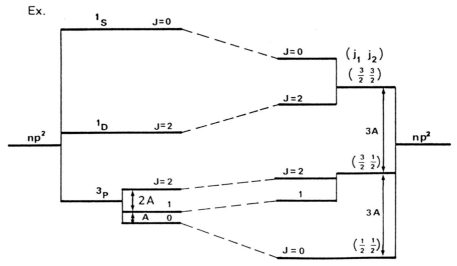

Fig.2.9. Coupling conditions for two equivalent p electrons

2.5 The Influence of External Fields

Since the particles making up the atoms have magnetic moments, and since the charged electrons describe "orbits" around the nucleus, the atom as a whole has magnetic properties and is influenced by external magnetic fields. In the same way, the negatively charged electron cloud must be displaced with regard to the positive nuclear charge under the influence of an external electric field.

2.5.1 Magnetic Fields

First consider an atom with only one electron outside the closed shells. The atom is situated in a homogeneous magnetic field **B**, which is *weak*. The meaning of the word "weak" will be clarified later. Associated with the orbital and spin motions there are magnetic moments, see (2.9), i.e.,

$$\boldsymbol{\mu}_s = -g_s \mu_B \mathbf{s} , \qquad (2.23)$$

$$\boldsymbol{\mu}_\ell = -g_\ell \mu_B \mathbf{l} . \qquad (2.24)$$

It has been found that $g_s = 2$ (apart from a 0.1% correction due to quantum electrodynamics) and $g_\ell = 1$. The spin-orbit interaction causes a precession of s and l about j, which results in a magnetic moment $\boldsymbol{\mu}_j$ precessing around **j** with a time-averaged value $\langle\boldsymbol{\mu}_j\rangle$ and in the opposite direction to **j**. As shown in Fig.2.10, the averaged moment will now precess slowly around an external field **B** which may be generated, for example, by fixed coils. In the vector model we can project the contribution of $\boldsymbol{\mu}_j$ in the **j** direction

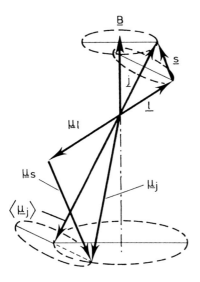

Fig.2.10. Vector model for the Zeeman effect

$$\langle \boldsymbol{\mu}_j \rangle = - \mu_B \frac{3j^2 - l^2 + s^2}{2j^2} \mathbf{j} = - \mu_B g_j \mathbf{j} \tag{2.25}$$

with

$$g_j = 1 + \frac{j(j+1) - \ell(\ell+1) + s(s+1)}{2j(j+1)} \quad \text{(Landé formula)}. \tag{2.26}$$

To obtain these expressions we have utilized the relations

$$\mathbf{l} \cdot \mathbf{j} = (j^2 + l^2 - s^2)/2 \tag{2.27}$$

and

$$\mathbf{s} \cdot \mathbf{j} = (j^2 - l^2 + s^2)/2 \tag{2.28}$$

and have inserted the "quantum-mechanical squares". For the orientational energy E_m in the external field we then obtain

$$E_m = - \langle \boldsymbol{\mu}_j \rangle \cdot \mathbf{B} = \mu_B B g_j m_j , \tag{2.29}$$

where the effect is seen of the space quantization m_j ($m_j = j, j-1,...-j$). The equidistant splitting of the m_j sublevels described by (2.29) is called the *Zeeman effect*. The Zeeman effect is obtained in magnetic fields of sufficiently low strengths that the orientational energy E_m of the atom in the external field is negligibly small compared with the internal orientational energy, the spin-orbit interaction energy. In such cases the coupling between l and s remains strong. According to (2.10 and 29) the precession velocity is proportional to the coupling energy. Another way of expressing the Zeeman effect condition is then to state that the precession movement about the external field is very slow compared with the internal (**j** = **l**+**s**) precession velocity. If we have the opposite condition, i.e. a *strong* field,

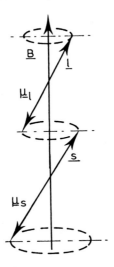

Fig.2.11. Vector model for the Paschen-Back effect

more energy is associated with the individual coupling of μ_ℓ and μ_s to the external field **B** than with the internal coupling. The coupling between l and s therefore breaks up and l and s precess independently about the field with separate space quantizations (Fig.2.11). This is the *Paschen-Back effect*. For this condition we have

$$E_m = \mu_B B(m_\ell + g_s m_s) + \zeta m_\ell m_s , \qquad (2.30)$$

where ζ is the fine-structure interaction constant from (2.13). Of course, all intermediate situations are also possible and the mathematical treatment then becomes more complicated. Still restricting ourselves to cases with only one outer electron, a general expression for the magnetic field dependence of the sublevels of a fine-structure doublet can be given, which describes the Zeeman, Paschen-Back and the intermediate field regions (the *Breit-Rabi formula* for the fine structure)

$$E_m(J, m_j) = - \frac{\Delta E_{so}}{2(2\ell + 1)} + \mu_B B m_j \pm \frac{\Delta E_{so}}{2} \sqrt{1 + \frac{4 m_j x}{2\ell + 1} + x^2} \qquad (2.31)$$

with

$$x = (g_s - 1) \frac{\mu_B B}{\Delta E_{so}} \quad \text{and} \quad \Delta E_{so} = (\ell + 1/2)\zeta .$$

The plus sign corresponds to the sublevels originating in the higher J level. In Fig.2.12 the magnetic field behaviour of a ^2P state with a fine-structure splitting ΔE of 110 cm^{-1} is shown. In the Zeeman region ($E_m \ll 110 \text{cm}^{-1}$) the energies are linear functions of the field B which is also the case in the Paschen-Back region ($E_m \gg 110 \text{cm}^{-1}$) which is not fully reached in the figure. For the quantum numbers m_j in the Zeeman region and m_ℓ and m_s in the Paschen-Back region we have $m_j = m_\ell + m_s$ (as $j_z = \ell_z + s_z$ is a

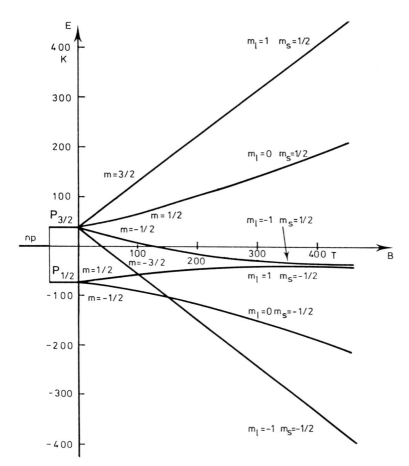

Fig.2.12. Energy level diagram for a ^2P state with fine-structure splitting 110 cm^{-1}

constant of motion). This relation in connection with the *non-crossing rule* (levels with the same $m_j = m_\ell + m_s$ cannot cross) prescribes which Zeeman level should be connected to which Paschen-Back level. (The non-crossing rule can be understood considering the signs of the second-order perturbation terms). Eq.(2.31) describes how sublevels with the same m_j value mix resulting in nonlinear field dependences. As the m = 3/2 and −3/2 levels have no mixing partner they remain linear in the magnetic field. (The expression under the square root becomes $(1\pm x)^2$).

For many-electron systems a similar description can be given for the magnetic field influence. For the Zeeman effect we have

$$E_m = g_J \mu_B m_J B \ . \tag{2.32}$$

The Landé formula for g_J, (2.26), in the many-electron case is only valid for pure LS coupling. For jj coupling of two electrons the following expression for the g-factor can be derived

$$g = \frac{J(J+1) + j_1(j_1+1) - j_2(j_2+1)}{2J(J+1)} g_1$$
$$+ \frac{J(J+1) + j_2(j_2+1) - j_1(j_1+1)}{2J(J+1)} g_2 \, . \qquad (2.33)$$

Here g_1 and g_2 are g-factors for electrons 1 and 2 according to the Landé formula (2.26). By measuring the g-factors for all states of a given configuration it is possible to determine the degree of intermediate coupling, i.e. to determine the coefficients c_i in the expansion $\psi' = \Sigma c_i \psi_i(LS,(jj))$ of a real state ψ' in pure LS or jj states.

2.5.2 Electric Fields

In a homogeneous electric field $\mathcal{E}$ an atom becomes *polarized*, i.e. the centre of the electronic cloud no longer coincides with the nucleus. The field induces an electric dipole moment **d**, which is proportional to the field

$$\mathbf{d} = \alpha \mathcal{E} \, . \qquad (2.34)$$

The orientation energy of the induced dipole is given by $E_e = -\mathbf{d} \cdot \mathcal{E}$. This so-called *Stark effect* thus generally increases as the square of the applied electric field strength. For hydrogen and hydrogen-like systems special conditions prevail due to the level degeneracy and hence a *linear* Stark effect is obtained. For the normal quadratic case it can be shown, that a common displacement of all sublevels (the *scalar* effect) is obtained as well as a differential displacement, depending on the m_J^2 value (the *tensor* effect). For a fine-structure level with quantum number J we have

$$E_s = -\frac{1}{2} \left[\alpha_0 + \alpha_2 \frac{3m_J^2 - J(J+1)}{J(2J-1)} \right] \mathcal{E}^2 \, , \qquad (2.35)$$

α_0 and α_2 being the *scalar* and *tensor polarizability constants*, respectively, can be determined experimentally and calculated theoretically. For J = 0 or 1/2, for which the formula breaks down, there is only a scalar effect. The Stark effect can be seen as an admixture of other states into the state under study. Perturbing states are those for which there are allowed electric dipole transitions (Sect.4.2) to the state under study. Energetically close-lying states have the greatest influence. A theoretical calculation of the constants α_0 and α_2 involves an evaluation of the matrix elements of the electric dipole operator (Chap.4). Investigations of the Stark effect are therefore, from a theoretical point of view, closely related to studies of transition probabilities and lifetimes of excited states. (Sects. 4.1 and 9.4.5). In Fig.2.13 an example of the Stark effect is given; different aspects of this phenomenon have been treated in [2.31].

Highly excited atoms with large principal quantum numbers n (*Rydberg atoms*) are very sensitive to electric fields since they have large α_0 and α_2 values. Clearly, the loosely bound outer electron is strongly influenced by an electric field. If the field strength is sufficiently high the electron

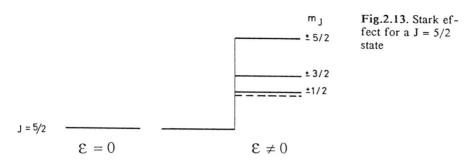

Fig.2.13. Stark effect for a $J = 5/2$ state

can be swept away (*field ionization*). For Rydberg atoms the critical field for field ionization is proportional to n^{-4}. Whereas ground-state atoms can be field-ionized only at field strengths of the order of some MV/cm, an atom in an $n = 30$ state only requires a few V/cm. The phenomenon of field ionization is illustrated by a potential diagram in Fig.2.14, showing the possibility of electron tunnelling through the potential barrier and the conversion from bound to unbound states. For a more detailed discussion of field ionization the reader is referred to [2.32,33]. It should be noted, that (2.35) is valid only for low electric field strengths far from the region of field ionization. Stronger fields must be handled with a different mathematical approach [2.34].

2.6 Hyperfine Structure

Hyperfine structure (hfs) in optical spectra was discovered independently by A. Michelson, and Ch. Fabry and A. Pérot at the end of the 19th century. The effect is explained by the presence of nuclear magnetic and electric moments, interacting with the electronic shell.

2.6.1 Magnetic Hyperfine Structure

Associated with its spin angular momentum **I**, a nucleus has a magnetic dipole moment μ_I

$$\mu_I = g_I \mu_N \mathbf{I} = g_I' \mu_B \mathbf{I} \, . \tag{2.36}$$

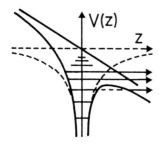

Fig.2.14. Atomic field ionization

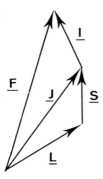

Fig.2.15. Coupling of angular momentum vectors

In this expression, which is analogous to (2.9,23,24), the *nuclear magneton* $\mu_N = \mu_B/1836$ and g_I (g_I') is a constant of proportionality. At the site of the nucleus the electron shell exhibits an effective magnetic field $\mathbf{B_J}$ directed along the atomic "axis of rotation" $\mathbf{J}$, i.e.,

$$\mathbf{B_J} = k\mathbf{J} \ . \tag{2.37}$$

There will be a magnetic coupling (orientational energy) between the nucleus and the electron shell and the interaction is described by

$$H_{mhfs} = a\mathbf{I}\cdot\mathbf{J} \tag{2.38}$$

with

$$a = -kg_I\mu_N \ . \tag{2.39}$$

Note that the *magnetic dipole interaction constant* a is a product of a nuclear quantity g_I which is proportional to the moment, and an electronic quantity k which is proportional to the internal magnetic field strength. $\mathbf{I}$ and $\mathbf{J}$ precess about their resultant $\mathbf{F}$, as shown in Fig.2.15. We have

$$\mathbf{F} = \mathbf{I} + \mathbf{J}, \quad F = I+J, I+J-1, \ldots |I-J| \ . \tag{2.40}$$

The energy contribution is calculated from (2.38) in the same way as the spin-orbit interaction (2.15)

$$E_{mhfs} = \frac{a}{2}[F(F+1) - I(I+1) - J(J+1)] \ . \tag{2.41}$$

As for the fine stucture, the Landé interval rule is also valid for the magnetic hyperfine structure

$$\Delta E_{mhfs(F,F-1)} = aF \ . \tag{2.42}$$

In Fig.2.16 hyperfine structures for two cases are displayed, one for the case of a positive a value ("*normal*" structure) and one for the case of a

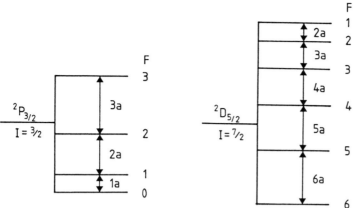

Fig.2.16. Examples of hyperfine structure

negative a value (*inverted* structure). Clearly, both signs can occur since the sign of the g_I factor varies for different nuclei. (The sign of k can also vary).

The magnetic dipole interaction constant a can frequently be determined accurately with precise methods which will be discussed in Chaps. 7 and 9. As we have mentioned, it represents a quantity in the field between atomic and nuclear physics. If the nuclear moment is known, the measurement yields an experimental value of the magnetic field at the nucleus which can be compared with the results of atomic calculations. If, on the other hand, the field can be calculated reliably, information on unknown nuclear moments can be obtained. The hyperfine structure will be particularly large if the atom contains an unpaired s electron, giving rise to the *Fermi contact interaction*, caused by the large probability of the s electron being found inside the nucleus.

The field strength at the nucleus is of the same order of magnitude as that exerted on the electron in its spin-orbit interaction. Since typical magnetic moments for nuclei are about 1000 times smaller than corresponding moments for the electron shell ($\mu_B/\mu_N = 1836$) the magnetic hyperfine structure is correspondingly smaller compared with the fine structure. As mentioned in Chap. 1, typical splittings are of the order of 10^{-6} eV ($\sim 10^{-2}$ cm^{-1}).

2.6.2 Electric Hyperfine Structure

In the same way as a magnetic dipole acquires an orientational energy in a magnetic field, a non-spherically symmetric charge distribution will acquire such an energy in an electric field gradient (Fig.2.17).

Atomic nuclei can be stretched like cigars (prolate shape) or compressed like discs (oblate shape). The deformation is described by the *electric quadrupole moment* Q (prolate: Q>0; oblate: Q<0). The principal interaction is, of course, the normal electrostatic (Coulomb) force on the charged nucleus (*monopole* interaction). The differential interaction, which depends

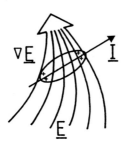

Fig.2.17. Orientation energy of an electric quadrupole in an electric field gradient

on the structure of the nucleus and on the variation of the field across its finite extension, is of course very much smaller (*quadrupole* interaction). It gives rise to an electric hyperfine structure. The energy contribution depends on the direction of the nuclear spin in relation to the electric field gradient. For the electric hyperfine interaction one obtains

$$E_{ehfs} = b \frac{(3C/4)(C+1) - I(I+1)J(J+1)}{2I(2I-1)J(2J-1)} \qquad (2.43)$$

with

$$C = F(F+1) - I(I+1) - J(J+1) \quad \text{and} \quad b = \frac{e^2}{4\pi\epsilon_0} q_J Q \ . \qquad (2.44)$$

In analogy with the magnetic dipole interaction constant a the *electric quadrupole interaction constant* b is a product of a nuclear quantity Q, the electric quadrupole moment, and an electronic quantity q_J, which is proportional to the electric field gradient. Thus, with the b factor experimentally determined, information on the nucleus *or* the electronic shell can be obtained. The electric hyperfine structure is of the same order of magnitude as the magnetic one, but generally somewhat smaller. It exhibits itself as a deviation from the Landé interval rule. In Fig.2.18 two examples of the combined action of magnetic and electric hyperfine structure are shown.

If the nucleus has no spin, i.e. I = 0, there is neither a magnetic nor an electric hyperfine structure. For I = 1/2 only a magnetic interaction is possible whereas the occurrence of electrical hyperfine structure requires I ≥ 1

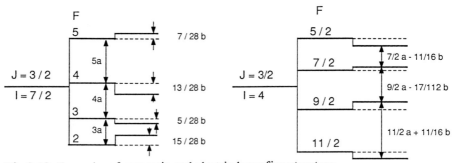

Fig.2.18. Examples of magnetic and electric hyperfine structure

and $J \geq 1$. Hyperfine structure and the determination of nuclear moments have been discussed in [2.35]. Extensive data on nuclear moments have been listed in [2.36,37]; hfs data for the extensively studied alkali atoms have been compiled in [2.38] and the theoretical aspects of atomic hyperfine interactions have been covered in [2.39-42].

2.7 The Influence of External Fields (hfs)

In a weak external magnetic field **B** there will be a splitting of magnetic sublevels for the hyperfine structure (hfs) case as for the previously discussed case, where the influence of the nucleus was omitted. However, also in the hyperfine structure case the electronic shell is responsible for the interaction. Because of the coupling between **I** and **J**, these vectors precess about **F**. Thus, the direction in relation to **B** of the electronic magnetic moment, which is associated with **J**, is influenced. The situation is illustrated in Fig.2.19.

As the nuclear magnetic moment is negligibly small in comparison with that of the electronic shell, the direct interaction of the nuclear moment with the field is normally negligible. The coupling in the inner, extremely strong field (the magnetic hyperfine coupling) is, however, much larger and, as a matter of fact, the *Zeeman effect for the hyperfine structure* demands that the external interaction be negligible compared with the internal one. Using the vector model, we obtain, as in the fine structure case

$$E_m = \mu_B B g_J \frac{F(F+1) - I(I+1) + J(J+1)}{2F(F+1)} M_F = \mu_B B g_F M_F \ . \tag{2.45}$$

In the presence of hyperfine structure, g_F now takes the place of g_J in describing the "gear ratio" between the magnetic energy contribution and

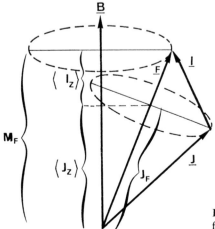

Fig.2.19. Vector diagram for the Zeeman effect of the hyperfine structure

field strength. Although again there is an *electronic* magnetic moment, the influence is modified by the nuclear coupling resulting in a different angle of $\langle \mu_J \rangle$ with respect to the magnetic field. If J and g_J are known a measurement of g_F gives information on the nuclear spin I.

In strong magnetic fields, the *Paschen-Back effect for the hyperfine structure* will result. As for the fine structure case the mathematical expression is simple, i.e.,

$$E_m = \mu_B B(g_J M_J - g_I' M_I) + a M_I M_J . \qquad (2.46)$$

In *intermediate fields* the calculation is more complicated. For I or J = 1/2, the *Breit-Rabi formula* can be expressed in analogy with the fine structure case, see (2.31). For J = 1/2 we have

$$E(F, M_F) = -\frac{\Delta E}{2(2I+1)} - g_I' \mu_B B M_F \pm \frac{\Delta E}{2} \sqrt{1 + \frac{4 M_F x}{2I+1} + x^2}$$

with $\quad x = (g_J + g_I') \dfrac{\mu_B B}{\Delta E} \quad$ and $\quad \Delta E = (I + \tfrac{1}{2})a$. $\qquad (2.47)$

Here the plus sign refers to the higher F value. In Fig.2.20 the dependence of the sublevels on the magnetic field, for a J = 1/2, I = 3/2 state is shown.

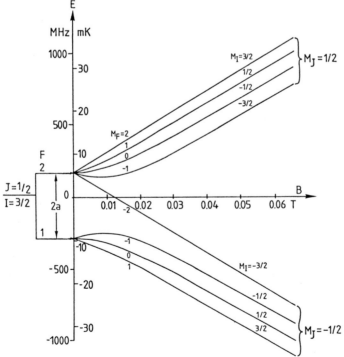

Fig.2.20. Breit-Rabi diagram for an alkali state with J = 1/2 and I = 3/2. (Ground state of ^{39}K)

The Zeeman and Paschen-Back regions can be identified and the relevant quantum numbers M_F, M_I, and M_J, are indicated ($M_F = M_I + M_J$). The $M_F = \pm 2$ levels have no mixing partner and proceed linearly into the Paschen-Back region.

For the general case a computer calculation, in which the quantum-mechanical energy matrix is *diagonalized*, is required. Atoms with hfs in external magnetic fields have been discussed in [2.35]. Hyperfine structures are normally given in MHz. Using this unit for energy and expressing B in Gauss (10^{-4} Tesla) μ_B/h has the numerical value of 1.400 (MHz/Gauss).

In the presence of an external electric field a Stark effect for the hyperfine structure occurs. The theory which applies to this situation has been given in [2.43, 44].

2.8 Isotopic Shifts

Isotopes with $I = 0$ have no hyperfine structure, but in transitions between energy levels in a mixture of $I = 0$ isotopes of the same element, a line structure may still be obtained. This effect is called the *isotopic shift*. It has two origins and a distinction is made between the *mass effect* and the *volume effect*. The mass effect can be divided up into the *normal* and the *specific* mass effects. The normal mass effect is due to the movement of the nucleus, which is due to the fact that it is not infinitely heavy. For hydrogenic systems it is possible to take this into account by using the *reduced mass* μ instead of m

$$\mu = \frac{mM}{m + M} \tag{2.48}$$

(m: electron mass, M: nuclear mass). For a transition in hydrogenic systems one finds that the mass shift between two isotopes of mass M and M+1 decreases with M according to

$$\Delta\nu \propto M^{-2} . \tag{2.49}$$

The specific mass effect is due to the interactions (correlations) between the different outer electrons. The mass effect is very prominent for hydrogen/deuterium but is quickly reduced for heavier elements. For such elements the volume effect becomes important. It is particularly prominent when the electron configuration contains unpaired s electrons. The nucleus has a charge density ρ_n over a finite volume. The s electron with a charge probability distribution $\rho_e(r)$ can penetrate the nucleus and is then no longer under the influence of the pure Coulomb field. A nucleus of mass M has a smaller radius r than one with the mass M+ΔM, and thus the potential begins to deviate from a Coulombic one at smaller values of r. The situation is illustrated in Fig.2.21. There will be an energy shift described by

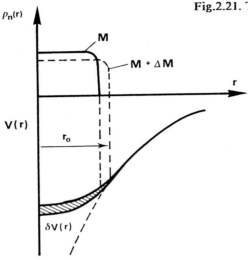

Fig.2.21. The origin of the volume isotopic shift

$$\Delta E = \int_0^{r_0} \rho_e(r) 4\pi r^2 \delta V(r) dr \ . \tag{2.50}$$

Studies of the volume effect yield information on the charge distribution in the nucleus. Hyperfine structure and isotopic shifts are of the same order of magnitude. Isotopic shifts can be studied in the visible region as well as in the X-ray region. Particularly prominent isotopic shifts are obtained for muonic atoms, in which, for example, a μ meson ($m_\mu = 209 m_e$) has taken the place of an electron. The classical radius of the orbit is reduced by a factor of 209, and thus the nuclear influences are much greater than those pertaining to the electrons. Isotope shifts and their interpretation have been discussed in [2.45-47]. Different aspects of current atomic physics research are covered in the proceedings of a series of international atomic physics conferences [2.48-57]. Further reviews may be found in [2.58, 59].

3. Molecular Structure

A molecule is formed by the binding of two or more atoms in such a way that the total energy is lower than the sum of the energies of the constituents. The bonds are normally of *ionic* or *covalent* nature. Particularly weak bonds occur in van der Waals molecules. The energy-level diagrams of molecules are significantly more complicated than those of atoms since, apart from energy levels corresponding to different electronic arrangements, there are also different states corresponding to vibrational and rotational motion. The structure is schematically shown in Fig.3.1. This chapter will mainly be concerned with *diatomic* molecules.

For a more detailed description of molecular structure the reader is referred to [3.1-17].

3.1 Electronic Levels

In the mathematical treatment of atoms it has been found that the interactions, especially electron-electron interactions, are rather complicated. Theoretical studies of molecules are also made more complicated than those of atoms because we do not have any given centre but many centres, one for

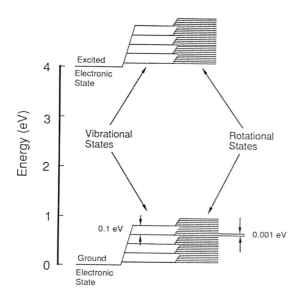

Fig.3.1. Schematic molecular energy level diagram with electronic, vibrational and rotational levels

each atom in the molecule. As for atoms, the theoretical treatment of molecules starts with the Schrödinger equation and, as in the case of atoms, it is useful to consider the independent-particle model. In this approximation one assumes that each electron moves independently of the others in the field generated by the nuclei and the other electrons. For an atom the central field approximation is then a natural further step. However, as there are many centres in a molecule such a model does not apply here. In order to facilitate the calculations one generally tries to utilize other symmetry properties of the molecule. In the treatment of diatomic molecules we have *cylindrical symmetry* and it is then possible to separate the Schrödinger equation into equations corresponding to different absolute values of the angular momentum projection onto the symmetry axis. Electronic orbitals denoted by σ, π, δ, ... are obtained corresponding to the m_ℓ values 0, 1, 2, ... (m_ℓ is the quantum number for the orbital angular momentum projection). In molecules with two atoms of the same kind (homo-nuclear molecules) there is a centre of symmetry. For such molecules the states are further classified by indicating the parity of the wavefunction: even, g (German "gerade") or odd, u (German "ungerade"), e.g., σ_g, σ_u, π_g, π_u, ...

Electrons are much lighter than nuclei and move much faster. Thus the electrons can adjust to the movement of the nuclei which means that the electronic states are at any moment essentially the same as if the nuclei were fixed. This is the basis of the *Born-Oppenheimer approximation*, which assumes fixed nuclei. The wave function can be expressed as a product of an electronic wavefunction with the nuclei assumed fixed and a nuclear wavefunction describing the relative nuclear motion. Energy eigenvalues for the electronic Schrödinger equation, solved for different nuclear separations, form a potential, that is inserted into the nuclear Schrödinger equation together with the nuclear repulsion term.

In order to further describe the molecular wavefunctions or the molecular orbitals, Linear Combinations of Atomic Orbitals (LCAO) are normally used (*LCAO method*). Such a method of solution is possible since the directional dependence of the spherical-harmonic functions for the atomic orbitals can be used. The Pauli principle can be applied to the single-electron molecular orbitals and by filling the states with the available electrons the molecular electron configurations are attained. Coupling of the angular momenta of the open shell then gives rise to molecular terms.

The total angular momentum of the electrons in a molecule is not a constant of motion since the electrons do not move in a central field. The coupling of angular momenta for the electrons will therefore be different from the atomic case.

Let us consider a diatomic molecule which has an axial symmetry, with regard to the axis through the two nuclei (i.e., cylindrical symmetry). Only the component of the electron orbital angular momentum L_z along the symmetry axis will be a constant of motion. The total angular momentum L will precess about the symmetry axis, as shown in Fig. 3.2.

The projected component is characterized by the quantum number M_L

$$M_L = L, L-1, \ldots -L . \qquad (3.1)$$

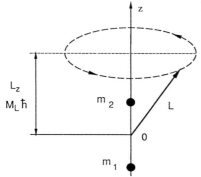

Fig.3.2. Precession of L in a diatomic molecule

However, since the internuclear field is of electric nature rather than magnetic, the energy is not changed for the exchange $M_L \rightarrow -M_L$. (Compare the Stark effect with the Zeeman effect). Since the field is very strong the energy separation between states with different M_L values is quite substantial. The absolute value for M_L is designated Λ

$$\Lambda = |M_L|, \quad \Lambda = 0, 1, 2, .., L. \tag{3.2}$$

The states are given the following symbols

$$\begin{array}{ccccc} \Lambda = & 0 & 1 & 2 & 3 \\ & \Sigma & \Pi & \Delta & \Phi. \end{array}$$

The states are doubly degenerate apart from Σ states because of the $M_L \leftrightarrow -M_L$ symmetry.

The resulting spin quantum number S of the electrons is also needed to characterize the molecular states. In diatomic molecules with $\Lambda>0$ S precesses about the internuclear axis and can have 2S+1 well-defined projections. The quantum number for S_z is called Σ. As for the atomic case the *multiplicity*, 2S+1 is placed as an index of the Λ symbol, e.g. $^3\Pi$, $^1\Delta$.

The total electronic angular momentum along the internuclear axis is designated Ω and is obtained from Λ and Σ

$$\Omega = \Lambda+\Sigma, \Lambda+\Sigma-1, ..., |\Lambda-\Sigma|. \tag{3.3}$$

For $\Lambda > 0$ the (2S+1) values of Ω with different energies are obtained. In Fig.3.3 the possible states for $\Lambda = 2$ and S = 1 are shown.

The strength and type of bonding between two atoms depend on the tendency of the participating atoms to donate, attract and share electrons. The variation of the electronic energy with the bond length r between the nuclei is schematically shown in Fig.3.4 for the OH (hetero-nuclear) and O_2 (homo-nuclear) molecules. The curve with the lowest energy corresponds to the ground state while the other curves represent different electronic states, which may also be unbound. The ground state is designated X, and the excited states are conventionally called A, B, C etc.

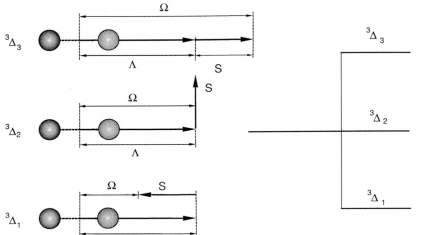

Fig.3.3. Term splitting for molecules

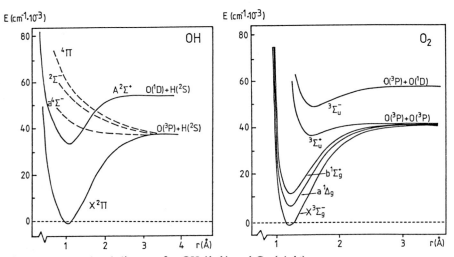

Fig.3.4. Energy level diagram for OH (*left*) and O_2 (*right*)

3.2 Rotational Energy

Consider a rotating diatomic molecule, as shown in Fig.3.5, with the atomic masses m_1 and m_2 at distances r_1 and r_2 from the centre of gravity. The moment of inertia with respect to the rotational axis is I. We have

$r = r_1 + r_2$,

$m_1 r_1 = m_2 r_2$,

$I = m_1 r_1^2 + m_2 r_2^2$.

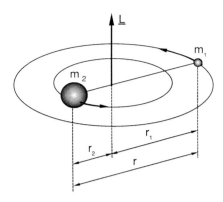

Fig.3.5. Rotation of a diatomic molecule

We then obtain

$$I = \frac{m_1 m_2}{m_1 + m_2}(r_1 + r_2)^2 = \mu r^2 , \quad (3.4)$$

where μ is the reduced mass. Classically, the angular momentum **L** and the energy E are given by

$$\left. \begin{array}{l} \mathbf{L} = I\omega/\hbar \\ \\ E = I\omega^2/2 \end{array} \right\} \Rightarrow E = \frac{\mathbf{L}^2 \hbar^2}{2I}$$

where ω is the angular frequency vector. Quantum mechanically **L** is given by

$$|\mathbf{L}| = \sqrt{J(J + 1)} , \quad J = 0, 1, 2, ... \quad (3.5)$$

and thus the quantized energy of the rotator is given by

$$E_J = J(J + 1)\hbar^2/2I = BJ(J + 1) . \quad (3.6)$$

This energy expression leads to energy levels such as those in Fig.3.6.

If the rotator is not completely rigid it is slightly extended in higher rotational states. Then I increases and E_J will decrease. This results in a successive downward movement of the upper energy levels. The elastic rotator is described by

$$E_J = BJ(J + 1) - D^+ J^2(J + 1)^2 . \quad (3.7)$$

Here D^+ is a positive constant and $D^+/B \ll 1$. In Fig.3.6 the modified rotational levels are also included. The energy separation between rotational levels is of the order of 10^{-3} eV.

In molecules with an open electron shell there is a coupling between the angular momentum of the electrons and the molecular rotation. The

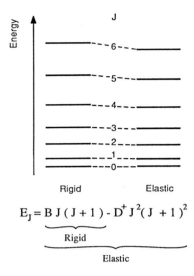

Fig.3.6. Rotational levels for a rigid rotator (*left*) and an elastic rotator (*right*)

situation is described by the *Hund coupling cases* a, b, c or d. (This complication, which sometimes leads to the absence of the first rotational levels, will not be treated in this book). The coupling between **L** and **J** can lead to a splitting of each J level into two states (Λ doubling) breaking the M_L degeneracy.

3.3 Vibrational Energy

The potential energy of a diatomic molecule depends on the internuclear distance r. In Fig.3.7 a typical potential curve is shown. The so-called *Morse potential* is often used

$$V(r) = D(1 - e^{-\alpha(r-r_0)})^2 \ . \tag{3.8}$$

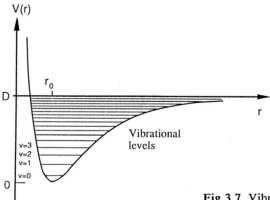

Fig.3.7. Vibrational potential and energy levels

Here r_0 corresponds to the bottom of the potential curve, and D is the dissociation energy. One atom oscillates with regard to the other in this potential. A Taylor expansion of V(r) close to $r = r_0$

$$V(r) = D\alpha^2(r-r_0)^2 + ... \qquad (3.9)$$

shows that the bottom of the potential curve may be approximated by a parabola and the molecule is then called a *harmonic oscillator*. The classical vibrational frequency ν_c is given by

$$\nu_c = \frac{1}{2\pi}\sqrt{\frac{k}{\mu}} \qquad (3.10)$$

where μ is the reduced mass and k is the force constant (defined by the force expression $F = -k(r-r_0)$. Thus $k = 2D\alpha^2$ according to (3.9)). Quantum mechanically

$$E_v = (v + 1/2)h\nu_c, \quad v = 0, 1, 2, ... \qquad (3.11)$$

leading to *equidistant* vibrational levels and the presence of a *zero-point energy* $h\nu_c/2$. For higher-lying vibrational levels the harmonic oscillator model is obviously not valid, since the higher terms of V(r) become important. For a non-harmonic oscillator the energy eigenvalues are given approximately by

$$E_v = (v + 1/2)h\nu_c - (v + 1/2)^2 x_e h\nu_c . \qquad (3.12)$$

Here x_e is a small positive constant. Increasingly closer-lying energy levels are obtained, developing into a continuum at $E = D$. The separation between low-lying vibrational levels is typically 0.1 eV.

3.4 Polyatomic Molecules

While the energy-level structure of diatomic molecules can be divided up reasonably easily, the degree of complexity is greatly increased for polyatomic molecules. Such molecules have several nuclear distances, several force constants, several dissociation energies etc., which must normally be determined simultaneously. While for diatomic molecules it is possible to start from empirical regularities in the spectra and arrive at a theoretical interpretation such a procedure is difficult for polyatomic molecules. Instead it is more advantageous to first develop the theory and then use the theory to interpret the observed spectra. An important point to consider is the *shape* of the molecule, i.e. the internal arrangement of the atoms, as certain qualitative features are associated with a certain shape. The study of *symmetry properties* is thus very important in the understanding of complicated molecules. The symmetry is described in group theory in terms

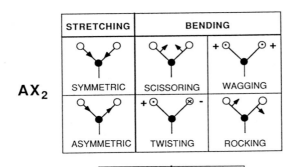

Fig.3.8. Stretching and bending modes for the molecular groups

of *point groups*. E.g., carbon tetrachloride CCl_4, which has a tetrahedral structure, belongs to a point group called T_d.

The vibrational motion of a polyatomic molecule is described by *fundamental frequencies*, (*modes*) corresponding to different types of vibration. Every atom has 3 degrees of freedom (the 3 dimensions of space) leading to 3N degrees of freedom for an N-atomic molecule. Of these degrees of freedom, three describe rotation around a centre of mass, which does not correspond to any vibrational motion. A further three degrees of freedom describe a translational motion which does not either give rise to any vibrational movement. Thus, in general an N-atomic molecule has 3N-6 vibrational modes. For a *linear* molecule there is one more mode (3N-5) because only two independent rotational movements exist. The different vibrational modes in the molecule represent stretching, where the distance between the atoms in the molecule is changed, or bending, where the angle between the atoms in the molecule is changed. In Fig.3.8 examples of stretching and bending modes for AX_2 and AX_3 molecular groups are given.

3.5 Other Molecular Structures

Apart from the energy-level structures discussed above, molecules exhibit both the Zeeman and Stark effects. Further, hyperfine structure and isotopic shifts also occur. The occurrence of isotopic shifts is particularly simple to understand considering the substantially altered values of the reduced mass found in the vibrational and rotational energy expressions.

4. Radiation and Scattering Processes

After the brief survey of atomic and molecular energy structures in Chapters 2 and 3 we will now consider radiation and scattering processes by which atoms and molecules change their energy state. The processes are accompanied by the absorption or release of radiation giving rise to spectra. These spectra, can be used to clarify the structure of atoms and molecules and for a wealth of analytical purposes. We will first consider the case of transitions at a frequency corresponding to given energy separations (resonance radiation) and then discuss Rayleigh, Raman and Mie scattering. A detailed presentation of the theory of radiation and scattering processes can be found in [4.1-11]. Several of the books on atomic, molecular and quantum mechanics, earlier cited, also discuss this topic in more detail.

4.1 Resonance Radiation

Consider an atomic or molecular system with energy levels E_k subject to a time-dependent perturbation $\mathcal{H}'$, e.g. the oscillating electric field of incoming monochromatic light (Fig.4.1). In this section we will consider in what state a system, initially in a given state, will be found after having interacted with an electromagnetic field for a time t. We start with the time-dependent Schrödinger equation

$$i\hbar \frac{\partial \Psi(t)}{\partial t} = (\mathcal{H}_0 + \mathcal{H}') \Psi(t) \; . \tag{4.1}$$

The time-independent eigenfunctions of $\mathcal{H}_0$ (describing the non-perturbed system) and the eigenvalues E_k are assumed to be known

$$\mathcal{H}_0 \psi_k^0 = E_k \psi_k^0 \; . \tag{4.2}$$

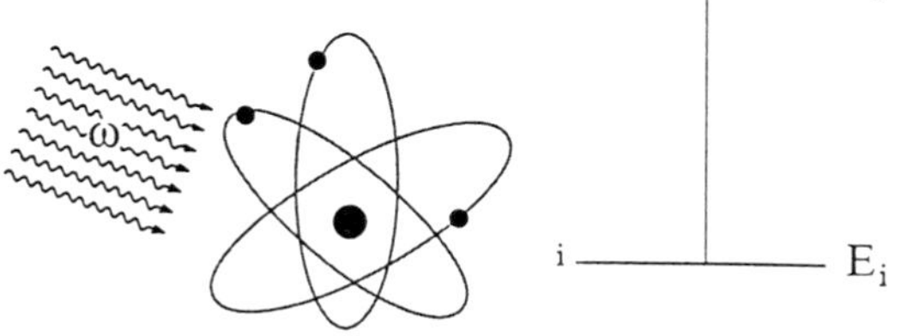

$\hbar \omega_{ni} \approx E_n - E_i$

Fig.4.1. Interaction between electromagnetic radiation and an atom

The eigenfunctions are supposed to be normalized and those with the same eigenvalues have been orthogonalized. These eigenfunctions have the time dependence

$$\Psi_k^0(t) = \psi_k^0 \exp(-i\omega_k t) \tag{4.3}$$

with $\omega_k = E_k/\hbar$. Since the functions constitute a complete system, an arbitrary time-dependent function can be expanded according to

$$\Psi(t) = \sum_k c_k(t) \Psi_k^0(t) = \sum_k c_k(t) \psi_k^0 \exp(-i\omega_k t) . \tag{4.4}$$

Inserting this expression into the Schrödinger equation (4.1) gives

$$i\hbar \sum_k \frac{dc_k}{dt} \exp(-i\omega_k t) \psi_k^0 = \mathcal{H}' \sum_k c_k \exp(-i\omega_k t) \psi_k^0 .$$

Multiplication by ψ_n^{0*} and integration yield

$$\frac{dc_n}{dt} = -\frac{i}{\hbar} \sum_k c_k \exp(i\omega_{nk} t) \langle \psi_n^0 | \mathcal{H}' | \psi_k^0 \rangle \tag{4.5}$$

with

$$\omega_{nk} = \omega_n - \omega_k = \frac{1}{\hbar}(E_n - E_k) .$$

The set of equations (4.5), written for the various values of n, constitutes a set of coupled linear differential equations. The coupling between these equations arises solely from the existence of the perturbation $\mathcal{H}'$, which has non-zero off-diagonal matrix elements. Assume that the system is in an eigenstate of $\mathcal{H}_0$, say E_i, at t=0. Thus, with the probability interpretation of the wavefunction we have

$$c_i(0) = 1 ,$$
$$c_k(0) = 0 , \quad k \neq i . \tag{4.6}$$

The first-order result is now obtained by integration of (4.5), using the condition (4.6) in the integral also for t > 0, i.e.,

$$c_n(t) = -\frac{i}{\hbar} \int_0^t \exp(i\omega_{ni} t') \langle n | \mathcal{H}' | i \rangle dt', \quad n \neq i , \tag{4.7}$$

where $|i\rangle$ abbreviates $|\psi_i^0\rangle$ and correspondingly for $\langle n|$. The transition probability from state i to state n is then $|c_n(t)|^2$. In order to calculate this quantity we must consider the time-dependent perturbation $\mathcal{H}'$ more explicitly.

We start by considering a one-electron system influenced by an electromagnetic field. In a source-free region this field can be expressed by the magnetic vector potential **A**, fulfilling the condition $\nabla \cdot \mathbf{A} = 0$ according to the Lorentz condition $\nabla \cdot \mathbf{A} + c^{-1} \partial \phi / \partial t = 0$. The Hamiltonian of the system is

$$\mathcal{H} = \frac{1}{2m}(\mathbf{p} + e\mathbf{A})^2 + V(\mathbf{r}) ,$$

$$\mathcal{H} = \frac{1}{2m}(-i\hbar\nabla + e\mathbf{A})^2 + V(\mathbf{r}) ,\qquad(4.8)$$

where $V(\mathbf{r})$ is the static potential. But $\nabla \cdot (\mathbf{A}f) = (\nabla \cdot \mathbf{A})f + \mathbf{A} \cdot (\nabla f) = \mathbf{A} \cdot (\nabla f)$ since $\nabla \cdot \mathbf{A} = 0$ according to the Lorentz condition, i.e., **A** and ∇ commute. Thus

$$\mathcal{H} = \underbrace{-\frac{\hbar^2}{2m}\nabla^2 + V(\mathbf{r})}_{\mathcal{H}_0} \underbrace{- i\frac{\hbar e}{m}\mathbf{A}\cdot\nabla + \frac{e^2}{2m}\mathbf{A}^2}_{\mathcal{H}'} .\qquad(4.9)$$

For weak fields the last quadratic term can be neglected (it is important for two-photon transitions, see Sect.9.1.3). If the radiation is of frequency ω, $\mathbf{A} = \mathbf{A}_0 \cos\omega t$ and the quadratic term is neglected then

$$\mathcal{H}' = -i\frac{\hbar e}{m}\mathbf{A}\cdot\nabla = C\cos\omega t \qquad(4.10)$$

with

$$C = -i\frac{\hbar e}{m}\mathbf{A}_0\cdot\nabla . \qquad(4.11)$$

Eq.(4.7) now becomes

$$c_n(t) = -\frac{1}{2\hbar}\langle n|C|i\rangle \left[\frac{\exp[i(\omega_{ni}-\omega)t] - 1}{\omega_{ni}-\omega} + \frac{\exp[i(\omega_{ni}+\omega)t] - 1}{\omega_{ni}+\omega}\right]. \qquad(4.12)$$

The case when $\Delta\omega = \omega_{ni} - \omega$ is close to zero, is of special interest. This is when the irradiation frequency ω is close to the energy difference between states i and n. Then the second term in (4.12) can be neglected and, using the Eulerian formulae, the first term becomes

$$c_n(t) = -\frac{i}{2\hbar}\langle n|C|i\rangle e^{i\Delta\omega t/2}\frac{\sin(\Delta\omega t/2)}{\Delta\omega t/2} t . \qquad(4.13)$$

The time-dependent probability of an i→n transition is thus

$$|c_n(t)|^2 \propto \left[\frac{\sin(\Delta\omega t/2)}{\Delta\omega t/2}\right]^2 t^2 . \qquad(4.14)$$

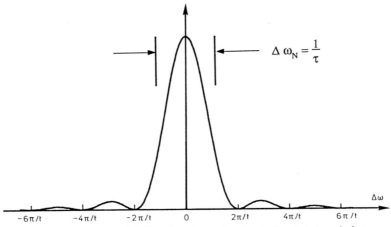
Fig.4.2. Transition probability for absorption and stimulated emission

The transition probability is plotted in Fig.4.2 as a function of $\Delta\omega$ for a given value of t. The function has a prominent maximum for $\Delta\omega = 0$, i.e. when the photon energy of the radiation field exactly matches the energy difference between the final and initial states

$$\hbar\omega = \hbar\omega_{ni} = E_n - E_i \ . \tag{4.15}$$

This condition is called *resonance*. We have assumed that a transition to a state with higher energy occurs, i.e. *absorption* of radiation. However, exactly the same result is obtained if

$$\hbar\omega = \hbar\omega_{in} = E_i - E_n \ , \tag{4.16}$$

where the final state n is below the initial state. In this case the first term in (4.12) is neglected instead of the last. The situation where the perturbation causes the system to emit a photon of the same energy as that of the incoming photons is called *stimulated emission*. From completely equivalent mathematics it follows that the probabilities for absorption and stimulated emission are the same. Furthermore, a stimulated photon is emitted in the same direction and the wave has the same phase as that of the incoming photon (coherence). This can be seen in a semi-classical picture of these radiative processes where the two processes are considered essentially equivalent. The coherence properties of the stimulated photon result in a strengthening of the incoming beam, a process that is the exact counterpart of the attenuation of a well-defined beam by absorption processes under other circumstances. We will consider later which process dominates.

According to (4.14) the maximum transition probability ($\Delta\omega = 0$) is proportional to t^2, where t is the time during which the system is subject to the perturbation. This seems unphysical; one would expect the transition probability to increase linearly with t. However, we must consider that the energy levels are not infinitely sharp, but have a certain width, associated with the natural lifetime (τ) of the state. This point will be considered later in this chapter. If the perturbation is applied for a time t $\gg \tau$ the resonance curve of Fig.4.2 will be much narrower than this level width. The maximum value of the curve then has no significance; instead the area below the curve yields the transition probability. Since the half-width of the curve is proportional to t^{-1} and its maximum value to t^2 (the internal shape being independent of t) the area under the curve is proportional to t, as expected. This is called the *Fermi Golden Rule*.

At this point we should recall that a prerequisite for any process to take place at all is that the matrix element $\langle n|C|i\rangle$ in (4.12) is non-zero. We will now consider this matrix element more closely, namely

$$\langle n|C|i\rangle = -i\frac{\hbar e}{m}\langle n|\mathbf{A}_0\cdot\nabla|i\rangle . \tag{4.17}$$

Transitions involving outer electrons normally occur in the optical or UV region, which means that the wavelength of the radiation ($\lambda > 100$ nm) is much greater than the dimensions of the atom (~ 0.1 nm). Thus, the spatial variation of the amplitude over the atom can be neglected and we can write

$$\langle n|C|i\rangle = -i\frac{\hbar e}{m}\mathbf{A}_0\cdot\langle n|\nabla|i\rangle . \tag{4.18}$$

Using the relations

$$i\frac{\hbar(d\mathbf{r})}{dt} = [\mathbf{r},\mathcal{H}_0]$$

and

$$\mathbf{p} = m\frac{d\mathbf{r}}{dt} = -i\hbar\nabla$$

we obtain

$$\nabla = -\frac{m}{\hbar^2}[\mathcal{H}_0,\mathbf{r}]$$

and thus

$$\langle n|\nabla|i\rangle = -\frac{m}{\hbar^2}\langle n|\mathcal{H}_0\mathbf{r} - \mathbf{r}\mathcal{H}_0|i\rangle = -\frac{m}{\hbar^2}(E_n - E_i)\langle n|\mathbf{r}|i\rangle .$$

Here the Hamiltonian operator $\mathcal{H}_0$, being Hermitian, acts to the left in the first term and to the right in the second term. We then obtain

$$\langle n|C|i\rangle = ie\omega_{ni}\mathbf{A}_0\cdot\langle n|\mathbf{r}|i\rangle . \tag{4.19}$$

Thus we have shown that the transition probability $|c_n(t)|^2$ is proportional to the square of the matrix element between the initial and final state formed with the atomic *electric dipole operator*

$$\mathbf{p} = -e\mathbf{r} . \tag{4.20}$$

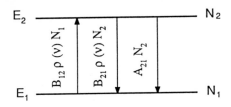

Fig.4.3. Radiative processes connecting energy levels E_1 and E_2

So far we have treated absorption and stimulated emission of radiation. However, it is well known that an atom can emit radiation even when it is not externally perturbed, i.e. *spontaneous emission*. It is not possible to treat this process fully here, since consideration of the quantization of the electromagnetic field as described by *Quantum ElectroDynamics* (QED) is necessary. According to QED a coupling between the atom and the "vacuum state" of the field is responsible for the emission.

A simple relation will be derived between the probabilities for spontaneous and stimulated emission and absorption of radiation using well-known statistical distribution laws. Consider a system such as that illustrated in Fig.4.3 with two energy levels, E_1 and E_2, populated by N_1 and N_2 atoms, respectively. Three radiative processes can occur between the levels, as discussed above. In the figure the processes are expressed using the so-called *Einstein coefficients*, B_{12}, B_{21} and A_{21}, which are defined such that the rate of change in the population numbers is

$$\frac{dN_1}{dt} = -\frac{dN_2}{dt} = -B_{12}\rho(\nu)N_1 + B_{21}\rho(\nu)N_2 + A_{21}N_2 \qquad (4.21)$$

where $\rho(\nu)$ is the energy density of the radiation field per frequency interval, and $\nu = (E_2 - E_1)/h$. At equilibrium we have

$$\frac{dN_1}{dt} = \frac{dN_2}{dt} = 0 \qquad (4.22)$$

yielding

$$\rho(\nu) = \frac{A_{21}}{B_{12}(N_1/N_2) - B_{21}}. \qquad (4.23)$$

We now assume the system to be in thermodynamic equilibrium with the radiation field. The distribution of the atoms is governed by Boltzmann's law

$$\frac{N_1}{N_2} = \exp\left(\frac{h\nu}{kT}\right), \qquad (4.24)$$

where T is the absolute temperature of the system and k is Boltzmann's constant. By identifying the expression for ρ given above with the Planck radiation law

$$\rho(\nu) = \frac{16\pi^2 \hbar \nu^3}{c^3} \frac{1}{\exp(h\nu/kT) - 1} \tag{4.25}$$

we obtain the following relations between the three coefficients

$$B_{12} = B_{21}, \tag{4.26}$$

$$\frac{A_{21}}{B_{21}} = \frac{16\pi^2 \hbar \nu^3}{c^3}. \tag{4.27}$$

The first relation shows that the probabilities for absorption and stimulated emission are the same for a transition between states 1 and 2. This is in accordance with the result obtained above using first-order perturbation theory. Note, that the result (4.26) is independent of the strength of the radiation field. It was in discussions of this kind that A. Einstein, in 1917, found it necessary to introduce the concept of stimulated emission in order to obtain agreement with the statistical laws known at that time [4.12].

From (4.27, 13 and 19) the spontaneous transition probability between the states i and k is found to be

$$A_{ik} = \frac{32\pi^3}{3} \frac{\nu^3}{4\pi\epsilon_0 \hbar c^3} |\langle i|er|k\rangle|^2. \tag{4.28}$$

The most important result here is that this transition probability is determined by the same matrix elements as the induced transitions. Thus there are common *selection rules* for all three types of transitions.

So far we have only treated electric dipole radiation. In a more detailed treatment the radiation field can be described by electric and magnetic *"multipole fields"*, i.e. magnetic dipole radiation, electric quadrupole radiation etc. Magnetic dipole radiation is analogous to electric dipole radiation and it depends on the magnetic dipole moment of the atom

$$\boldsymbol{\mu}_J = -\mu_B \sum_i (l_i + 2s_i) = -\mu_B(\mathbf{L} + 2\mathbf{S}). \tag{4.29}$$

If electric dipole radiation is allowed, i.e. if the matrix element of **p** between the two states is non-zero, this type of radiation strongly dominates over the other types. If, however, an electric dipole transition is not allowed, other types of radiation become important.

The total spontaneous transition probability per unit time for an atom, in a specific state i, can be expressed as

$$A_i = \sum_k A_{ik} \tag{4.30}$$

where the summation is over all levels of the atom having energies less than E_i, and A_{ik} is the spontaneous transition probability for a single process. This means that the number of atoms, N, in a certain state i will decrease exponentially with time

$$N = N_0 \exp(-t/\tau_i) \quad (4.31)$$

with

$$\tau_i = 1/A_i . \quad (4.32)$$

Here we have assumed that the considered level is not re-populated by decay from higher-lying levels.

On average, the time elapsed before an atom in the upper state decays to another state is

$$\bar{t} = \frac{\int_0^\infty t \exp(-t/\tau) dt}{\int_0^\infty \exp(-t/\tau) dt} = \tau \quad (4.33)$$

and therefore τ is called the *mean lifetime*. It is also possible to calculate the *variance* Δt of the lifetime of the atom in the state of interest

$$(\Delta t)^2 = \frac{\int_0^\infty (t-\tau)^2 \exp(-t/\tau) dt}{\int_0^\infty \exp(-t/\tau) dt} = \tau^2 . \quad (4.34)$$

Thus, the "uncertainty" in the lifetime is also equal to τ. The corresponding uncertainty (ΔE) in an energy determination of the level must fulfil the Heisenberg uncertainty relation

$$\Delta E \cdot \Delta t \geq \hbar/2 . \quad (4.35)$$

Here, mathematically the uncertainties are *variances*. The minimum value of this product yields the minimum energy uncertainty which can never be surpassed

$$\Delta E_{min} = h \Delta \nu . \quad (4.36)$$

We find that the so-called *natural radiation width* (in frequency units) is given by

$$\Delta \nu_N = 2\Delta \nu = 1/2\pi\tau . \quad (4.37)$$

It is interesting to compare this "microscopic" treatment of radiative processes with the "macroscopic" treatment of classical physics. In the

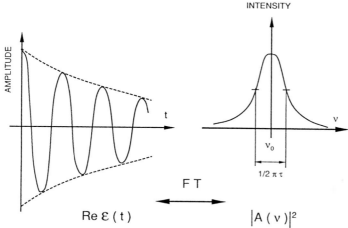

Fig.4.4. Semiclassical picture of a radiating atom

latter, the atom is considered as an exponentially damped harmonic oscillator with an amplitude (Fig.4.4)

$$\mathcal{E} = \mathcal{E}_0 e^{-t/2\tau} \cos(2\pi\nu_0 t) . \tag{4.38}$$

Because the oscillation is damped the frequency cannot be infinitely sharp. A Fourier analysis allows a determination of the frequency distribution giving rise to this special time dependence for the oscillation. In Fourier analysis it is convenient to work with the complex function

$$\mathcal{E}(t) = \mathcal{E}_0 e^{-t/2\tau} e^{i2\pi\nu_0 t} \tag{4.39}$$

the real part of which is the function given in (4.38). The *amplitude* distribution is described by the Fourier transform

$$A(\nu) = \frac{1}{2\pi} \int_0^\infty \mathcal{E}_0 e^{-t/2\tau} e^{i2\pi(\nu_0 - \nu)t} dt$$

$$= \frac{\mathcal{E}_0}{2\pi} \frac{-1}{i2\pi(\nu_0 - \nu) - 1/2\tau} .$$

The intensity distribution is $|A(\nu)|^2$

$$|A(\nu)|^2 = A(\nu) A^*(\nu) = \frac{\mathcal{E}_0^2}{4\pi^2} [4\pi^2 (\nu - \nu_0)^2 + 1/4\tau^2]^{-1}$$

$$= \frac{(\mathcal{E}_0^2 \tau^2)}{\pi^2} \left[1 + \left[\frac{\nu - \nu_0}{1/4\pi\tau} \right]^2 \right]^{-1} . \tag{4.40}$$

This is a Lorentzian curve with a full-width at half maximum of $\Delta \nu = (2\pi\tau)^{-1}$. This classical approach thus yields the same radiative width as quantum mechanics. The broadening of the transition due to the finite lifetime of the excited state is called the *homogeneous* broadening in contrast to *inhomogeneous* broadening, which will be discussed in Chap.6.

At the end of this section we would like to mention two kinds of recently investigated phenomena that can influence the natural radiative lifetime. The first is the *ambient black-body radiation* which, even at room temperature, can have a substantial effect on the lifetime of highly-excited, normally long-lived states. Such states are connected to nearby levels by microwave electric dipole transitions. The substantial black-body radiation in this wavelength region can induce transitions through absorption or stimulated emission effectively reducing the lifetime of the investigated state. The effects can be quite drastic and can only be eliminated by cooling the atomic environment to low temperatures [4.13, 14].

The second phenomenon is of a much more fundamental nature. As we have discussed above, spontaneous emission can be seen to be caused by a coupling to the vacuum state of the electromagnetic field. It has recently been shown [4.15, 16] that it is possible to manipulate this fundamental interaction by enclosing the atom in a cavity. If the cavity is so small that the modes of the vacuum field at the transition frequency cannot be supported by the cavity (dimensions smaller than the transition wavelength) the lifetime is prolonged. If, on the other hand, a cavity is tuned to resonance with a transition the spontaneous emission is enhanced. Clearly, the *Lamb shift*, which is a pure quantum electro-dynamical effect is also influenced by a cavity.

In a complete quantum theory of radiation several interesting phenomena regarding photon statistics are predicted. One such effect is *photon anti-bunching* for a two-level system. When an atom has just emitted a photon it cannot immediately radiate a second photon since it is in the lower state [4.17, 18]. This behaviour is experimentally observed in photon correlation experiments.

A more far-reaching phenomenon is the possibility of generating radiation in "squeezed" states [4.19]. Such radiation exhibits reduced noise below the quantum limit and could have important applications for optical communication and precision interferometric measurements of small displacements, e.g. in gravity-wave detection experiments. A considerable degree of "squeezing" has recently been experimentally demonstrated [4.20, 21]. Various aspects of modern quantum optics have been discussed in [4.22-25].

4.2 Spectra Generated by Dipole Transitions

A spectrum is generated by transitions between different energy states according to certain selection rules. The selection rules for allowed transitions essentially reflect the requirement of conservation of angular momentum for the atom/molecule-photon system. Considerations of symmetry for the

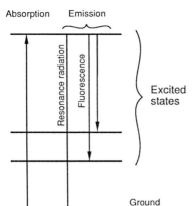

Fig.4.5. Terminology for radiative processes

wavefunctions describing the states involved are also important. For electric dipole transitions the establishing of selection rules is equivalent to a determination of the conditions under which the matrix element $\langle n|\mathbf{r}|i\rangle$ is non-zero. Since $\mathbf{r}$ is an odd operator we can immediately state that the two considered states must be of *opposite parity*.

The selection rules will not be derived here, but stated with some comments. We will then see how spectra are generated by allowed transitions between energy levels described in Chaps. 2 and 3. It is best to treat atoms and molecules separately, but first some general features will be discussed in connection with Fig.4.5.

Spectra of atoms and molecules resulting from absorption or emission can be studied. In *absorption*, a wavelength continuum is used, of which certain wavelengths are absorbed. *Emission* spectra may be generated in a discharge in a light source where the excited levels are populated by, for example, electron collisions. If atoms or molecules are irradiated by light of a wavelength that corresponds to the energy of an allowed transition from the ground state, there will be a *resonance absorption* of photons followed by the release of *resonance radiation* in the decay back to the original level. If the decay occurs to levels above the ground state the emitted light is called *fluorescence light*.

A special type of emission, *phosphorescence*, can be obtained from certain molecules that are excited from the ground state to a higher-lying state

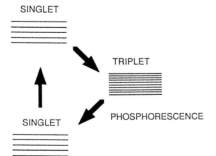

Fig.4.6. The phosphorescence phenomenon

in the normal way. If the molecules make transitions to a lower-lying state of another multiplicity, for example, through collisions, the molecules can accumulate here for a long time since the transition to the ground state is normally radiatively forbidden (Fig. 4.6). In phosphorescing substances the state is not depopulated through radiationless transitions and therefore an extremely long lifetime is obtained (seconds or more), and weak light will be emitted long after the excitation has been terminated.

4.2.1 Atoms

For one-electron systems we have the following selection rules

$$\Delta \ell = \pm 1 ,$$
$$\Delta j = 0, \pm 1 . \tag{4.41}$$

For many-electron systems we have

$$\Delta J = 0, \pm 1 \quad 0 \leftrightarrow 0 \text{ forbidden} . \tag{4.42}$$

In pure LS coupling we also have

$$\Delta L = 0, \pm 1 \quad 0 \leftrightarrow 0 \text{ forbidden} ,$$
$$\Delta S = 0 . \tag{4.43}$$

In Fig.4.7 allowed transitions between a p^2 and an sp configuration are indicated as an example.

Even at small deviations from LS coupling transitions between states of different multiplicity (*intercombination lines*) are observed.

In the presence of hyperfine structure there is a further selection rule

$$\Delta F = 0, \pm 1 \quad 0 \leftrightarrow 0 \text{ forbidden} . \tag{4.44}$$

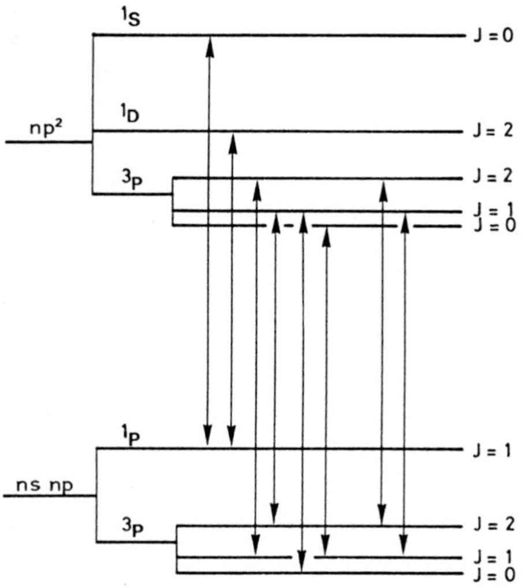

Fig.4.7. Allowed electric dipole transitions between a p^2 and an sp configuration

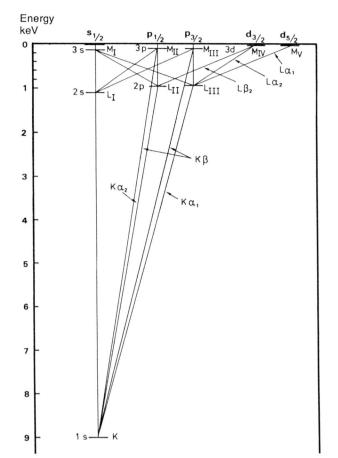

Fig.4.8. X-ray transitions in the copper atom

So far we have only considered transitions of outer electrons giving rise to lines in the IR, visible and UV spectral regions. If a vacancy is created in an inner shell, de-excitation occurs by successive transitions inwards by electrons from outer shells (Chap. 5). In Fig.4.8 level and transition designations for the inner shells of the copper atom are given.

So far we have not discussed the selection rules for the M quantum number. M describes projections of the *resulting* angular momentum of the atom, which can be **J** or **F**. The rule is

$$\Delta M = 0, \pm 1 \;. \tag{4.45}$$

The different values of ΔM correspond to different *angular distributions* of the radiation and to different *polarization conditions*. For $\Delta M = 0$ only the z component of $\mathbf{p} = -e\mathbf{r}$ contributes and the radiation can be compared to that of a classical dipole oscillating along the direction of the field (quantization axis). The radiation then has an intensity which is proportional to $\sin^2\theta$, θ being the angle between the field and the direction of radiation. As for the polarization, the electrical vector oscillates in the plane defined by the field and the direction of the radiation. (The magnetic vector is perpendicular to the electric one and both vectors are perpendicular to the direction of propagation). In a similar way, it is found that the x±iy components of **p** are responsible for $\Delta M = \pm 1$ transitions corresponding to an electric dipole rotating in the xy plane, i.e. perpendicularly to the field direction. The intensity is, in this case, proportional to $(1+\cos^2\theta)$, and the radiation is generally elliptically polarized. In the z-direction it is circularly polarized and in the xy plane linearly polarized. The radiation that corresponds to $\Delta M = 0$ is called π radiation and for $\Delta M = \pm 1$ the term σ radiation is used. The intensity and polarization distributions are illustrated in Fig.4.9.

Above we have discussed the distribution of the emitted radiation. Clearly, the process is reversible and the given angular distributions then give the relative probability of absorption of an incoming photon by an atom in the lower energy state.

Radiative transitions in atoms and selection rules have been discussed in more detail in the atomic physics books referred to in Chap.2.

4.2.2 Molecules

Classically, light is emitted by a system only if its electric dipole moment is changed. This rule is also valid quantum mechanically and is of special im-

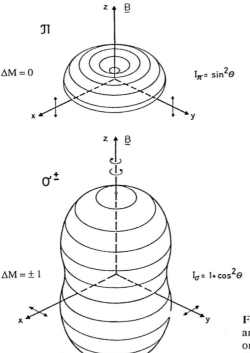

Fig.4.9. Intensity distributions for π and σ radiation. The marks on the coordinate axes indicate the oscillation direction of the electric field vector

portance in the consideration of the radiative properties of molecules. We will first treat rotational and vibrational transitions within a given electronic molecular state. Primarily, we only consider $^1\Sigma$ states for which the complications, discussed at the end of Sect.3.2, do not occur since the rotational angular momentum is the only form of angular momentum present.

a) Rotational Transitions

A diatomic molecule with two different kinds of atoms has a permanent dipole moment in the direction of the symmetry axis. If the molecule rotates the dipole should classically emit radiation. (See also Fig.4.9). Quantum mechanically, radiation occurs when the rotation is changed. For the rotational quantum number J, which was defined in Chap. 3 we have

$$\Delta J = \pm 1 . \tag{4.46}$$

Using the energy expression (3.6) for rotational levels we have

$$\nu_{J+1 \leftrightarrow J} = \frac{2B}{h}(J+1) , \quad J = 0, 1, 2, ... \tag{4.47}$$

Thus we obtain equidistant rotational lines in the far IR (~100 μm) as illustrated in Fig.4.10.

If we take the elasticity of the molecule into account (3.7) we obtain

$$\nu_{J+1 \leftrightarrow J} = \frac{2B}{h}(J+1) - \frac{4D^+}{h}(J+1)^3 . \tag{4.48}$$

b) Vibrational Transitions

If the molecule has a permanent dipole moment at the equilibrium distance (r_0 in Fig.3.7) this moment will vary periodically during vibration. Classic-

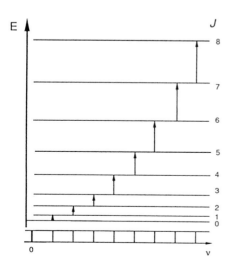

Fig.4.10. Rotational transitions in a diatomic molecule

ally, radiation will then be emitted. If a diatomic molecule has a harmonic oscillatory movement we obtain the quantum mechanical selection rule

$$\Delta v = \pm 1 \ . \tag{4.49}$$

Since the energy levels of the harmonic oscillator are equidistant only one IR vibrational line ($\sim 10\,\mu$m) is obtained. If the potential deviates from the harmonic oscillator, transitions with $\Delta v = \pm 2$, $\Delta v = \pm 3$, etc. can also occur. These transitions, which are generally weak, are called overtones (harmonics). As for rotational motion, molecules of the type O_2, N_2 (i.e., homo-nuclear molecules), do not exhibit electric-dipole vibrational lines. However, quadrupole- and pressure-induced transitions of homonuclear molecules can be observed faintly.

c) Vibrational-Rotational Spectra

If there were no interaction between rotational and vibrational motion the energy of a rotating vibrator would simply be the sum of its rotational and vibrational energies according to the above expressions. However, the moment of inertia of the molecule is influenced by molecular vibration during rotation. The vibrational energy is much greater than the rotational energy and we can use the mean value of r^2, see (3.4,6), during the vibration to calculate an effective rotational constant B′, which is slightly lower than the one corresponding to the equilibrium separation. Writing the energy of a molecular level as

$$E = (v + 1/2)h\nu_c + B'J(J + 1) \tag{4.50}$$

and using the selection rules $\Delta v = \pm 1$, $\Delta J = \pm 1$ we obtain the absorption spectrum

$$\Delta E = h\nu_c \begin{cases} +2B'(J + 1) & J \to J + 1 \quad J = 0, 1, \dots \text{ (R branch)} \\ -2B'J & J \to J - 1 \quad J = 1, 2, \dots \text{ (P branch)} \end{cases} \tag{4.51}$$

Lines corresponding to the upper expression form the so-called *R branch*, while the lower expression yields the *P branch*. In Fig.4.11 the transitions are indicated for a simple case. Note, that there is no line at $\Delta E = h\nu_c$, since $\Delta J = 0$ is forbidden for diatomic molecules. As a matter of fact, the value of B′ will decrease for higher vibrational states. Designating the rotational constants for the two states B′ and B″ (upper and lower states, respectively) we obtain

$$\Delta E = h\nu_c \begin{cases} +2B' + (3B'-B'')J + (B'-B'')J^2 & J \to J+1 \ \ J = 0, 1, \dots \\ & \text{(R branch)} \\ -(B'+B'')J + (B'-B'')J^2 & J \to J-1 \ \ J = 1, 2, \dots \\ & \text{(P branch)} \end{cases} \tag{4.52}$$

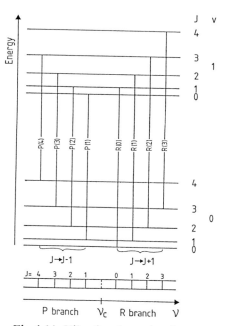
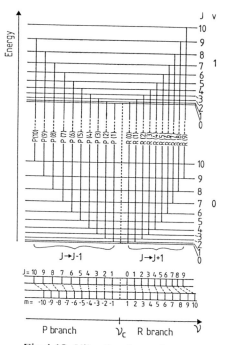

Fig.4.11. Vibrational-rotational spectrum when the rotational constant B is the same in both vibrational states

Fig.4.12. Vibrational-rotational spectrum when the rotational constant is smaller in the upper vibrational state

Both branches can be represented by a single formula

$$\Delta E = h\nu_c + (B'+B'')m + (B'-B'')m^2 \tag{4.53}$$

with

$m = 1, 2, 3 \ldots$ for the R branch

and

$m = -1, -2, -3 \ldots$ for the P branch.

Since $B' < B''$, because of the avaraging over r^2, see (3.4), this formula shows that the distance between lines becomes successively smaller in the R branch while there is a corresponding increase in the P branch, as illustrated in Fig.4.12.

For sufficiently high values of J there can be an inversion of the R branch, giving rise to the formation of a *band head*. Since the B values differ still more between states belonging to different electronic levels, band heads are more frequently observed in spectra obtained in electronic transitions. Eq.(4.53) represents a parabola and a diagram like the one given in Fig.4.13 is called a *Fortrat parabola*. Depending on the relative sizes of B' and B" the parabola has its vertex towards higher or lower frequencies. If $B' < B''$ the band is said to be *shaded to the red*, whereas if $B' > B''$ it is *shaded to the violet*.

The population of different vibrational levels is given by the Boltzmann distribution. Thus only the state with $v = 0$ is well populated for

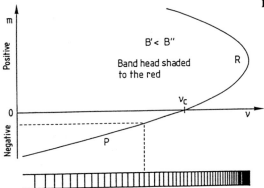

Fig.4.13. Fortrat parabola

molecules with a large separation between the vibrational levels, e.g. N_2 and O_2. The distribution for a molecule with a small value of $h\nu_c/kT$ is shown in Fig.4.14.

The intensity distribution *within* a vibrational-rotational band is also determined by the Boltzmann distribution. It is then necessary to take the 2J+1 magnetic sublevels of a rotational level into account. The distribution factor is

$$N_J \propto (2J + 1)e^{-BJ(J+1)/kT} . \qquad (4.54)$$

In the left part of Fig.4.15 this distribution for the HCl molecule (B = 10.44cm^{-1}) is shown for room temperature. This distribution leads to a somewhat higher intensity of the R branch compared with the P branch which is also illustrated in the figure for two molecules with large and small B values, respectively. This effect can be utilized for temperature measurements.

d) Electronic Transitions – The Franck-Condon Principle

Vibrational and rotational spectra arise from the movements of the atomic nuclei of the molecule. In transitions between different electronic configu-

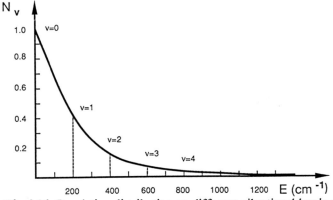

Fig.4.14. Population distribution on different vibrational levels of I_2 [4.26]

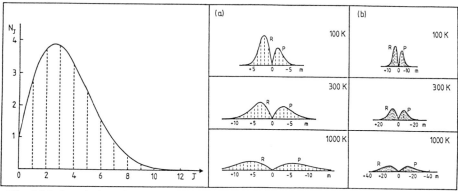

Fig.4.15. Population distribution on different rotational levels and resulting spectra for different temperatures. (a) $B = 10.44 \text{cm}^{-1}$, (b) $B = 2 \text{cm}^{-1}$ [4.26]

rations much higher energies are involved (~eV, visible or UV region). Normally a transition in the electronic shell is accompanied by transitions between different vibrational and rotational levels of the molecule. Therefore *bands* are also obtained in the visible region. The intensities of the components of the observed vibrational structure are explained by the *Franck-Condon principle*. Electronic transitions in molecules occur so quickly that the nuclei do not change their relative positions in vibrational motion. Some cases are illustrated in Fig.4.16. In part (a) of the figure the equilibrium distances of the two potential curves are the same. Primarily, we then obtain a $v=0 \rightarrow v=0$ transition in absorption since the relative positions and velocities of the nuclei are not changed. In (b) transitions occur primarily from the $v = 0$ level to a level with higher v, since the wavefunction in the excited state has a maximum at the classical turning point. In (c) dissociation primarily occurs, resulting in a continuum in the absorption spectrum. Typical spectra are also shown beside the potential curves.

The transition back to the ground state from an excited electronic state can occur in several ways. Transitions can occur by direct emission of a photon. De-excitation can also occur through successive transitions via the vibrational levels in the excited electron configuration and only after that be followed by a return to the ground state. The light is then shifted towards red compared with the excitation wavelength (*Stokes shifting*). This can be observed, for example, in organic dye molecules.

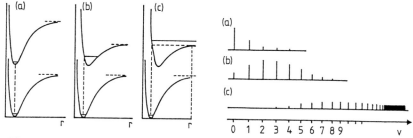

Fig.4.16. Illustration of the Franck-Condon principle and resulting spectra [4.26]

While $\Delta J = 0$ is forbidden in a pure rotational-vibrational transition, this type of transition is allowed in connection with electronic transitions (still not for $^1\Sigma$-$^1\Sigma$ transitions). The new selection rule gives rise to a Q branch, in addition to the R and P branches.

Molecular spectra have been discussed in more detail in the molecular-physics books referred to in Chap.3.

4.3 Rayleigh and Raman Scattering

As shown in Sect.4.1, resonant transitions can be obtained if an atomic or molecular system is irradiated by light with an energy corresponding to the energy separation between two levels. Even if the light is not of a resonant frequency, weak scattering effects are still obtained, so-called *Rayleigh* and *Raman scattering*. This scattering can be explained classically as well as quantum mechanically. Detailed descriptions of these phenomena and their utilization can be found in [4.10, 11].

If a molecule is subject to an electric field $\mathscr{E}$, an electric dipole moment **P** is induced in the molecule (see the Stark effect), i.e.

$$\mathbf{P} = \alpha \mathscr{E} . \qquad (4.55)$$

Generally, **P** is not directed along $\mathscr{E}$ since the molecule is frequently asymmetric and α is then replaced by a *polarization tensor*. For an oscillating field

$$\mathscr{E} = \mathscr{E}_0 \sin(2\pi\nu t) \qquad (4.56)$$

the polarization will vary at the frequency ν resulting in a re-radiation of light of the same frequency. This *Rayleigh scattering* is *elastic* in nature. For the total radiated energy I from an oscillating dipole

$$I = \frac{2}{3c^3} \overline{\left(\frac{d^2 \mathbf{P}}{dt^2}\right)^2} , \qquad (4.57)$$

where the bar denotes time averaging. Eqs.(4.57, 56) yield

$$I = \frac{16\pi^4 c}{3\lambda^4} \alpha^2 \mathscr{E}_0^2 . \qquad (4.58)$$

Here we have used the relation $\overline{(\sin^2 2\pi\nu t)} = 1/2$. As can be seen, Rayleigh scattering strongly increases with diminishing wavelength.

If a molecule vibrates, its polarizability varies. Further, the polarizability depends on the orientation of the molecule with regard to the field, as mentioned in connection with the introduction of the polarizability tensor. Thus the polarizability of the molecule varies as it rotates. We can then state

$$\alpha = \alpha_0 + \alpha_{1v}\sin(2\pi\nu_{vibr}t) \quad \alpha_{1v} \ll \alpha_0 \quad (4.59)$$

for the vibrational motion and

$$\alpha = \alpha_0 + \alpha_{1r}\sin(2\pi 2\nu_{rot}t) \quad \alpha_{1r} \ll \alpha_0 \quad (4.60)$$

for the rotational motion, where the variation occurs at twice the rotational frequency ν_{rot}. This is due to the fact that the polarizability is the same for opposite directions of the field (cf. the Stark effect). If an external oscillating field of frequency ν is applied there will be a coupling between the applied and the internal oscillation. This coupling becomes evident if (4.59 or 60) is inserted into $|P| = \alpha|\mathcal{E}|$ and simple trigonometric relations are employed.

For vibrational motion

$$P = \alpha_0 \mathcal{E}_0 \sin(2\pi\nu t) + (1/2)\alpha_{1v}\mathcal{E}_0[\cos 2\pi(\nu-\nu_{vibr})t - \cos 2\pi(\nu+\nu_{vibr})t] \quad (4.61)$$

and for rotational motion

$$P = \alpha_0 \mathcal{E}_0 \sin(2\pi\nu t) + (1/2)\alpha_{1r}\mathcal{E}_0[\cos 2\pi(\nu-2\nu_{rot})t - \cos 2\pi(\nu+2\nu_{rot})t] \quad . \quad (4.62)$$
$$\uparrow \qquad\qquad\qquad \uparrow \qquad\qquad \uparrow$$
$$\text{Rayleigh} \qquad\quad \text{Stokes} \qquad \text{Anti-Stokes}$$

Thus a sideband is obtained on both sides of the Rayleigh line. The sidebands are shifted from the Rayleigh frequency by the vibrational frequency and twice the rotational frequency, respectively. The down- and upshifted components are called the *Stokes* and the *Anti-Stokes* lines, respectively, and their strength is generally ~1/1000 of the strength of the Rayleigh line. The phenomenon is called the *Raman effect* and was first observed experimentally in 1928 by the Indian scientist C.V. Raman (Nobel Prize 1930). The Raman effect represents *inelastic scattering*. In the quantum mechanical theory of Raman scattering, *virtual* levels are introduced which mediate the scattering (Fig.4.17).

In the interaction with a molecule, an amount of energy can be emitted or absorbed by the molecule corresponding to a change in the fre-

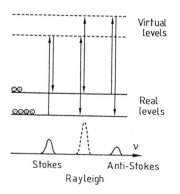

Fig.4.17. Raman scattering

quency of the scattered light quantum. From (4.61,62) it is evident that the Raman effect occurs only when the polarizability is changed in the process ($\alpha_1 \neq 0$). The Raman scattering increases for shorter wavelengths and also exhibits resonances when real levels are approached.

4.4 Raman Spectra

When a molecular gas is intensely irradiated with monochromatic light, usually from a laser, a spectrum of Raman lines is generated close to the strong elastic Rayleigh component.

4.4.1 Vibrational Raman Spectra

For the Raman effect vibration in a harmonic oscillator potential leads to the same selection rule as for the normal IR spectrum

$$\Delta v = \pm 1 \ . \tag{4.63}$$

Thus a Stokes and an anti-Stokes component are obtained, shifted by a frequency ν_c from the Rayleigh line. At normal temperatures, most of the molecules are in the lowest state (v=0) and few are in the v=1 state. The intensity of the Stokes line, corresponding to the transition 0→1, is thus much greater than that of the anti-Stokes line (1→0). In the quantum mechanical picture the intensities reflect the populations of the levels. On the other hand, the classical picture falsely predicts components of equal intensity. By comparing the intensities of the Stokes and the anti-Stokes components the temperature can be determined, for example, in a flame. Because of the nonharmonicity of the vibrational potential the spacing between the vibrational levels is not constant. This results in the occurrence of separated Stokes components originating from different vibrational levels. Utilizing the components from higher levels (*hot bands*), the temperature can conveniently be determined as illustrated in Fig.4.18.

Diatomic molecules with identical atoms do not exhibit any IR spectrum because of the absence of a permanent dipole moment. They are, however, *polarizable* and the Raman effect is then observed. In most di-

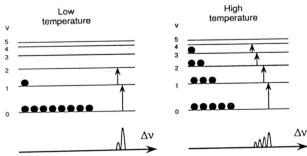

Fig.4.18. The occurrence of vibrational "hot bands" in Raman spectra

atomic molecules the separation between the lowest vibrational levels is so large that only the Stokes line can be observed at room temperature. However, anti-Stokes lines can be observed for most polyatomic molecules which exhibit lower vibrational frequencies.

4.4.2 Rotational Raman Spectra

While the vibrational Raman effect has the same selection rule as the IR transitions, different rules apply for the rotational Raman effect (compare with the classical picture!)

$$\Delta J = \pm 2 . \tag{4.64}$$

Since several rotational levels with different J's are normally populated the rotational Raman spectrum consists of several lines. $\Delta J = 0$ corresponds to the Rayleigh line. Stokes and anti-Stokes branches of equidistant lines corresponding to $\Delta J = \pm 2$ are obtained. The branches are called S and O for $\Delta J = +2$ and $\Delta J = -2$, respectively. Utilizing (3.6), we have

$$\nu_{J+2 \leftarrow J} = \frac{4B(J + 3/2)}{h} . \tag{4.65}$$

An example of the rotational Raman effect is shown in Fig.4.19.

4.4.3 Vibrational-Rotational Raman Spectra

As for dipole transitions, we obtain a combined vibrational-rotational spectrum for Raman scattering. Such a Raman band contains three branches, S, Q ($\Delta J=0$) and O. Since the rotational constants for the vibrational levels with v = 0 and 1 are almost identical the lines of the Q branch occur almost on top of each other and can frequently not be resolved. Because of this, a very strong central line occurs. The S and O branches are much

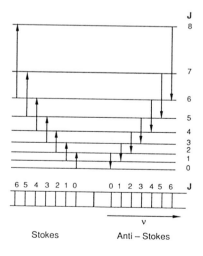

Fig.4.19. Rotational Raman spectrum

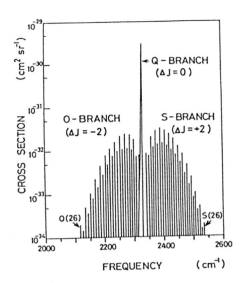

Fig.4.20. Vibrational-rotational Raman spectrum of N_2 [4.27]

weaker since the components are spread out. These branches correspond to the R and P branches in IR spectra with the exception that the separations in the Raman spectrum are twice as large. In Fig.4.20 an example of a Raman spectrum is given. The intensity distribution, previously discussed for IR transitions, is illustrated.

As we have seen, IR and Raman spectra frequently yield the same information. Raman spectra can, in many respects, be considered as IR spectra which have been moved into the visible region employing a visible excitation line. However, IR and Raman spectra also complement each other as different transitions can sometimes be observed.

4.5 Mie Scattering

In Sect.4.3 we discussed Rayleigh scattering, which occurs when a light wave induces a varying polarization in molecules. In this case the wavelength was larger than the molecular diameter. When light falls on particles of a size considerably exceeding the light wavelength, elastic scattering is observed. This type of scattering was investigated as early as 1908 for the case of spherical particles by G. Mie. *Mie scattering* has been discussed in detail in [4.7]. The probability (cross section) for Mie scattering is a complicated function of wavelength λ, particle radius r, index of refraction and absorption

$$\sigma_{Mie} = f(x, m_{rel}) \,, \tag{4.66}$$

where

$$x = \frac{2\pi r}{\lambda} \,, \quad m_{rel} = m_1/m_2 \,, \tag{4.67}$$

m_1 is the complex index of refraction ($m_1 = n - ik$) for the particles and m_2 is the corresponding quantity for the surrounding medium. For air, $m_2 \simeq 1$. σ_{Mie} oscillates rapidly as a function of the parameter x due to interference effects related to surface waves on the particles. For natural particle distributions in the atmosphere the oscillations are smeared out and the scattering intensity varies only slowly with the wavelength. The intensity increases towards shorter wavelengths with an approximate λ^{-2} dependence. In the atmosphere Mie scattering from particles is normally more important than Rayleigh scattering from molecules. The visibility is determined by Mie scattering which can be understood from the fact that in the absence of particles in the atmosphere the visibility would be hundreds of kilometres. A relatively simple relation between the visibility and the Mie scattering cross section exists.

Mie scattering can be used to monitor particles in ambient air and water (Sect. 10.2). It is also useful in the laboratory. Measurements in the direction of the probing beam are called *turbidimetry* while measurements in other directions are called *nephelometry*. (Greek nephele = cloud). Particle sizes and shape parameters can be determined from the angular and polarization distributions, at least under favourable conditions [4.28, 29].

4.6 Atmospheric Scattering Phenomena

The wavelength dependence of Mie and Rayleigh scattering is responsible for the blue of the clear sky and the red of the setting sun. The sun emits essentially "white" light. Blue light is scattered more effectively than red and so red light is transmitted better when the angle at which the sun shines is small (long path length through particle-rich layers). These well-known atmospheric scattering phenomena are illustrated in Fig. 4.21.

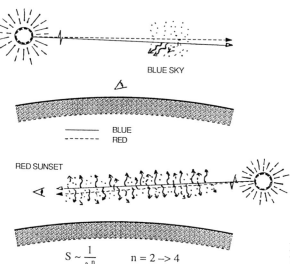

Fig. 4.21. The occurrence of blue sky and red sunset

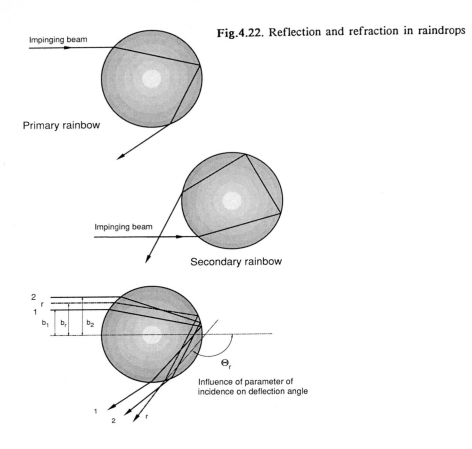

Fig.4.22. Reflection and refraction in raindrops

Another well-known atmospheric feature is the *rainbow*. This phenomenon can be explained in detail using Mie-scattering theory. However, the general principles of the rainbow can be described considerably more simply. As early as the 17th century R. Descartes explained the most important points of the rainbow phenomenon. The normal, most intense rainbow (the *primary* rainbow) is formed by the rays of the sun being reflected once in the interior of drops of moisture, as illustrated in Fig.4.22. The colours occur (starting at the outside) in the order red, orange, yellow, green, blue, indigo and violet. Often, a weaker, *secondary* rainbow can be observed outside the primary rainbow. This is formed by light which has been reflected twice in the interior of the droplets, and the colours occur in the reverse order. The angle of deflection for a ray impinging on a drop depends on the index of refraction and the parameter of incidence b, which is the vertical distance between the incoming ray and an axis drawn through the centre of the drop parallel to the incoming beam (Fig.4.22). Parallel light rays impinging on a drop are scattered in many directions because all parameters of incidence are represented and because of partial reflection from the drop surface. However, a strong concentration of intensity in a particular direction is obtained. Rays with zero parameter of

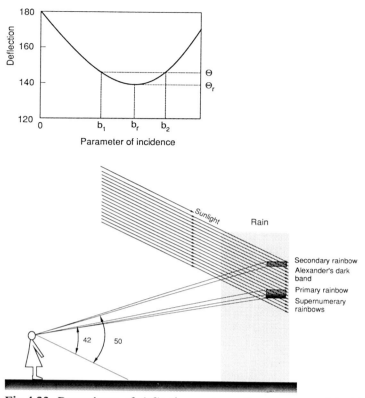

Fig.4.23. Dependence of deflection angle on the parameter of incidence (*above*) and observation angles for the rainbow (*below*)

incidence are reflected retracing their path giving a deflection angle of 180°. For increasing values of b, the angle of deflection θ decreases to a certain value, $\theta = \theta_r$, from which it slowly increases again for increasing values of b, as illustrated in Fig.4.23. Around θ_r the variation of θ with b is slow and all rays with a parameter of incidence close to b_r leave in a narrow angular range resulting in a high intensity. For water droplets θ_r (the *rainbow angle*) is 138° for red light and 140° for blue light. The secondary rainbow, corresponding to two inner reflections, can be explained in the same way. A high light intensity is obtained around 130°.

In the range of angles between the two rainbows there will be no rays corresponding to one or two reflections. Therefore the sky brightness is reduced. This darker region is called *Alexander's dark band* (Greek philosopher, 200 B.C.).

On the inner bright side of the primary rainbow (close to the violet primary band) a number of extra bands often occur, alternating in colour between green and pale red. These extra bands are caused by interference between rays with b values slightly higher and slightly lower than b_r being deflected in the same direction (cf. Young's double-slit experiment). Since the interference effect is due to differences in the path length through the drop the appearance of extra bands depends on the drop size. Frequently the bands are most visible below the top of the primary rainbow. The raindrops are smallest at high altitude and increase in size as they fall. The interference effect is smeared out for large drops.

The scattered light in rainbows is almost fully polarized since the angle of incidence for the internal reflection is close to Brewster's angle ($\tan\theta = n$).

The angle of observation for a rainbow is shown in Fig.4.23. A rainbow is circular due to the spherical symmetry of the drops. From an aeroplane, in principle, a complete rainbow can be seen centred around the shadow of the aeroplane. Common atmospheric scattering phenomena such as the rainbow have been discussed in [4.30]. Other phenomena such as atmospheric haloes, mirages, the "green flash" etc. have been treated in [4.31].

4.7 Comparison Between Different Radiation and Scattering Processes

In Table 4.1 the cross sections for different radiation and scattering processes are compared. Of course, the strength varies considerably for a particular process, but the numbers give an indication of the relative strengths. Resonance absorption and the associated fluorescence process (electric dipole radiation) are the strongest processes. Strong fluorescence only occurs at low pressures at which collisional processes can be neglected. At atmospheric pressure the return to the lower state usually occurs through collisional processes (*quenching*), see (10.1). The fluorescence intensity can then be reduced by a factor of 10^3 to 10^5. Rayleigh scattering is generally 10^{10} times weaker than resonance absorption. Raman scattering is a further factor of 10^3 weaker. The strength of Mie scattering varies strongly with the particle size.

Table 4.1. Comparison between different radiation and scattering processes

Process	Cross section σ [cm^2]
Resonance absorption	10^{-16}
Fluorescence	10^{-16}
Fluorescence (quenched)	10^{-20}
Rayleigh scattering	10^{-26}
Raman scattering	10^{-29}
Mie scattering	$10^{-26} - 10^{-8}$

4.8 Collision-Induced Processes

In this chapter, we have dealt with radiative transfer of atoms and molecules between different energy states rather extensively. However, transitions can also be induced by collisions. Extensive information on the static and dynamic properties of atoms and molecules can be obtained from collision physics. Although this book is centred on the spectroscopy of atoms and molecules, the importance of collisional physics should be clearly pointed out. For studies of these aspects the reader is referred to [4.32-37].

5. Spectroscopy of Inner Electrons

In this chapter we will discuss spectroscopic methods that involve inner electrons [5.1]. Such electrons are much more strongly bound than outer electrons and the interaction energies become correspondingly high. Two kinds of methods are used to study inner electrons, those that are based on absorbed or emitted X-ray radiation (*X-ray spectroscopy*) and those dealing with energy measurements on emitted photo-electrons (*photo-electron spectroscopy* (XPS or ESCA).

5.1 X-Ray Spectroscopy

When a solid is bombarded by electrons at an energy of few keV or more X-rays are emitted. The radiation consists of a continuous part (*Bremsstrahlung*) and a discrete (*characteristic*) part. The Bremsstrahlung is generated by the charged electrons that undergo deceleration and a change in direction of motion when interacting with the atoms of the sample. The maximum energy of the X-ray quanta corresponds to a full utilization of the kinetic energy of the incoming electrons. The discrete radiation is a line spectrum that is characteristic for the material and, as for optical radiation, it is caused by spontaneous transitions between atomic states. As a

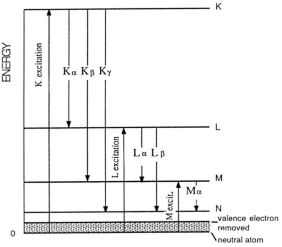

Fig.5.1. Inner electron energy levels and corresponding X-ray transitions

result of electron bombardment an electron can be knocked out of an inner shell of the atom. The atom is then in a state of high excitation and a transition to a state of lower energy quickly occurs through an electron from an outer shell falling into the vacancy to fill the hole. Thus the electron hole is seen to move outwards towards outer shells. A series of emission lines, corresponding to the successive atomic energy losses, is obtained. The processes are illustrated in Fig.5.1 in which energy levels corresponding to an electron vacancy in the K, L, M, N etc. shell are indicated, and emission lines corresponding to transitions between such states are shown. The transitions are denoted $K\alpha$, $K\beta$, $K\gamma$ etc., corresponding to the movement of the electron hole from the K shell to the L, M etc. shell. Correspondingly, $L\alpha$, $L\beta$ etc. lines are emitted when an L-shell vacancy is filled. If fine structure is also considered, energy levels are given a further index L_I, L_{II}, L_{III} etc. and emission lines a further specification: $K\alpha_1$, $K\alpha_2$ etc. as already indicated in Fig.4.8, in which the more commonly used convention of negative level energies was used. As an example, an X-ray emission spectrum from molybdenum excited by 35 keV electrons is shown in Fig. 5.2 with the continuum Bremsstrahlung and the most energetic characteristic lines.

X-rays were discovered by K. Röntgen in 1895, and were shown to be electromagnetic radiation with a wavelength comparable to the distance between crystal planes (~0.1nm) in investigations by K. von Laue and W.H. and W.L. Bragg, father and son (1913). More detailed studies of X-ray spectra were performed by H. Moseley and high-precision techniques were introduced by M. Siegbahn. The wavelengths of a large number of emission lines have now been very accurately determined. The atomic X-

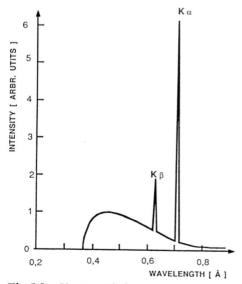

Fig.5.2. X-ray emission spectrum from molybdenum, obtained for an acceleration voltage of 35 kV. The continuum radiation (Bremsstrahlung) as well as characteristic lines is shown

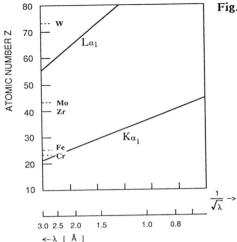

Fig.5.3. Diagram illustrating Moseley's law

ray investigations have resulted in a very thorough charting of the energy levels of inner electrons.

The energy levels in the inner electron shells are comparatively little affected by the chemical environment of the atoms. Thus, a spectral analysis of the characteristic X-ray emission is well suited for elemental analysis. The relation between the wavelength λ of a particular X-ray line and the nuclear charge Z of the corresponding atom is given by *Moseley's law*

$$1/\sqrt{\lambda} = C(Z - \sigma) \ . \tag{5.1}$$

Here C and σ are constants characterizing a particular spectral series. Moseley's law can be derived from the simple Bohr atomic model (Sect.2.1) taking shielding into account. A Moseley diagram for Kα and Lα emission is shown in Fig.5.3. Using such a diagram the identification of different elements in a sample is greatly facilitated.

5.1.1 X-Ray Emission Spectroscopy

X-ray emission can be induced in different ways. We have already mentioned the use of energetic electrons. Alternatively, heavier charged particles can be used. It is also possible to create an inner shell vacancy by irradiating the sample with X-ray radiation. We are then dealing with an inner shell photo-electric effect. The characteristic radiation following X-ray absorption is referred to as *X-ray fluorescence* following the terminology in the optical spectral region. The energy (wavelength) of the X-ray quanta can be determined in two basically different ways. Measurement systems are referred to as being *wavelength-dispersive* or *energy-dispersive*. In the former type of instrument an X-ray spectrometer using a crystal for the wavelength analysis is used. In Fig.5.4 a diagram of a wavelength-dispersive X-ray fluorescence system is shown. An X-ray tube with an anode (also called "anti-cathode") of tungsten or sometimes chromium, gold or

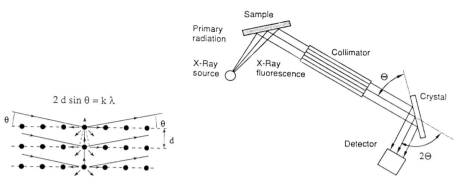

Fig.5.4. Illustration of Bragg diffraction (*left*) and a wavelength-dispersive X-ray fluorescence spectrometer (*right*)

rhodium, is used for exciting X-ray fluorescence in the sample. With the aid of a collimator, parallel beams from the sample are directed towards the flat crystal at an angle of incidence θ. The analysing crystal has been cut to have its crystal planes, separated by d, parallel to the crystal surface. The detector of the spectrometer is placed behind a collimator to receive radiation that has been deflected by 2θ from the original direction. When using the spectrometer the crystal is turned by a motor at a constant angular velocity at the same time as the detector arm is moved at twice the angular velocity. Radiation quanta can pass into the detector when the *Bragg relation* is fulfilled (Fig.5.4)

$$2d\sin\theta = k\lambda \ . \tag{5.2}$$

NaCl (d=0.56nm) or LiF (d=0.4nm) are frequently used as crystals. The X-rays bundles are collimated using systems of thin metal plates (thickness: 50μm) arranged parallel to each other at small separations (0.5mm). In this way the divergence can be limited to one degree or less. Alternatively, a curved crystal can be used: this will cause radiation diverging from an entrance slit to be focused towards an exit slit. Curved crystals of alkali halides or mica can be used.

A Geiger counter, a proportional counter or a scintillation counter can be used for X-ray detection. The two former types are gas-filled. The incoming X-rays cause the formation of ions which are then detected. Such detectors are mostly used for long-wavelength radiation ($\lambda>0.2$nm). For X-rays of shorter wavelengths scintillation counters are used, in which X-ray-induced light flashes in sodium iodide crystals are detected by a photomultiplier tube (Sect.6.3).

An example of an X-ray fluorescence spectrum of an alloy sample is shown in Fig.5.5.

Fluorescence spectrometers are widely used in the metal industry. Frequently parallel spectrometers are employed. Such an instrument actually consists of a number of crystal spectrometers, each set for a particular emission line. The spectrometers are arranged around the sample, which is irradiated by an X-ray tube. One of the spectrometers is set for a standard

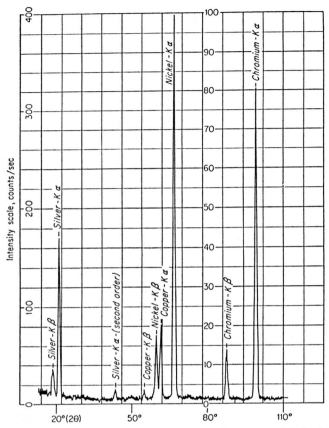

Fig.5.5. X-ray fluorescence spectrum of chrome-nickel plating on a silver-copper base. The recording was taken with a wavelength-dispersive spectrometer [5.2]

sample that is contained in the sample holder. In this way the intensity of the X-ray tube can be monitored. Frequently, a measurement is terminated when a preset number of counts for the reference sample has been obtained. The corresponding number of counts from the other detectors can then be directly used for a parallel assessment of the elemental composition of the sample. With a sequential spectrometer, a number of selected elements are measured sequentially by turning the crystal and the detector to preset positions. With computer steering the measurement process is automatic. This type of instrument is well suited for varying types of analysis whereas parallel spectrometers are more suited to continuous control of the operation of, e.g., a steel mill in near real time.

As was mentioned above, X-ray analysis can also be performed using an *energy-dispersive* system. In this case no analysing crystal is needed. Instead the radiation falls directly on an energy-dispersive detector. Normally a lithium-doped silicon detector (Si(Li)) is used, which yields voltage pulses that are proportional to the energy of the X-ray quanta. The pulses are sorted according to their amplitude (energy) in a multi-channel analyser using an Analogue-to-Digital Converter (ADC). Gradually the full spectrum emerges on the system display. The linewidth obtained with a Si(Li) detector is about 150 eV. Partly overlapping lines can be deconvoluted using computer analysis. It should be noted that an energy-dispersive system of this kind automatically permits multi-element analysis.

It might be expected that the intensity of a spectral line from a sample would be directly proportional to the amount of the corresponding element in the sample. In practice the intensity can deviate considerably from the expected linear relation due to absorption in the matrix material and multiple scattering processes. However, it is possible to correct for such effects and very reliable quantitative analyses can be performed. X-ray fluorescence measurements on alloys have an elemental sensitivity of about 10 ppm (ppm: parts per million, $1:10^6$). The typical penetration depth of the radiation in the metal is about 1 μm and thus, primarily, the surface is analysed. X-ray emission techniques have been discussed in [5.3,4].

As we have already mentioned characteristic X-rays can also be induced using accelerated heavy particles such as protons. This technique is called PIXE (Particle-Induced X-ray Emission) [5.5,6]. The cross-section for the creation of an inner shell vacancy is very high using protons at an energy of a few MeV and therefore a sensitivity much better than that typical for X-ray fluorescence is obtained. Most elements except the very light ones (Z<14) can be detected in concentrations below 1 ppb (parts per billion: $1:10^9$). With a specially focused proton beam (microbeam) quantities of an element as low as 10^{-18} g can be detected. The comparatively large number of small accelerators that are presently less suitable for nuclear physics work are generally very useful for PIXE. As an example of a PIXE spectrum, illustrating the use of an energy-dispersive Si(Li) detector, a spectrum of sea-water is shown in Fig.5.6. Some of the lines are identified.

X-ray analysis using different excitation and detection techniques has wide applications in fields ranging from biology and medicine to archaeology and forensic science. One interesting area of application is the measurement of heavy metals in particulate air pollution [5.7]. Airborne particles can be collected in different size fractions using special devices called cascade impactors. The particles deposited on the foils are then analysed. Small particles with diameters less than 2 μm are particularly important, since they follow inhaled air down into the alveoles of the lung. The human uptake of metals due to general air pollution or due to special working environments (e.g., inhalation of fumes by welders) can be investigated with samples of blood serum.

Our discussion of X-ray emission spectroscopy has so far mainly been focused on the analytical capabilities of systems with moderate resolution. Clearly, X-ray emission spectroscopy is also a field of active research, particularly in the soft X-ray region. Here dipole transitions between well defined inner orbitals and more diffuse (perturbed) valence orbitals are studied yielding valuable information on the latter orbitals. In the measurements it is clearly desirable to increase the resolution as much as possible. Wavelength-dispersive instruments are then mandatory. (Above 3nm two

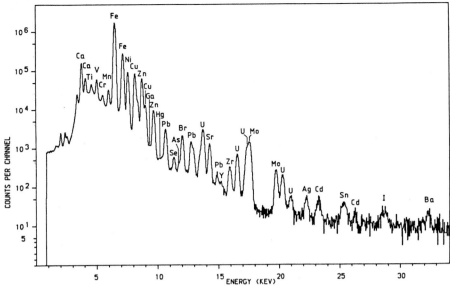

Fig.5.6. PIXE spectrum of Arctic sea water. An energy-dispersive detector system was utilized [5.5]

orders of magnitude better resolution than in energy-dispersive systems is attainable). For soft and ultra-soft X-ray wavelengths special organic crystal materials with a large crystal plane separation, such as potassium hydrogen ftalate (d = 2.7 nm) and lead stearate (d = 10 nm), can be used. If the wavelength is not too short the best choice is frequently to use a concave grating at grazing angle of incidence (see also Sect.6.2.2.). Ion-etched holographic X-ray gratings can have an efficiency exceeding 10%. A photographic plate provides parallel detection of many lines. Recently, electronic multichannel techniques for X-ray spectra recording have been introduced (see also Fig.6.38). If the entrance slit, the plate and a grating with a radius of curvature r are placed on a circle of radius r/2 (the *Rowland circle*), sharp spectral lines are obtained without using any collimators (Fig.6.25).

As an example of a spectrum obtained with a 10 m instrument (grating radius 10m) a recording of the carbon K emission line from the CO_2 molecule is shown in Fig.5.7. As can be seen, the high resolution reveals a clear structure due to molecular vibration. Through careful analysis of a spectrum of this kind it is possible to evaluate the C-O bond length very accurately in the core-ionized molecule. It turns out that the bond length is shortened by about 2% when the 1s core vacancy has been formed in the carbon atom. From the linewidth it is also possible to evaluate the natural lifetime of the C 1s state (Sect.9.4.5). The lifetime is of the order 10^{-14} s. Atomic structure research using X-ray emission spectroscopy has been discussed in [5.9-13].

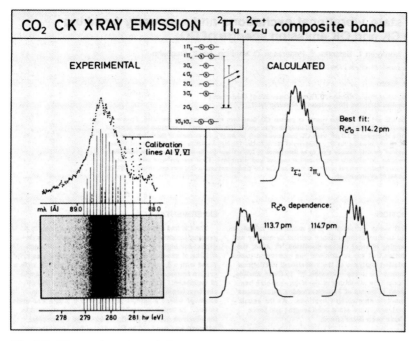

Fig.5.7. The carbon K emission spectrum from the CO_2 molecule. To the left the photographic plate and a corresponding densitometer trace are shown. Calculated spectra are shown to the right, illustrating the sensitivity of the spectrum to the C-O bond length [5.8]

5.1.2 X-Ray Absorption Spectroscopy

We have seen how measurements of X-ray emission lines have an important analytical potential and also yield fundamental information on the structure of atoms and molecules. An emission line yields information on the *difference* in binding energy between two electronic states. However, from a theoretical point of view the *absolute* binding energies are of even greater interest. In order to determine such energies *absorption* of X-rays, rather than emission, has to be studied. For such purposes an X-ray continuum is used and the sample absorption as a function of the wavelength is measured. If the energy of the X-ray photons is sufficiently high an inner electron can be excited to an unoccupied state, either in the discrete or in the continuous part of the spectrum. In the latter case an electron has been released from the system (photoemission). The available empty discrete levels (the excited valence electron states) are closely spaced in a region of a few eV and converge at the ionization limit. For increasing X-ray energies the absorption is strongly increased at the energy at which an additional, more strongly bound electron is released. An *absorption edge* is observed, from which the approximate binding energy is obtained.

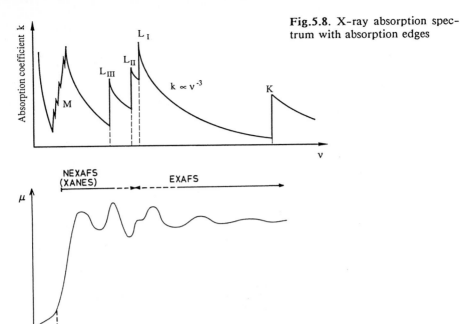

Fig.5.8. X-ray absorption spectrum with absorption edges

Fig.5.9. Fine structure at a K absorption edge, observed at high resolution (EXAFS spectrum). [Adapted from a figure by C. Nordling]

For increasing photon energy the thresholds for photo-emission from deeper and deeper shells are reached and additional contributions to the total absorption are obtained resulting in a number of edges, as illustrated in Fig.5.8. Apart from the edges there is a general fall-off in absorption due to a ν^{-3} dependence of the absorption coefficient. The absorption edges exhibit fine structure corresponding to the fine structure of the core states (Sect.5.1). However, even the K edge exhibits a structure, as shown in Fig.5.9, which is due to discrete levels close to the series limit. The position of the edges also depends slightly on the chemical composition of the studied material (chemical shifts, Sect.5.2.2). The shifts are of the order of a few eV, and the difference in position of the edge for a solid and free atoms can be expected to be of the same order of magnitude. A measurement on a solid is in relation to the Fermi level and thus a correction for the work function must be applied in order to allow a comparison with measurements or calculations for free atoms. The different effects discussed above result in an uncertainty of at least 5 eV for atomic binding energies obtained from X-ray absorption data. However, since binding energies are frequently of the order of keV or larger this uncertainty is frequently of less importance. X-ray absorption data are available for most element in the solid state. During recent years high-resolution measurements of the edge structures have become possible using synchrotron radiation (Sect. 6.1.3). Such measurements are referred to as EXAFS (Extended X-ray Absorption Fine Structure) studies or NEXAFS (Near-Edge X-ray Absorption Fine Structure) - also termed XANES for X-ray Absorption, Near-Edge Structure - depending on how far from the edge one records and analyses the X-ray absorption spectrum. Information on distances to neighbouring atoms from the studied atoms is obtained from EXAFS spectra.

X-ray absorption and EXAFS spectroscopy has been discussed in [5. 14-18].

5.2 Photo-Electron Spectroscopy

In the previous section we described how X-ray absorption measurements can be used to obtain information on energy levels in atoms. In the absorption process an electron can be released and an alternative way of obtaining energy-level information is therefore to study the energies of the emitted electrons (photo-electrons). This technique was applied as early as the beginning of the 20th century, but the instruments used had such a low resolution that little information was obtained. At the same time the optical and X-ray methods of spectroscopy were developed to higher and higher precision and there was little interest in electron spectroscopy for a long time.

While atomic physics developed significantly during the beginning of the 20th century interest was gradually diverted to nuclear physics. In this field of research new techniques were introduced to allow high precision and some of the techniques developed turned out to be very well suited to atomic physics experiments. This was the case for precision beta-particle spectroscopy that was developed by K. Siegbahn and co-workers. Their doubly focusing iron-free spectrometer with two coaxial coils was found to be very well suited to the analysis of X-ray excited photo-electron spectra and the first results were reported in 1957. This was the starting point for a new type of spectroscopy called XPS (X-ray Photo-emission Spectroscopy) or ESCA (Electron Spectroscopy for Chemical Analysis). This field of research has since developed very rapidly.

The principle of XPS is very simple, as illustrated in Fig.5.10. Using a characteristic X-ray line from an X-ray tube an electron is ejected from an inner shell in a sample atom, and its kinetic energy E_{kin} is then the difference between the photon energy $h\nu$ and the binding energy E_B of the electron:

$$E_{kin} = h\nu - E_B . \tag{5.3}$$

(The recoil energy of the atom can normally be disregarded). The binding energy of the electron is defined as the energy difference between the final and initial states

$$E_B = E_{ion} - E_{atom} . \tag{5.4}$$

By analysing the emitted electrons it is possible to determine the binding energies of electrons in different shells. This has been done for most

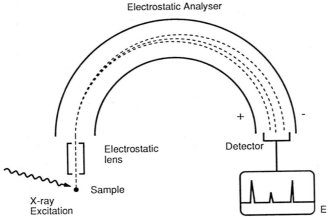

Fig.5.10. The principles of XPS. The photoelectrons emitted from the sample are analysed for energy in an electrostatic spectrometer

elements and the results are in general more reliable than those obtained by X-ray absorption. In Fig.5.11 the different peaks obtained in measurements on a complex organic molecule are shown. From the recording the binding energies of 1s and 2p electrons in several light atoms can be deduced. Initially, electron spectroscopy measurements were only made on solids, and uncertainties similar to those discussed in connection with the X-ray absorption technique occurred. Since some 20 years it has become possible to perform measurements on free atoms and molecules with a corresponding increase in accuracy in binding energy determinations. This is important for comparisons with theoretical calculations, which can presently be performed very accurately.

Electron spectroscopy has been discussed in detail in [5.20-26].

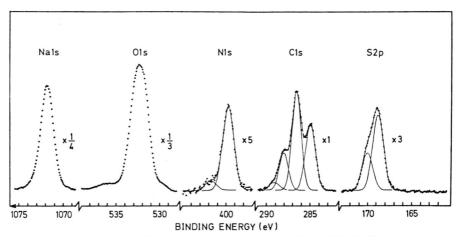

Fig.5.11. XPS spectrum of a complex organic molecule (heparin) [5.19]

5.2.1 XPS Techniques and Results

Presently, electrostatic energy analysers have fully replaced magnetic ones, since the former have many practical advantages. An electrostatic analyser consists essentially of two concentric cylindrical capacitor electrodes, that produce a well-defined radial electric field.

The resolution of an instrument will be limited by the monochromaticity of the exciting X-ray radiation. Frequently, the K emission from aluminium at 1486 eV is used. The natural radiation width of the line is about 1 eV and the line is broadened by unresolved fine structure (the overlapping $K\alpha_1$ and $K\alpha_2$ components). *X-ray monochromatization* by filtering out a narrow spectral region can be accomplished by placing the X-ray anode, a spherical single crystal analyser and the sample on a *Rowland circle* (Fig.6.25). For a certain angle between the three objects the Bragg relation will be fulfilled only for a specific wavelength. Since both the anode and the sample have a finite size the monochromatization has limitations. A typical effect of monochromatization is shown in Fig.5.12. Although a linewidth below the natural one has been achieved this does not violate the Heisenberg uncertainty relation, since it is still not possible to measure the $K\alpha$ photon energy of Al with a higher precision than before.

For solid samples it is possible to increase the resolution further by compensating for the finite extension of anode and sample by constructing the spectrometer with a certain dispersion. This is illustrated in Fig. 5.13. If the anode has a certain size the Bragg relation will be fulfilled at slightly different wavelengths for different parts of its surface. These different wavelengths will be focussed at slightly different parts on the sample, resulting in slightly different emitted electron energies from different parts of the sample. By making modifications in the design of the spectrometer, these electrons can be focussed to the same point in the detector plane for a given setting of the spectrometer. Unfortunately, this technique does not work for gases which are of particular interest, since the surface of the sample is no longer restrained to the Rowland circle. Instead a *fine focusing technique* has been developed in which the extension of the X-ray source has been strongly reduced by focusing the exciting electron beam onto a very small surface. In order to avoid overheating the anode must then be rotated at high speed. The fine focusing technique provides the most efficient monochromatization in work with gaseous, liquid and solid samples. Monochromatized radiation from synchrotrons is presently widely used in XPS studies (Fig.6.13) [5.27]. Outer shells can be investigated using

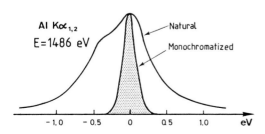

Fig.5.12. Line profile of the $K\alpha$ line of Al, with and without monochromatization [5.25]

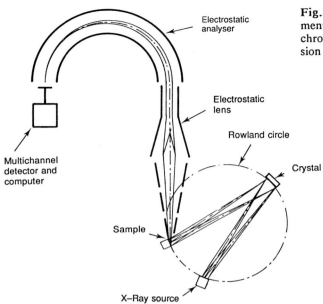

Fig.5.13. An ESCA instrument with X-ray monochromatization and dispersion compensation

special short-wavelength lamps, e.g. strong He lamps emitting the 1s2p-1s^2 resonance line at 58.4 nm (21.2eV) [5.28].

In order to be able to measure many electron lines simultaneously a multichannel detector is frequently placed in the detector plane instead of using a detector behind an exit slit. The electrons then impinge on a micro-channel plate (see also Fig.6.38), in which electron multiplication occurs due to secondary emission by the inner wall material in the densely packed tubes in the plate. The original electron line image in the focal plane of the spectrometer is amplified and, by using two channel plates in series an electron multiplication of 10^8 can be obtained. The electron showers are converted into optical signals on a phosphor screen which is viewed by a diode array or vidicon (TV camera) (Fig.6.38).

As an example of an atomic photo-electron spectrum a recording for mercury is shown in Fig.5.14, in which all populated levels in the N, O and P shells (n = 4, 5 and 6) can be observed. The common level designations from X-ray spectroscopy (Sect.5.1; Fig.4.8) have been used. Thus N_I corresponds to the 4s level, N_{II} and N_{III} are the $4p_{1/2}$ and $4p_{3/2}$ fine structure levels, N_{IV} and N_V are the $4d_{3/2}$ and $4d_{5/2}$ levels etc. It can be clearly seen how the spin-orbit interaction decreases with increasing principal quantum number, as for alkali atom valence-electron excited states (Table 2.1). An enlarged section of the spectral region to the left of the N_{VI} and N_{VII} lines (the $4f_{5/2}$ and $4f_{7/2}$ levels) is also included, revealing many weak components. Some of these are referred to as *shake-up lines*. At the same time as a photo-electron is ejected an outer electron can be excited resulting in a lower photo-electron energy than otherwise, i.e. corresponding to a line for a higher binding energy. The presence of shake-up lines is a fundamental quantum-mechanical phenomenon. The wavefunction

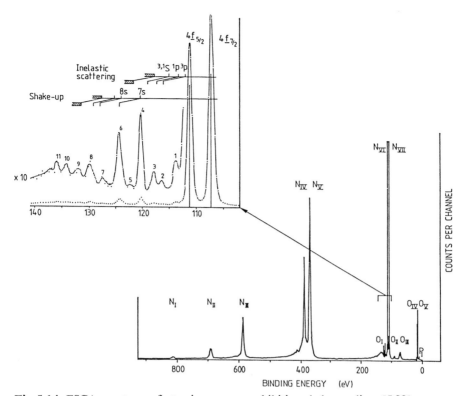

Fig.5.14. ESCA spectrum of atomic mercury exhibiting shake-up lines [5.29]

$|\psi_{N-1}^o\rangle$ for the system with a newly created inner shell vacancy can be expanded in the wavefunction system for the ordinary ion system that is obtained after the relaxation of the electronic shell. The intensity I_i of a shake-up line, corresponding to the state $|\psi_{N-1}^i\rangle$ of the ion is given by

$$I_i \propto |\langle\psi_{N-1}^o|\psi_{N-1}^i\rangle|^2 \ . \tag{5.5}$$

A sulphur L-shell XPS spectrum showing valence orbitals of the SF_6 molecule is shown in Fig.5.15. An ultra-soft X-ray emission spectrum for the same molecule is also included. The figure illustrates how the different selection rules for X-ray emission and XPS lead to complementary information for the interpretation of the valence orbital structure of the molecule.

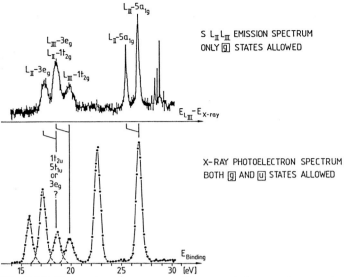

Fig.5.15. Comparison between an X-ray emission spectrum and an XPS spectrum for SF_6, illustrating different selection rules [5.12]

5.2.2 Chemical Shifts

In this section we will discuss an aspect of electron spectroscopy that is of great importance because of its applications: the presence of *chemical shifts* in the binding energies. Such shifts, which have already been mentioned in connection with X-ray absorption spectroscopy (Sect.5.1.2), constitute a source of error in the determination of binding energies, but also provide rich information on the atomic chemical environment. It is found that when the binding energy of a particular inner orbital is studied in different chemical environments a variation (chemical shift) of typically a few eV is found. This is illustrated in Fig.5.16 for the case of acetone. Whereas the 1s electron for oxygen only shows one distinct binding energy, the corresponding electronic state for carbon gives rise to a split line with a 1:2 intensity relation between the two components. The double line reflects the different positions of the three carbon atoms in the acetone molecule; one related to a double bond to oxygen and two equivalent positions in CH_3 groups. For comparison, the corresponding lines in the CO_2 molecule are included in the figure showing still other shifts. Chemical shifts are evident in the already shown Fig.5.11, where different positions of, e.g. the carbon atom in the molecule are reflected in the spectrum. A particularly illustrative example is given for the case of the ethyltrifluoroacetate molecule in Fig.5.17, featuring 4 equally strong carbon 1s lines corresponding to 4 different chemical environments. By comparing experimentally determined chemical shifts with theoretical calculations, valuable information on the molecular structure is obtained at the same time as the reliability of different theoretical models can be tested. The shift of the inner energy levels is

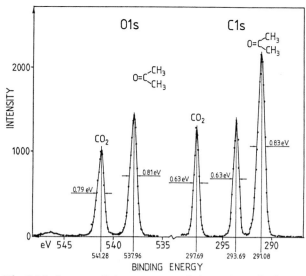

Fig.5.16. Spectra of the acetone and the carbon dioxide molecules, illustrating chemical shifts [5.30]

not due to a deformation of the corresponding orbitals but rather to a change in the electrical potential in the inner part due to the change in the outer orbitals. Detailed calculations using molecular self-consistent-field approaches yield good agreement with experiments. In a simpler approach, the molecule is represented by ions with specific "effective" charges. Using atomic calculations the shift of inner levels can be obtained as a function of this charge. Comparisons with experimentally determined shifts give a description of the charge distribution in the molecule.

As we have pointed out, the presence of chemical shifts has resulted in important applications for electron spectroscopy. Since the positions of the lines depend on the chemical composition, the method can be used for qualitative as well as quantitative chemical analysis. One of the commonly

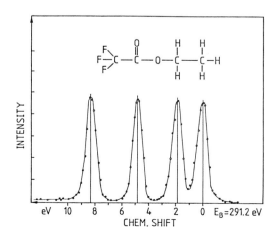

Fig.5.17. 1s electron lines for carbon in ethyltrifluoroacetate [5.30]

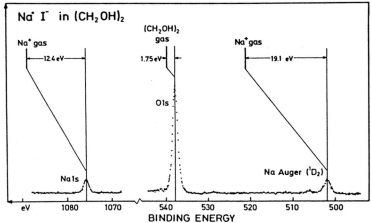

Fig.5.18. Electron spectrum of a solution of Na^+I^- in glycole. Solvation shifts with respect to the gas phase line positions are observed [5.31]

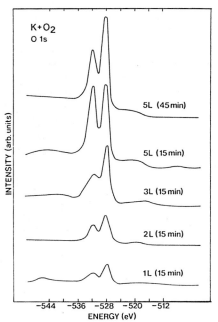

Fig.5.19. ESCA surface sensitivity illustrated by progressing oxidation of a potassium surface. 1L corresponds to 10^{-6} torr·s of O_2 exposure [5.32]

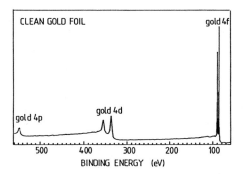

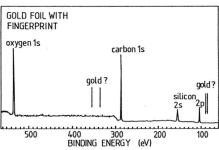

Fig.5.20. ESCA surface sensitivity illustrated by spectra from a clean and a contaminated gold surface [5.26]

used names for the XPS method, ESCA (Electron Spectroscopy for Chemical Analysis), focuses on this aspect.

One can also use the shift between free atomic or molecular species and the condensed phase to study cohesive properties. An example for a liquid phase system is shown in Fig.5.18, viz. the photoelectron spectrum of a NaI solution. In this case the free-ion/dissolved-ion shift can be related to the energy of dissolution of the ion in the solution (solvation energy) providing information on the solution structure [5.31].

In the study of condensed phase samples it is the outermost layers which are probed in the photoelectron experiments (down to 1nm below the surface). The *surface sensitivity* is illustrated in Fig.5.19 for the case of slow oxidation of a potassium surface. Peaks due to potassium oxide and physisorbed O_2 occur. A further example of surface studies is shown in Fig.5.20 which shows ESCA spectra for a clean gold foil and for a gold foil contaminated by a faint fingerprint from a finger that had previously been in contact with silicone oil. Surface monitoring has been discussed in [5.33, 34].

5.3 Auger Electron Spectroscopy

In normal photo-electron spectroscopy as described above, the energy of the ejected electrons varies according to the energy of the exciting X-ray quanta, (5.3). However, an emission of electrons with a kinetic energy that does not depend on the excitation energy is also observed. This process is called the *Auger effect* (P. Auger, 1925). The energy that would normally be emitted as, e.g., a $K\alpha$ quantum is instead used to release an outer electron, which emerges with a kinetic energy E_{kin} determined by the $K\alpha$ energy $E_{K\alpha}$ and the binding energy of the outer electron E_B

$$\Delta E_{kin} = E_{K\alpha} - E_B . \qquad (5.6)$$

The process is illustrated in Fig.5.21 for the case of a KLL Auger process, corresponding to the primary removal of a K electron, the infall of an L electron into the hole and the ejection of a further L electron.

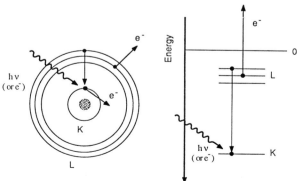

Fig.5.21. Illustration of the Auger process (KLL Auger process)

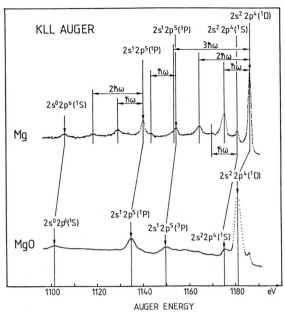

Fig.5.22. Auger electron spectra for Mg and MgO (solids) illustrating the influence of chemical shifts and electron hole coupling. Additional peaks $n\hbar\omega$ are due to volume plasmons [5.35]

Clearly, the Auger electron energy must also depend on the coupling of the two holes in the final state. Corresponding Auger spectra from Mg and MgO are shown in Fig.5.22. The influence of the hole coupling is shown as well as the chemical shift between the metal and the oxide. Additional peaks in the Mg spectrum are due to volume plasmons (a solid-state phenomenon).

Auger electrons can be emitted if the inner-shell vacancy is created by the absorption of an X-ray quantum or by electron bombardment. However, in the latter case an electron continuum due to the scattered electrons tends to mask the frequently weak Auger lines in the case of solid samples. Spectral derivation techniques can then be used to enhance the signals. For gases the background is weak and causes no problems.

The probability of Auger-electron emission dominates over the emission of characteristic X-rays for light elements, while the opposite conditions prevail for heavy elements.

The Auger technique is comparatively simple and it has the same surface sensitivity as XPS. Because of this, Auger spectroscopy is frequently used as a diagnostic tool, e.g. for controlling the cleanliness of a surface in connection with surface physics investigations. A particularly valuable technique is scanning Auger spectroscopy, that can provide surface images with a resolution of 200 nm [5.34].

6. Optical Spectroscopy

The method of spectroscopy, using a suitable light source and spectral apparatus for radiation analysis, has its natural field of application in the determination of the general energy-level structure in the energy range corresponding to UV, visible and IR light. The energy-level scheme for atoms and ions of many different charge states has been established from spectral analysis in different wavelength regions, as discussed in Chap.2. Many of the observed spectral lines are listed in standard monographs [6.1-5]. Hyperfine structure can also be studied in many cases using high-resolution instruments. The first observations of hyperfine structure in optical spectra were made at the end of the 19th century by A. Michelson (1891), and by Ch. Fabry and A. Pérot (1897). An interpretation of the structure was put forward at the end of the 1920's. The optical method for studies of hyperfine structure is particularly suitable when unpaired s-electrons are present (large hyperfine structure). A large number of nuclei have been studied with regard to nuclear spin and moments through the years. Many radioactive isotopes have also been studied using very small samples. Although the classical optical method has low accuracy, compared with resonance methods (Chap.7) or laser techniques (Chap.9), its field of application is wide. A very large number of excited levels can be studied through the structure in the large number of lines emitted by a light source. The structure in spectral lines, connecting a ground state or a well-populated metastable state with higher-lying, short-lived states can also be studied in absorption experiments, in which the atomic absorption in a continuous spectral distribution is recorded. The techniques of classical optical spectroscopy have been covered in [6.6-8].

6.1 Light Sources

Many different types of light sources (lamps) have been developed for atomic spectroscopy investigations. If the radiation from the source is to be analysed directly, the lamp should, of course, contain the atomic species for which the emission spectrum is to be studied. Such light sources are designated *line light sources*. If the light source is to be used for excitation of atoms that are to be studied with some spectroscopic method, any light source yielding a sufficient intensity at the excitation wavelength can be used: accidental coincidences between lines from different elements can, for example, be used. Close-lying lines can cause confusion in spectral analysis and have therefore been listed [6.9]. However, such perfect coincidences are rare, considering that the lines normally have to overlap within 0.001 nm. In certain experiments (e.g., in absorption experiments) *continuum light sources*, that do not produce any characteristic line spectrum, can be used. However, such light sources generally yield too low an intensity per spectral interval to be useful for excitation purposes, unless the absorption bands are broad. The opposite is true for the third main group of light sources, *lasers*. Here, the intensity per spectral interval is many orders of magnitude higher than that which is obtained from line sources. For quite some time only laser transitions in fixed-frequency lasers [6.10], or accidentally coincident lines could be studied. However, tunable lasers are now available, making more general applications of the unique properties of laser light possible in spectroscopy. Lasers will be discussed in Chap. 8.

6.1.1 Line Light Sources

The accuracy of a spectroscopic measurement is determined by the sharpness of the lines to be measured. The experimentally recorded width of the lines is due to two contributions: the primary width (from the source) and the instrumental width of the spectral apparatus used. We will first consider the factors contributing to the primary width of the light source. These broadening factors are also relevant in absorption experiments. We have already dealt with the *natural radiation width* (*homogeneous broadening*) of energy levels in Chap. 4. For most light sources this width is small compared with the *Doppler broadening*. Because of the varying velocities and directions of movement for the atoms under study, light within a certain frequency interval can be emitted or absorbed, as illustrated in Fig. 6.1.

Doppler broadening is an example of *inhomogeneous* broadening, in which different atoms contribute to different parts of the line profile. Light of frequency ν_0, emitted from a source moving with a thermal velocity v, towards (v>0) or away (v<0) from an observer is recorded with a frequency shift $\Delta \nu$ given by

$$\Delta \nu = \nu - \nu_0 = \nu_0 v/c \ . \tag{6.1}$$

This expresses the *Doppler effect*.

In a light source or in an absorption cell the velocities of the atoms follow a Maxwell distribution and the number of atoms with velocity components between v and v+dv in the direction of observation is

$$dN = Nf(v)dv \ , \tag{6.2}$$

where f(v) is the distribution function for the particular velocity component v:

$$f(v) = \sqrt{\frac{M}{2\pi RT}} \exp\left(-\frac{M}{2RT}v^2\right) . \tag{6.3}$$

Here N is the total number of atoms, M is the atomic weight and R is the general gas constant. The light power emitted in the frequency interval ν to

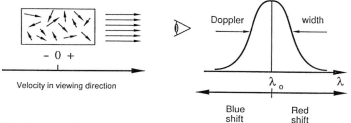

Fig.6.1. The origin of Doppler broadening of a spectral line

$\nu+\Delta\nu$ is denoted by $P_\nu d\nu$. This power is proportional to the number of light-emitting atoms in this band, i.e. the atoms with a velocity component v between v and v+dv, hence

$$v = c\frac{\nu - \nu_0}{\nu_0} \quad \text{and} \quad dv = c\frac{d\nu}{\nu_0} \tag{6.4}$$

and

$$P_\nu d\nu = KNf(c\frac{\nu - \nu_0}{\nu_0})c\frac{d\nu}{\nu_0} \tag{6.5}$$

and

$$P_\nu = K'f(c\frac{\nu - \nu_0}{\nu_0}) = K''\exp\left[-\frac{M}{2RT}c^2\frac{(\nu - \nu_0)^2}{\nu_0^2}\right]. \tag{6.6}$$

K, K' and K'' are constants of proportionality. Eq.(6.6) describes a Gaussian with a half-intensity at $e^{-x} = 1/2$, i.e. when

$$\delta\nu = \nu - \nu_o = \frac{\nu_0}{c}\sqrt{\frac{2RT}{M}\ln 2} \ . \tag{6.7}$$

The full width at half maximum (FWHM), the *Doppler width*, is then

$$\Delta\nu_D = 2\delta\nu = \frac{2\sqrt{2R\ln 2}}{c}\nu_0\sqrt{\frac{T}{M}} \ . \tag{6.8}$$

The value of the constant is $7.16\cdot 10^{-7}$ $K^{-1/2}$. As an example, for λ = 500 nm, T = 500 K and M = 100 the Doppler broadening is

$$\Delta\nu_D = 7.16\cdot 10^{-7}\frac{3\cdot 10^8}{5\cdot 10^{-7}}\sqrt{5} = 960\,\text{MHz} \leftrightarrow 0.001\,\text{nm} \ .$$

A general rule, which may easily be memorized, is that the typical Doppler width in the visible range for typical light sources is about 0.001 nm or 1000 MHz. In the IR region the width is, of course, proportionally smaller, i.e. about 50 MHz at 10 μm.

If the atoms under study undergo frequent *collisions*, the spectral lines will be broadened. With increasing pressure *Lorentz broadening* (due to collisions between different kinds of atoms and *Holtsmark broadening* (due to collisions between the same kind of atoms) will be more and more prominent. These two broadening mechanisms are generally treated together as pressure broadening. The collisional width, $\Delta\nu_{coll}(P_0,T_0)$, of the resulting Lorentzian curve is about 0.5-5 GHz at atmospheric pressure and room temperature. The width can be calculated for any temperature and pressure using the relation

$$\Delta\nu_{coll}(P,T) = \Delta\nu_{coll}(P_0,T_0)\frac{P}{P_0}\sqrt{\frac{T_0}{T}} \ . \tag{6.9}$$

If collisions with electrons and ions occur, as in discharge lamps, *Stark broadening*, due to the strong electrical fields experienced by the atoms during collisions, will also contribute to the total line broadening.

Line broadening mechanisms and resulting lineshapes have been discussed in [6.11-14].

In Fig.6.2 a number of line light sources have been schematically depicted. With a *dc gas discharge* between two electrodes the spectra of most gases can be conveniently produced. The discharge current is generally low, about 100 mA. Considerably stronger currents are used in *dc arcs*. The cathode contains the atomic species for which the spectrum is to be produced. Atoms are brought into the arc, which is burning between the anode and the cathode in an inert-gas atmosphere through sputtering. This type of lamp has been extensively used for studies of the strengths of spectral lines (transition probabilities) [6.15,16].

The *hollow cathode* is a very useful light source, which produces comparatively narrow spectral lines. A dc discharge of typically 10 to 500 mA runs in an inert-gas atmosphere (pressure: 1-5 torr) between a ring-shaped anode and the cathode, which is made of the element to be studied. A hole, with a typical depth and diameter of 5 mm is drilled in the cathode. Because of the field distribution around the hole the discharge will run down into the hole. Atoms are sputtered into the discharge through the ion bombardment, and the atoms are excited in the discharge. Intense light emission is obtained from the hole. By cooling the hollow cathode with liquid nitrogen, the Doppler width, which is the dominating source of broadening, can be further reduced (Schüler hollow cathode).

In an electric *spark discharge* considerably higher excitation energies are obtained than in the earlier described light sources. A large capacitor (~0.5 μF) is discharged through a spark gap between two electrodes of the element to be studied. Sometimes a piece of insulating material is placed between the electrodes to guide the spark (*sliding spark*). Highly ionized atoms with more than 10 electrons removed can be obtained and the corresponding spectra studied. The intermittent light emission from the spark can best be studied with spectrographs with photographic recording, since the photographic plate is a perfect light integrator. Photoelectric recording, which is otherwise frequently employed, is not suitable when working with this kind of light source. The blackening of photographically recorded plates can conveniently be recorded with photoelectrical methods (microdensitometer).

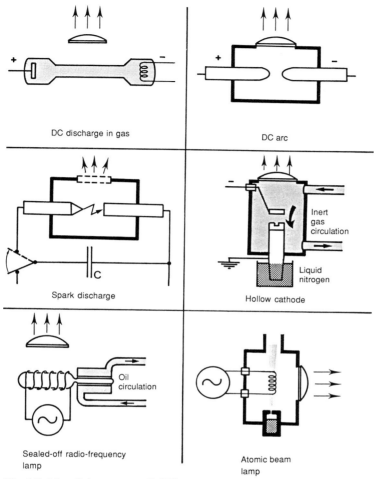

Fig.6.2. Line light sources of different constructions

A sealed *radio-frequency discharge lamp* can be used for most elements which develop a sufficient vapour pressure ($>10^{-3}$ torr) within the temperature range in which glass or quartz vessels can be employed (below 1000°C). The lamp vessel is sealed under high vacuum after a small amount of the element of interest has been brought into it by heating a lump of the material that is kept in a sidearm of the vacuum manifold. The sealed lamp vessel is placed in an RF coil inside an oven, and an RF power of typically 100 W is applied. This kind of lamp is frequently used for alkali atoms. At the low temperatures needed in these cases, the metal vapour pressure can be accurately regulated by circulating oil around a cold finger on the lamp vessel, which is kept at a slightly higher temperature. The coldest point on the vessel will determine the metal vapour pressure.

In the *atomic-beam lamp*, an atomic beam is excited in an RF coil, and light emitted at right angles to the atomic beam is collected. In this way the

Doppler broadening of the emission lines will be much reduced. The same principle for linewidth reduction is used in absorption measurements, in which absorption from a continuous wavelength distribution by atoms is recorded at right angles to the collimated beam. The resulting Doppler width when atoms with a preferred direction of motion are used can be calculated in a similar way as for the case with evenly distributed directions of motion. In practical cases, linewidths of tens of MHz are obtained in optical absorption measurements on atomic beams at thermal velocities (a few hundred m/s).

A special type of light emission is obtained from a *laser-produced plasma* [6.17]. Here the beam from a high-power laser is focused onto a small spot on the surface of a metallic substrate. Typically a Nd:Glass laser (Chap.8) is used with a pulse energy of about 10 J and a pulse length of a few ns. Plasma temperatures as high as 10^6 to 10^7 K can be obtained. The ionization occurs very quickly and the low states of ionization do not have time to radiate. The maximum state of ionization (up to 60 electrons removed) can be varied by changing the pulse energy and focusing. The laser-produced plasma spectrum is easier to interpret than the spectrum from a spark, where a large number of ionization stages simultaneously yield spectral lines. Particularly if a high-Z target material is used a strong soft X-ray continuum will also be obtained [6.18-21]. This radiation is comparable to that produced by an electron synchrotron (Sect.6.1.3). Such short-wavelength radiation is very useful both for spectroscopic investigations and for technical applications such as X-ray microscopy and lithography in connection with semiconductor circuit production [6.22,23].

In laser-driven fusion research high-power beams are made to impinge on small targets of deuterium and tritium [6.24]. It is also possible to study highly ionized spectra from other plasma generators used in fusion research, such as *Tokamak* machines. A smaller plasma light source is the *theta pinch*, where a strong current pulse in a one-turn coil excites and contracts the plasma. Spectral analysis of the plasma light serves as an important diagnostic tool in fusion research machines [6.25].

Finally, the *beam-foil light source* and its applications will be described. These techniques were introduced in 1963 by *Bashkin* and *Kay* [6.26,27]. The principles of beam-foil spectroscopy are illustrated in Fig. 6.3. Ions of the element to be studied are produced with a well-defined energy (velocity) in an ion accelerator. The ions (with a typical velocity of 10^6 m/s) pass a 50 nm thick carbon foil, in which very abrupt excitation occurs (the passage time is $\sim 10^{-14}$ s). Ions of different charges are formed. An electron can also be picked up and a neutral atom is then formed. The excited states will decay after passing through the foil and the corresponding spectrum can be recorded with a spectrometer. Because of a certain lateral scattering in the foil and since a finite solid angle must be used for the detection of the photons the Doppler broadening for the fast particles will be large; typically a few 0.1 nm. The linewidth can, however, be reduced using special imaging techniques (refocusing) [6.28]. A typical beam-foil spectrum is shown in Fig.6.4.

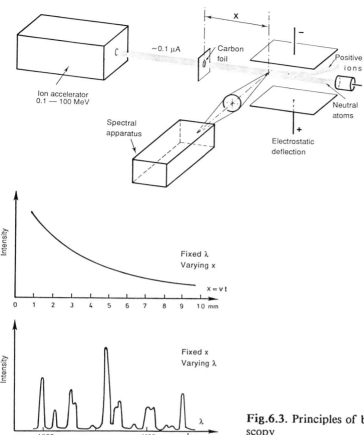

Fig.6.3. Principles of beam-foil spectroscopy

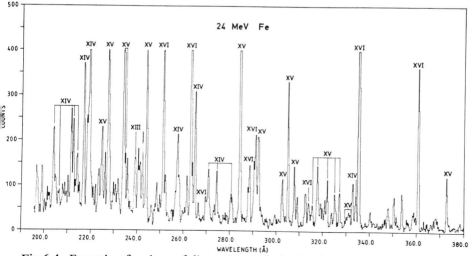

Fig.6.4. Example of a beam-foil spectrum obtained using an accelerated iron-ion beam [6.29]

Since the velocity of the particles is known, a well-defined time scale is obtained after the foil and the decay can be observed directly. In such measurements the foil is normally moved back and forth while the detection system remains fixed. A certain energy loss and velocity spread (straggling) is obtained as the ions pass through the foil and these small effects can be studied if an electrostatic velocity analyser is placed after the foil, as is illustrated in the figure. Lifetimes of excited states of atoms and ions can be conveniently measured with the beam-foil method. Since the excitation process is not selective, the decay curve must be carefully analysed so that cascade decays from higher-lying states can be identified and corrected for [6.30, 31]. Lifetime measurements using the beam-foil technique are illustrated in Fig.6.5. The beam-foil method yields low signal strengths, since the particle density in the beam is typically 10^5 ions/cm^3. Ion accelerators in different energy ranges can be used; from mass separators yielding 100 keV energy to heavy-ion accelerators yielding particle energies up to 300 GeV. Atoms ionized 90 times have been observed. In order to determine which lines belong to a certain ionization state, a strong electric field can be applied which deflects the beam according to the charge states of the ionic beam constituents.

As a result of the abrupt foil excitation atoms and ions can be excited into a quantum-mechnical state, that is a coherent superposition of eigenstates. By "coherent" we mean in this context that a well-defined relation

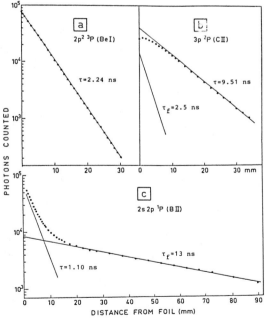

Fig.6.5. Examples of decay curves obtained with beam-foil spectroscopy [6.32]. Curve *a* presents the decay of a BeI level unaffected by cascades. In (*b*) a level in CII is repopulated by a short-lived level, while in (*c*) a level in BII is repopulated by a longer-lived level. τ_f denotes the half-life of the level feeding the level of interest

exists between the phase factors of the different eigenstates. If the sublevels are not equally populated but a certain degree of "alignment" is brought about, some components of the light from the decaying, abruptly excited state will be modulated at frequencies that correspond to the energy splittings between the substates. Such an alignment can be obtained in the beam-foil excitation process, especially if the symmetry is broken by inclining the foil with respect to the ion beam [6.33]. This so-called *quantum-beat spectroscopy*, which was treated theoretically for the beam-foil case in [6.34], is very useful for measuring fine and hyperfine structure separations. The technique is illustrated by a fine-structure determination for ^{4}He in Fig.6.6. An example of hyperfine structure quantum beats is given in Fig.

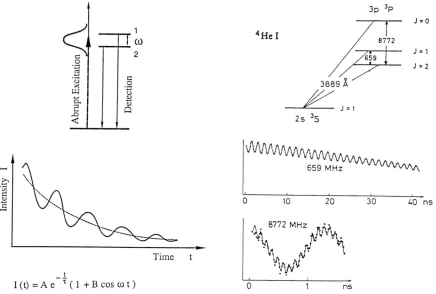

Fig.6.6. Principle of quantum beats (*left*) and illustration of fine-structure quantum beats in ^{4}He (*right*) [6.35]

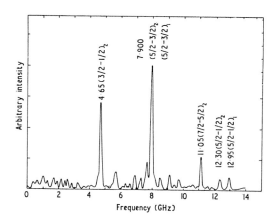

Fig.6.7. Fourier analysis of hyperfine-structure quantum beats in the 1s2p $^3P_{2,1}$ → 1s2s $^3S_1$ decay of Be III [I(^{9}Be) = 3/2] [6.36]

6.7. We will return to quantum-beat experiments in Sect.9.4.6, where the more well-defined case of optical (laser) excitation is discussed.

As we have seen the beam-foil light source offers many possibilities for studies of spectra, lifetimes and level separations and extensive data have been collected during the last 20 years [6.37-41]. Comprehensive reviews of the field of beam-foil spectroscopy have been written [6.42-48].

6.1.2 Continuum Light Sources

The *black-body emitter* is the simplest type of continuum light source. The radiation from a heated tungsten ribbon approaches that of a Planck emitter. However, for a given temperature only about 30% of the power emitted by a truly black body is obtained (emissivity: 30%). The highest temperature attainable in this way is 3400 K. Normally, the working temperature for a tungsten lamp is about 2900 K. Such a lamp is useful in the wavelength region 320 nm to 2.5 μm. Planck-radiation diagrams for certain characteristic temperatures are shown in Fig.6.8, which also illustrates the displacement of the wavelength of maximum emission towards shorter wavelengths with increasing temperature (the Wien displacement law: $\lambda_{max} \propto T^{-1}$). For the IR region a so-called *Nernst glower* is frequently used. This consists of a heated rod of sintered cerium and zirconium oxides (1500-2000 K).

In the UV and visible regions lamps utilizing a gas discharge are frequently used. In a *deuterium lamp* at a pressure of some tens of torrs, the

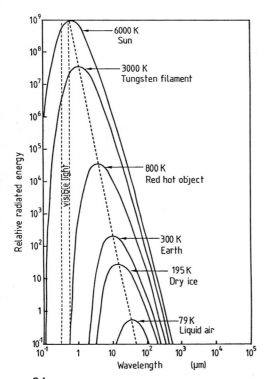

Fig.6.8. Planck radiation diagrams for different temperatures

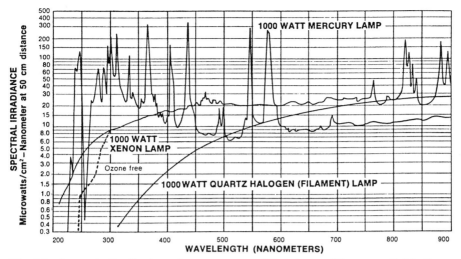

Fig.6.9. Spectral distributions from some gas-filled lamps (Courtesy: Oriel Corp., Stratford, CT, USA)

spectral lines are broadened to a continuum-like spectrum with only certain peaks remaining. The practical region of use is 180 to 380 nm. Lamps with much higher pressures can also be used. In *xenon* and *mercury lamps* pressures of tens of atmospheres are often used. Spectral distributions of some gas-filled lamps are presented in Fig.6.9. Although the total output power from a continuum source of this kind can be very high (>1kW) the power within a region corresponding to an individual atomic absorption line (0.001-0.005nm) is always small.

6.1.3 Synchrotron Radiation

During recent years *electron synchrotrons* have been increasingly used as sources of continuum radiation. Originally, the radiation was obtained "parasitically" from machines that were constructed for nuclear physics experiments. Lately, many machines dedicated for light generation have been built. Because of the strong centrifugal acceleration ($\sim 10^{16}$ g for electrons with a velocity close to c in an orbit of 1m radius), the electrons will emit radiation. At low velocities the emission pattern is a normal $\sin^2\theta$ dipole distribution with the orbit radius as the symmetry axis (Fig.6.10). Because

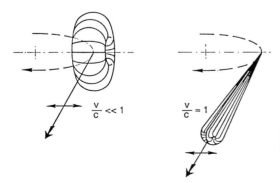

Fig.6.10. Generation of synchrotron radiation [6.50]

of relativistic effects the emission characteristics will be strongly peaked in the momentary flight direction of a high-speed electron [6.49,50]. The opening angle of the emission cone is inversely proportional to the electron energy and is typically a small fraction of a degree. The light is linearly polarized. The total power P of the synchrotron radiation increases as the fourth power of the electron energy E:

$$P[W] = 88.5 \frac{I[mA](E[GeV])^4}{R[m]}, \qquad (6.10)$$

where I is the electron beam current and R is the radius of the electron orbit. Classically, the frequency of the radiation emitted by a circulating electron would be the inverse of the period of revolution and would be typically 10 MHz. However, because of relativistic effects overtones of very high orders are emitted at high powers. Because of the short time for the radiation lobe of an individual electron to sweep over the detector and fluctuations in energy and orbit position the frequency distribution is smeared out into a continuum which extends down to very short wavelengths. A synchrotron radiation source is normally classified by its *characteristic wavelength* λ_c

$$\lambda_c[\text{Å}] = 5.6 \frac{r[m]}{(E[GeV])^3}. \qquad (6.11)$$

The maximum of the intensity distribution is close to λ_c. The characteristic wavelength for typical, large machines is around 1 nm. It is possible to express the distribution of any electron synchrotron radiation in normalized intensity units using the characteristic wavelength as the wavelength unit. The universal distribution function is shown in Fig.6.11.

In order to achieve high radiation intensities *electron storage rings* are used. The lay-out of such a ring is shown in Fig.6.12. In these rings, which are pumped by (filled from) an electron accelerator, circulating currents of the order of 1 A can be obtained. Using radio-frequency fields sufficient energy is delivered to the electrons to compensate for the radiative losses. (Compare with the Bohr atomic model!) In the ring the electrons are arranged in circulating bunches. Thus the emission as seen by a user who is placed tangentially to the orbit will be pulsed with a pulse width in the subnanosecond region and a repetition rate of the order of MHz to GHz.

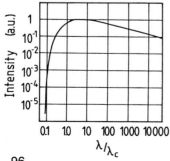

Fig.6.11. Universal intensity distribution function for synchrotron radiation

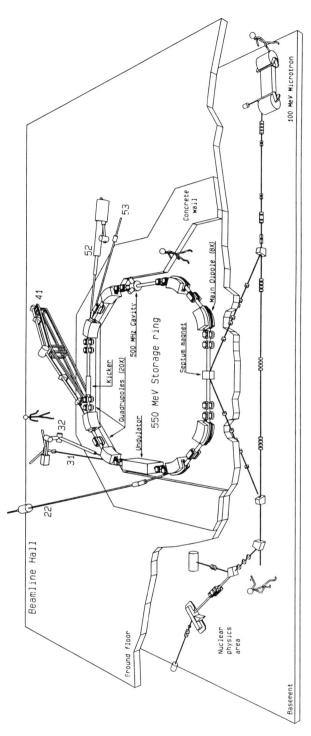

Fig.6.12. Lay-out of the MAX storage ring (Lund) for the generation of synchrotron radiation (Courtesy: P. Röjsel, MAX Synchrotron Radiation Laboratory)

Table 6.1. Data for some synchrotron radiation facilities (Courtesy: P.O. Nilsson, CTH, Göteborg)

VUV Facilities	E [MeV]	I [mA]	λ_c [Å]
SOR (Tokyo, Japan)	400	250	95
MAX (Lund, Sweden)	550	250	40
NSLS I (Brookhaven, USA)	750	500	30
UVSOR (Okazaki, Japan)	750	500	30
SUPERACO (Orsay, France)	800	400	20
BESSY (Berlin, FRG)	800	500	20
ALADDIN (Wisconsin, USA)	1300	500	12
ADONE (Frascati, Italy)	1500	60	8

X-Ray Facilities	E [GeV]	I [mA]	λ_c [Å]
DCI (Orsay, France)	1.8	300	1.9
SRS (Daresbury, UK)	2.0	500	3.9
PHOTON FACTORY (Tsukuba, Japan)	2.5	500	2.7
NSLS II (Brookhaven, USA)	2.5	500	3.0
SPEAR (Stanford, USA)	4.0	100	1.1
DORIS II (Hamburg, FRG)	5.0	50	0.5

Thus time-resolved experiments can be performed [6.51, 52]. Data for a number of facilities producing synchrotron radiation are given in Table 6.1. With synchrotrons of resonable sizes the intensity per Doppler width that is achievable is comparable to that which is obtained from efficent line light sources. However, the intensity increases towards the extreme UV (XUV) and X-ray regions, where no comparable continuum light sources exist.

The performance of syncrotrons and storage rings can be further enhanced by using *wigglers* or *undulators* [6.53]. In the former type of system a number of local, sharp bends in the electron beam are made using strong magnets. In this way a local emission of still shorter wavelength can be obtained, see (6.11). An undulator is a periodical structure of magnets on a straight part of the electron path between the large bending magnets of the machine, where the electrons are bent back and forth resulting in particularly strong radiation in narrow wavelength bands (fundamental band and high overtone bands) and in a well defined direction. By introducing suitable mirrors laser action can be achieved in a so-called *Free-Electron Laser* (FEL) [6.54-56]. Tunable radiation with extremely high powers should ultimately be feasible. FEL operation in the visible region has also been demonstrated at several facilities. Coherent light generation from an undulator is possible without using laser mirrors employing the *optical klystron* process [6.57]. The Inverse Free-Electron Laser (IFEL) action [6.58] can be used to accelerate electrons using intense laser beams.

Synchrotron radiation has many applications in a large number of scientific and industrial areas: physics, chemistry, biology, medicine and electronics. With regard to atomic and molecular spectroscopy synchrotron

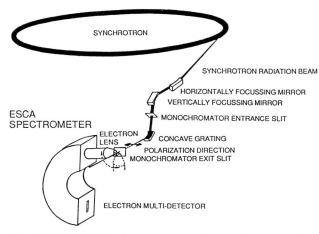

Fig.6.13. XPS (ESCA) arrangement at a synchrotron radiation source [6.59]

radiation is particularly applicable for X-ray and photo-electron studies. In most experiments it is necessary to have monochromatized photons, which can be obtained using specially designed monochromators capable of handling the intense radiation. An arrangement for ESCA studies (Sect.5.2) using synchrotron radiation is shown in Fig.6.13. Another important area in which synchrotron radiation can be applied is EXAFS (Extended X-ray Absorption Edge Spectroscopy, Sect.5.1.2). A spectrum can be recorded in typically 1/1000 of the time needed when a conventional X-ray source is used. Synchrotron radiation and its applications have been discussed in [6.60-63].

6.1.4 Natural Radiation Sources

In connection with continuum light sources we will also consider some natural radiation sources. The sun is clearly our most powerful radiation source, and apart from a large number of absorption and emission lines it essentially radiates like a Planck radiation source at a temperature of 6000 K (Fig.6.14). The moon exhibits a similar distribution although about 10^6 times weaker. The clear sky also yields a continuum-like distribution (Fig. 6.15). The distribution from the sun is enhanced towards the blue region because of the strong wavelength dependence of Rayleigh scattering. Further, there is a general ~300 K radiation in the IR region due to the thermal emission of the atmosphere. The spectral distribution of the night sky is determined by scattered starlight, galactic light and zodiacal light (Fig.6.16).

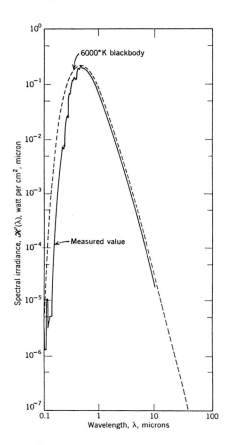

Fig.6.14. Spectral distribution of sunlight outside the atmosphere [6.64]

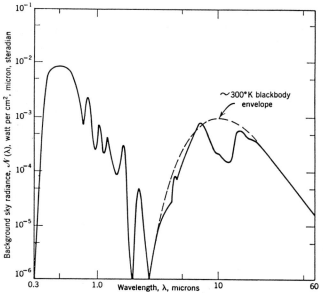

Fig.6.15. Spectral distribution of the blue sky [6.65]

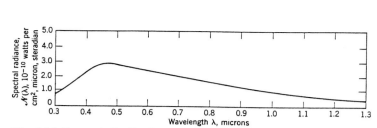

Fig.6.16. Spectral distribution of the night sky [6.65]

6.2 Spectral Resolution Instruments

Spectral resolution instruments of different kinds are used for resolving the different wavelength components of light. Important factors of merit for such devices are resolving power and light transmission. The resolving power $\mathscr{R}$ is defined by

$$\mathscr{R} = \frac{\lambda}{\delta\lambda}, \qquad (6.12)$$

where $\delta\lambda$ is the resulting linewidth of the spectral apparatus when using monochromatic light of wavelength λ.

We will describe four types of instruments: prism and grating instruments, the Fabry-Pérot interferometer and the Fourier transform spectrometer. A large number of varieties of these different types are used in spectroscopic research and various applications. Spectroscopic instruments have been discussed in [6.6, 7, 66].

6.2.1 Prism Spectrometers

Names such as spectrometer, spectroscope, spectrograph or monochromator are used for basically the same types of instruments, which are, however, applied in different ways. The principle of a prism spectrograph is illustrated in Fig.6.17. The resolving power $\mathscr{R}$ is determined by the dispersion, $dn/d\lambda$, and the length of the prism base b

$$\mathscr{R} = \frac{\lambda}{\delta\lambda} = b\frac{dn}{d\lambda}. \qquad (6.13)$$

If the prism is made of flint glass ($dn/d\lambda = 1200 \text{cm}^{-1}$ at $\lambda \sim 500\text{nm}$) and has a base length of 5 cm, then the resolving power $\mathscr{R}$ will be 6000. For

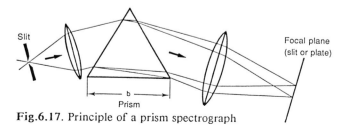

Fig.6.17. Principle of a prism spectrograph

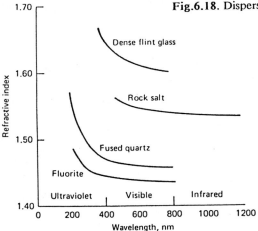

Fig.6.18. Dispersion of various prism materials [6.67]

practical reasons the size of the prism must be limited and thus this type of spectrograph will necessarily have a relatively low resolution. The finite width of the slits used will clearly reduce the theoretical resolving power given above. Since the dispersion of a prism strongly increases towards shorter wavelengths the resolution obtainable with a prism instrument will vary with the wavelength. In Fig.6.18 the dispersions of different prism materials are shown. Figure 6.19 displays the slit width necessary to achieve a 1 nm band pass for a typical prism spectrometer.

Commercial instruments frequently use a so-called *Littrow* mount. Here a Littrow prism with a 30° angle and one surface silvered, as shown in

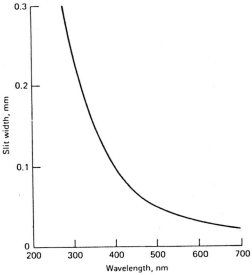

Fig.6.19. Diagram showing the slit width required to achieve a constant band pass of 1 nm from a prism spectrograph [6.67]

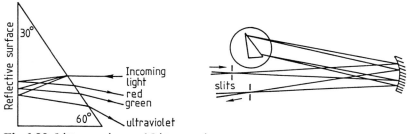

Fig.6.20. Littrow prism and Littrow prism mount

Fig.6.20, is used for increased dispersion. Arrangements with lenses or mirrors can be used for collimating the incoming light onto the prism and for focusing the refracted beams towards the photographic plate or the exit slit.

Prisms of different kinds can be used to deflect or deviate a beam of light. They can invert or rotate images and they can be used for separation of different states of polarization. Some prisms that are frequently used in optical and spectroscopic systems will be discussed here. Illustrations are found in Fig.6.21.

A *right-angle prism* using total internal reflection off the uncoated hypotenuse deviates a beam by 90°, if it impinges perpendicular to one of the entrance surfaces. This type of prism is very useful for deviating high-power laser beams.

In the *Amici (roof) prism*, the hypotenuse in the right-angle prism is replaced by two internal reflector surfaces oriented at 90° to each other (a

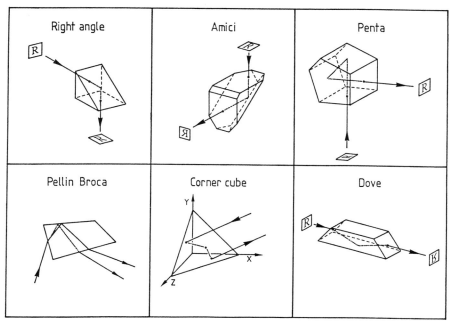

Fig.6.21. Prisms frequently used in optical and spectroscopic systems (After material from Melles Griot)

"roof"). This prism also deflects a beam by 90° but at the same time rotates the image by 180°. The *penta-prism* deflects the incoming beams by 90° *regardless* of the angle of incidence and does not invert or reverse the image. This prism does not operate on total internal reflection but requires a mirror coating on two faces.

The *Pellin-Broca* prism also deviates beams by roughly 90° but is a dispersive prism, frequently used to separate laser beams of different colours after nonlinear frequency conversion (Sect.8.6). By combining 4 Pellin-Broca prisms, beam separation with maintained direction of propagation can be achieved. In the *straight-view prism* colour dispersion in the forward direction is instead accomplished. The *dove prism* also transmits light in the forward direction but without dispersion. It has the important property of rotating the image (at twice the angular rate of the prism). A *corner-cube prism*, with three orthogonal total internal reflection faces acts as a *retroreflector*. A beam entering the corner cube will, after 3 reflections, be sent in the opposite direction regardless of the angle of incidence. Frequently, corner cube reflectors are made from 3 first-surface mirrors rather than from solid glass or quartz.

6.2.2 Grating Spectrometers

A reflection grating is used for the spectral separation in a grating spectrometer. The basic arrangement is shown in Fig.6.22. Constructive interference is obtained when the optical path difference is an integer number (m) of wavelengths for diffraction at adjacent lines, as expressed by the grating equation

$$m\lambda = d(\sin\alpha + \sin\beta) . \qquad (6.14)$$

Here d is the line separation and α and β the angles of incidence and reflection, respectively. The resolving power $\mathscr{R}$ of the grating is determined by the total number of illuminated lines N and by the diffraction order m, i.e.,

$$\mathscr{R} = \frac{\lambda}{\delta\lambda} = N\cdot m . \qquad (6.15)$$

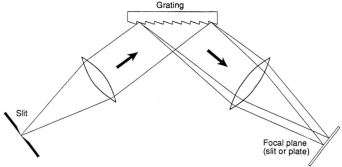

Fig.6.22. Principle of a grating spectrograph

Thus a 10 cm grating with 3,000 lines/cm has a resolving power of 30,000 in the first order.

Gratings were first made by J. Fraunhofer in 1823. Normally, a grating is ruled in a surface layer of aluminum on a substrate by means of a diamond tip. The ruling is performed with high-precision ruling machines, which are interferometrically controlled. *Replica gratings* which are the ones that are marketed, are manufactured by a casting procedure. During recent years *holographic gratings* have been much used. These gratings are produced by recording interference fringes from two crossed laser beams. An Ar^+ laser acting on photoresist can be used. A density of up to 6,000 lines/mm can be attained.

The intensity that is diffracted at a certain wavelength depends on the shape of the lines. Ruled gratings are made with a certain *"blaze" angle*, chosen according to which wavelength region is to be enhanced through reflective action (Fig.6.23). The efficiency of a grating can be up to 70% at the blaze angle for a certain order. Recently, it has also become possible to produce gratings with a blaze by the holographic process. From mechanically ruled gratings so-called ghost lines can appear due to periodical errors in the mechanical feeding of the machine. Such lines do not result from holographic gratings, which also have a lower level of diffusely scattered light (stray light). While the first diffraction order is normally used in small spectrometers, higher diffraction orders are frequently used in large research instruments. One drawback of grating spectrometers is that for a certain grating setting a series of wavelengths (λ_0, $\lambda_0/2$, $\lambda_0/3$, ...) is diffracted in the same direction (overlapping orders). The problem can be eliminated using filters or a premonochromator. An especially high resolution, but also many overlapping orders, is obtained with an echelle grating (Fig.6.24). Such gratings operate at such a high angle that the steep side of the line is utilized. Echelle gratings have comparatively few lines/mm but operate at a very high diffraction order. A resolving power approaching 10^6 can be achieved with grating instruments. Grating characteristics are discussed in [6.68].

Grating spectrometers are, in general, equipped with mirrors instead of lenses. Some common arrangements are shown in Fig.6.25. Instruments for the visible region frequently make use of the Ebert or the Czerny-Turner arrangement. A grating in a Littrow mount is frequently used for tuning pulsed dye lasers (Sect.8.5.1). By using concave gratings the need for col-

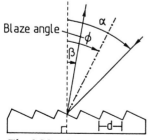

Fig.6.23. Blaze angle of a grating

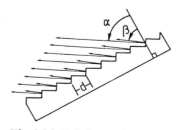

Fig.6.24. Echelle grating

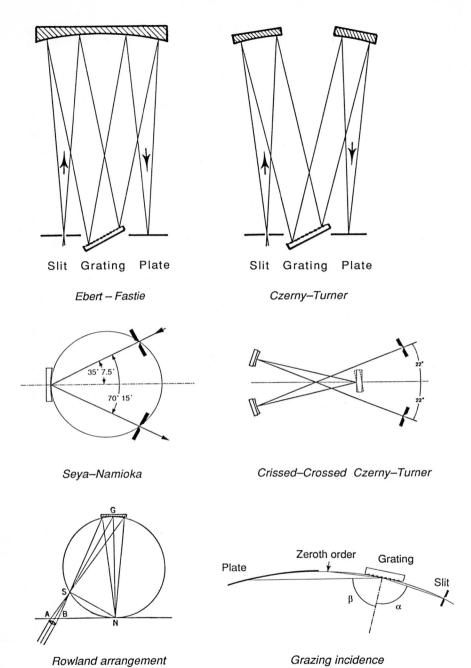

Fig.6.25. Some common arrangements for grating spectrometers

limating and focussing mirrors is eliminated, which is particularly valuable in the VUV (vacuum ultraviolet) and XUV (extreme UV) regions where conventional mirrors are ineffective.

Frequently an arrangement utilizing the *Rowland circle* is used: The entrance slit, the grating and the spectral image are all on a circle which has a diameter equal to the radius of curvature of the concave grating. For the very short wavelengths that are obtained in spectra from highly ionized atoms or in X-ray spectra, a *grazing angle of incidence* is used to minimize the absorption losses in the grating.

The beam leaving a monochromator always contains small amounts of radiation at wavelengths other than the selected one. This *stray light* is due to reflections and scattering from different parts of the monochromator. The amount of stray light can be much reduced by using a *double monochromator* which consists of two adjacent single monochromators connected only through a common intermediate (exit/entrance) slit.

6.2.3 The Fabry-Pérot Interferometer

The Fabry-Pérot interferometer was introduced by C. Fabry and A. Pérot in 1896. The interferometer consists of two flat, parallel mirrors with high reflectivity and low absorption. Light impinging on the interferometer will undergo multiple reflections between the mirrors whereby part of the light is transmitted (Fig.6.26). The different components of the transmitted light will interfere at infinity, but the interference pattern can be imaged on a screen using a lens. According to the Fermat principle the relative phases of the different rays will not be changed by their passage through the lens. For the analysis we introduce the following symbols pertaining to light *intensity*:

R : reflectivity of a mirror layer,
T : transmission of a mirror layer,
A : absorption of a mirror layer

with the relationship

$$R + T + A = 1 \ . \tag{6.16}$$

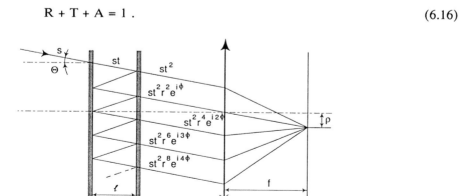

Fig.6.26. Ray tracing in a Fabry-Pérot interferometer

For the *amplitudes* r, t, and a are the corresponding quantities, i.e.

$$r^2 = R, \quad t^2 = T, \quad a^2 = A. \tag{6.17}$$

Further, we introduce

- ℓ : distance between the layers,
- n : index of refraction between the layers,
- θ : angle of incidence of the light,
- f : focal length of the lens,
- λ : wavelength of the incoming monochromatic light,
- ϕ : phase shift between two successively emerging rays, and
- s,S : amplitudes.

We will first calculate the optical path difference Δ (the *retardation*) between two successive rays making use of Fig.6.27, namely

$$\Delta = 2n\ell\cos\theta. \tag{6.18}$$

Constructive interference requires

$$\Delta = m\lambda; \tag{6.19}$$

m is normally a large integer (m $\simeq 10^4$). All rays that are incident on a conical surface are equivalent. Thus we will obtain a system of rings on the screen. The radius of a ring is denoted ρ. For small angles we have $\cos\theta = 1-\theta^2/2$ and $\theta = \rho/f$, which together with (6.18, 19) yields $2n\ell(1-\rho^2/2f^2) = m\lambda$ and

$$\rho = f\sqrt{m\lambda/n\ell}. \tag{6.20}$$

Thus, for a case when constructive interference occurs on the symmetry axis (the square root is zero for m=m_0) rings with m = m_0+1, m_0+2, ... will have radii proportional to $\sqrt{1}$, $\sqrt{2}$, $\sqrt{3}$ etc.

We shall now study the intensity distribution in more detail. Using Fig.6.26 we find that the total transmitted amplitude S for the case of a = 0, i.e. $t^2 = 1-r^2$, is given by the geometrical series

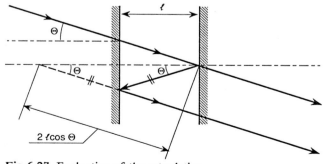

Fig.6.27. Evaluation of the retardation

$$S = st^2(1 + r^2 e^{i\phi} + r^4 e^{i2\phi} + r^6 e^{i3\phi} + ...) = \frac{s(1 - r^2)}{1 - r^2 e^{i\phi}} , \quad (6.21)$$

where the phase shift ϕ is given by

$$\phi = \frac{\Delta}{\lambda} 2\pi = \frac{\Delta}{c} 2\pi\nu . \quad (6.22)$$

From the amplitude S we obtain the corresponding intensity I

$$I = |S|^2 = \frac{s^2(1 - r^2)^2}{1 - r^2(e^{i\phi} + e^{-i\phi}) + r^4} .$$

But $(e^{i\phi} + e^{-i\phi})/2 = \cos\phi = 1 - 2\sin^2(\phi/2)$. With $s^2 = I_0$ we then have

$$I = I_0 \left[1 + \frac{4r^2}{(1 - r^2)^2} \sin^2\phi/2 \right]^{-1} .$$

Finally, if the absorption in the layers is considered we obtain

$$I = \left(\frac{T}{1-R}\right)^2 I_0 \left[1 + \frac{4R}{(1-R)^2} \sin^2\phi/2 \right]^{-1} . \quad (6.23)$$

This is called the *Airy distribution*, and is illustrated in Fig.6.28. The maximum intensity

$$I_{max} = I_0 \left(\frac{T}{1-R}\right)^2 \quad (6.24)$$

is obtained for $\phi = 0, 2\pi, ...$ while the minimum intensity

$$I_{min} = I_0 \left(\frac{T}{1+R}\right)^2 \quad (6.25)$$

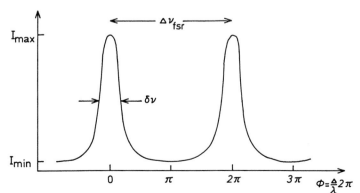

Fig.6.28. Airy distribution from a Fabry-Pérot interferometer

is obtained for $\phi = \pi, 3\pi, \ldots$ etc.

The separation between maxima is called the *free spectral range*. In frequency units it is given by

$$\Delta\nu_{fsr} = \frac{c}{\Delta} = \frac{c}{2n\ell\cos\theta} = \frac{c}{2n\ell} . \tag{6.26}$$

The halfwidth of a transmission peak is

$$\delta\nu = \frac{c}{\pi} \frac{1-R}{\Delta\sqrt{R}} \quad \text{(for large values of R)} . \tag{6.27}$$

The ratio between these two quantities is called the *finesse* N and is determined only by the reflectivity R of the layers

$$N = \frac{\Delta\nu_{frs}}{\delta\nu} = \frac{\pi\sqrt{R}}{1-R} . \tag{6.28}$$

The resolution of the Fabry-Pérot interferometer is

$$\mathscr{R} = \frac{\lambda}{\delta\lambda} = \frac{\nu}{\delta\nu} = \frac{\nu}{\Delta\nu_{frs}}\frac{\Delta\nu_{fsr}}{\delta\nu} = \nu\frac{2n\ell}{c}N . \tag{6.29}$$

The variable ϕ in the Airy function is

$$\phi = \frac{2n\ell\cos\theta}{\lambda}\cdot 2\pi . \tag{6.30}$$

Thus, for a given wavelength λ we can run through the transmission curve in Fig.6.28 by varying θ, n or ℓ.

In a photographic recording the ring system is recorded as a function of θ. For a photoelectric recording $\theta = 0°$ is chosen by placing a small hole on the symmetry axis, and the intensity of the central spot is recorded with a photomultiplier tube. Now either ℓ or n is changed. The Fabry-Pérot interferometer normally consists of two flat quartz substrates with reflective coatings and the substrates are precisely separated by Invar spacers. n can then be varied by changing the gas pressure in the Fabry-Pérot housing (pressure scanning). Alternatively, the spacing can be varied by using piezo-electric crystals in the mounts of one of the plates.

As an example, we will consider a Fabry-Pérot interferometer which has a mirror separation of 1 cm (air). The coatings have a reflectivity of 98%. We then have $\Delta\nu_{fsr} = c/2\ell = 15$ GHz, $N \simeq 150$ and $\delta\nu = 100$ MHz. If the interferometer is used at 500 nm $\leftrightarrow 6\cdot 10^{14}$ Hz then $\mathscr{R} = 6\cdot 10^6$. Clearly, for such an interferometer the width of the spectral line from the light source is a limiting factor.

Because of small imperfections and the finite size of the circular aperture the practically achievable resolution is frequently reduced. Because of its high resolution the Fabry-Pérot interferometer has been much used for

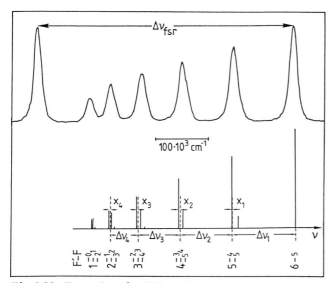

Fig.6.29. Examples of a Fabry-Pérot recording obtained by pressure scanning. The hyperfine structure of ^{55}Mn (I=5/2) in a J = 7/2 ↔ 5/2 transition at 5395 Å has been recorded with an instrumental free spectral range of 15 GHz [6.69]

measuring hyperfine structure and isotope shifts. The spectral line of interest is then first selected by a monochromator or an interference filter. First a free spectral range of sufficient width to accomodate all the spectral components of the line is chosen to allow the correct order of components to be determined. Then the plates are moved further apart resulting in an increased resolution but also the inclusion of overlapping orders.

All hyperfine components have their own Airy functions, which are shifted with respect to each other. The Fabry-Pérot pattern repeats itself with a period of the free spectral range, which is used for the frequency calibration. An example of a Fabry-Pérot recording is illustrated in Fig. 6.29. The Fabry-Pérot interferometer was discussed in [6.70].

Interferometers with spherical mirrors can also be used. A frequently used arrangement is used in the *confocal* interferometer where the mirror separation ℓ equals the radius r of the mirrors (Fig.6.30). The naming of this interferometer is due to the fact that the focal length f of a mirror of radius r is f = r/2. In a confocal interferometer a light ray makes 2 double passes between the mirrors before the primary and the reflected beams again coalesce at the second mirror and can interfere.

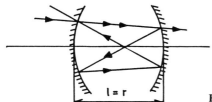

Fig.6.30. Confocal interferometer [6.71]

Since this corresponds to a planar interferometer with double the mirror separation the free spectral range for a confocal interferometer is given by

$$\Delta\nu_{fsr} = \frac{c}{4n\ell}. \qquad (6.31)$$

This interferometer is a special case of the *multi-pass interferometer*, where N double passes between the mirrors occur before interference. For such an interferometer we have

$$\Delta\nu_{fsr} = \frac{c}{N \cdot 2n\ell}. \qquad (6.32)$$

It can be shown that the mirror radius and the mirror separation must fulfil the relation

$$r = \ell[1 - \cos(\pi/N)]^{-1}. \qquad (6.33)$$

For $N = 2$ we have the special case of the confocal interferometer. As an example, we can choose a multi-pass interferometer with a mirror separation of 0.75 m operating with $N = 4$ (Fig.6.31). We find $r = 0.75/(1-1/\sqrt{2})$ = 2.56 m. The free spectral range for this device is $\Delta\nu_{fsr} = c/4 \cdot 2 \cdot 0.75$ m = 50 MHz.

Multi-pass interferometers are frequently used for monitoring laser beams of a very sharp frequency. A small free spectral range can be obtained without using a very long interferometer. Using special multilayer techniques it has recently become possible to produce coatings resulting in a finesse of the order of 10^3. Clearly, such instruments have important applications for ultra-stable lasers and precision metrology.

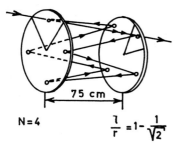

Fig.6.31. Multi-pass interferometer for N=4 [6.71]

6.2.4 The Fourier Transform Spectrometer

The Fourier transform spectrometer (FTS) is a dual-beam interferometer, which is most frequently of the Michelson type, as shown in Fig.6.32. If the arms of the interferometer have equal lengths the path difference between the two interferring beams would be 0. If the mirror is moved $\Delta/2$ an optical path difference of Δ is introduced. For the case of monochromatic radiation and equally intense beams the intensity at the detector will be

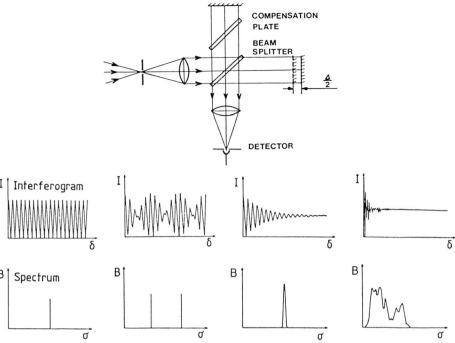

Fig.6.32. Fourier transform spectrometer and basic signals (Courtesy: Ulf Litzén)

$$I(\Delta) = I_0 \cos^2 \phi/2 \qquad (6.34)$$

with

$$\phi = \frac{\Delta}{\lambda} 2\pi = \frac{\Delta}{c} 2\pi\nu . \qquad (6.35)$$

If the light source emits a spectrum $B(\nu)$ we then obtain

$$I(\Delta) = \int_0^\infty B(\nu) \cos^2\left(\frac{\Delta}{c}\pi\nu\right) d\nu$$

$$= \frac{1}{2} \int_0^\infty B(\nu) \left[1 + \cos\left(\frac{\Delta}{c} 2\pi\nu\right)\right] d\nu . \qquad (6.36)$$

The part of the above expression that depends on Δ is called the *interferogram*

$$J(\Delta) = \frac{1}{2} \int_0^\infty B(\nu) \cos\left(\frac{\Delta}{c} 2\pi\nu\right) d\nu . \qquad (6.37)$$

The spectrum B(ν) can be calculated from the interferogram J(Δ) as its Fourier cosine transform

$$B(\nu) \propto \int_0^\infty J(\Delta)\cos\left[\frac{\Delta}{c}2\pi\nu\right]d\Delta \ . \tag{6.38}$$

Such a calculation is conveniently performed on a computer [6.72] and special, very fast processors have been constructed. In practice, the movable mirror can only be moved a limited distance (<1 m) and therefore the integration must be performed over a finite interval. Such a limited integration gives rise to side maxima on the spectral lines. By using a mathematical trick, which involves the multiplication of the integrand in (6.37) by a particular function, the unwanted maxima can be suppressed. However, this procedure, which is called *apodization*, gives rise to broader spectral lines. The interferogram has many similarities to a *hologram*. Every part of it contains information on the whole structure, but a high resolution is obtained only when a large part of it is utilized. One of the great advantages of Fourier transform spectroscopy is that all spectral lines are recorded at the same time (the *multiplex* or *Felgett advantage*). Unless photographic or array detector recording is used the earlier mentioned types of spectral apparatus only collect information at a particular wavelength at any time. In the IR region, where photographic plates are not available, Fourier instruments are particularly valuable. Further, a Fourier transform spectrometer has a comparatively high light collecting efficiency, since it does not require narrow slits for the above-mentioned reasons (the *Jaquinot advantage*). Instruments of this kind have a very good signal-to-noise ratio and instruments with a resolution of $\mathscr{R} = 10^7$ can be constructed. The movement of the mirror is normally controlled by fringes from a HeNe laser. Examples of FTS spectra are shown in Fig. 6.33. Interferometers and interferometry have been discussed in [6.74-77].

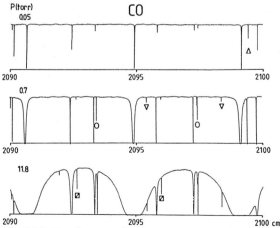

Fig.6.33. Fourier-transform spectrometer recording for a gas containing different isotopic species of a CO at increasing pressures [6.73]

6.3 Detectors

Historically, the photographic plate is an important detector of light that has been spectrally decomposed by a spectral apparatus. Photographic emulsions are still important for recording the emission lines from ionized atoms. As we have already mentioned, *photographic material* has the unique property of being an integrator over the intensity fluctuations of a light source. There are many types of films for spectroscopic use. In Fig.6.34 regions of sensitivity for some emulsions are shown. The information stored on a plate can be sensed by a *microdensitometer* and transferred to a recorder. For the evaluation of emission spectra reference lines from a different light source are frequently exposed on the same plate. A *comparator* is used to determine the relative positions of the lines on the plate.

In modern spectroscopy in the visible or the UV region a *Photo-Multiplier Tube* (PMT) is normally used. A PMT is an amplifying photocell. The light falls on a photocathode from which electrons are released due to the photoelectric effect. The electrons are accelerated through a chain of secondary electrodes, dynodes, that are held at increasingly higher potential. The dynodes, which are normally covered with CsSb, release secondary electrons and an avalanche amplification is obtained. A current that can be more than 10^6 times stronger than the primary photocurrent is obtained from the collecting anode. In Fig.6.35 a PMT and its electrical supply are shown. A number of cathode materials with different spectral sensitivities can be chosen. The sensitivity is frequently expressed as the *quantum efficiency*, which is the number of photo-electrons (<1) that can be obtained per incoming absorbed photon. Two common materials are Na_2KSbCs ("trialkali") and AgOCs. In Fig.6.36 curves for a number of materials are shown. S designations for the materials are frequently used (e.g., S20).

In many cases it is important to minimize the *dark current* of the PMT. The number of thermally released electrons can be reduced by cooling. For

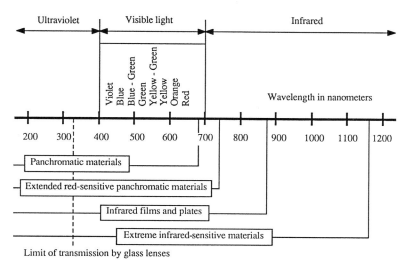

Fig.6.34. Sensitivity regions for photographic emulsions [6.78]

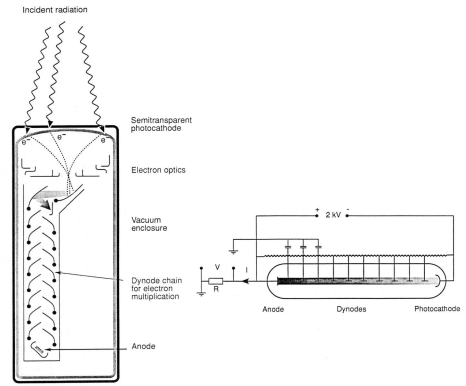

Fig.6.35. Construction of a photomultiplier tube and its high-voltage supply

most materials no further improvement is obtained below -30°C, except for red-sensitive materials, with e.g. an S1 response, for which dry ice (222K) or liquid nitrogen (77K) cooling is advisable. The dark current in PMT's can be reduced by a factor of 10 or more by cooling.

No detectors of the PMT type exist for the IR region. Other types of detectors are instead used. A *thermocouple* can be a useful detector at long wavelengths. A *bolometer* is a resistance thermometer with a platinum wire or a thermistor as a sensor. Certain crystals are *pyroelectric*, i.e. they give rise to an electrical potential between opposite surfaces when heated. Lithium niobate is used for such purposes. *Photoconducting detectors* are particularly useful in the IR region. Especially common is the *lead sulphide (PbS) cell*, which operates at room temperature, but with a rather slow (~ms) response time. Detectors made of *indium arsenide* (InAs) or *indium antimonide* (InSb) are considerably faster and are frequently cooled to liquid nitrogen temperatures (77K). Further down in the IR region, e.g. around 10 μm, cooled *mercury cadmium telluride* (HgCdTe) detectors are used. Sensitivity curves for different detector materials are shown in Fig.6.37.

Since the internal low-noise amplification of PMT:s has no counterpart in the IR region (except possibly for the case of *avalanche diodes*) one has

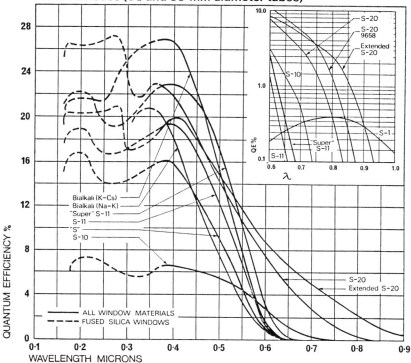

Fig.6.36. Quantum efficiency curves for various photocathode materials (Courtesy: EMI)

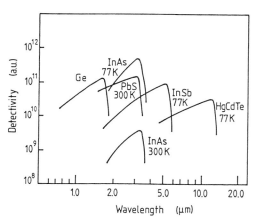

Fig.6.37. Sensitivity curves for infrared detector materials

to rely on low-noise electronic amplification. Noise is normally a greater problem in the IR region than in the visible region. One remedy in the IR region is to use *heterodyne detection*, as discussed in Sect. 10.2.1. Optical detectors have been discussed in [6.79-83].

We shall conclude this section on detectors by briefly describing *array detectors* which combine the advantages of the integration and multiplexing characteristics of the photographic plate, as well as the electronic read-out of the PMT. In these solid state devices, a large number of light-sensitive elements (256, 512 or 1024) are closely arranged in a row. Each one is connected to one channel of a multichannel analyser, in which a certain number of counts are stored, proportional to the intensity of the impinging light. If the array of detectors is arranged in the focal plane of a spectrograph the spectrum is immediately obtained in the multichannel analyser (Optical Multichannel Analyser, OMA). A linear diode array can consist of, e.g. 512 optical diodes, each 50 μm wide and 2.5 mm high. The elements then form an "electronic photographic plate" of the size 25×2.5 mm². In order to attain a higher light sensitivity the array can be placed behind an *image intensifier tube*, which is based on a microchannel plate [6.84]. The arrangement is shown in Fig.6.38. The spectrum illuminates a photocathode in which electrons are released. The electrons are sucked into the narrow channels of the microchannel plate. The channels are covered with a material emitting secondary electrons and multiplication occurs as the electrons pass along the channel, propelled by an applied electric field. The showers

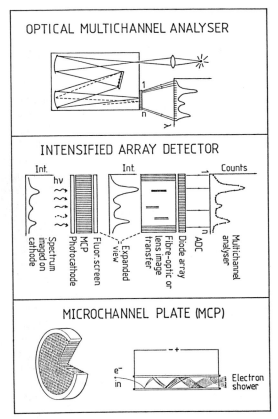

Fig.6.38. Optical multichannel analyzer with microchannel-plate image intensifier

of electrons, that are spatially arranged corresponding to the primary spectrum, impinge on a phosphor screen that produces an amplified image of the original spectrum. The light is then transferred to the diode array with retained spatial information, sometimes employing bunches of optical fibres. An amplification of more than 10^4 can be obtained. The high voltage for the channel-plate operation can be pulsed rapidly, and thus the whole detector assembly can be gated to be sensitive during time intervals down to 5 ns. This type of operation is of great interest, e.g. for background rejection in connection with spectroscopy using short laser pulses.

Array detectors can also be two-dimensional and can be used for imaging. We will discuss spectroscopic applications of such imaging later. The field of solid state array detectors is rapidly developing driven by TV applications. The whole area of CCD (Charge Coupled Device) technology will undoubtedly yield interesting new possibilities for advanced detection of spectroscopic information. Normal TV camera techniques using vidicon tubes are also in use for optical multichannel analysers, as discussed above. Also here image intensifier techniques are used as in low light level TV applications. Imaging electro-optic detectors have been discussed in [6.85].

6.4 Optical Components and Materials

6.4.1 Interference Filters and Mirrors

Filters of different kinds are employed to select a certain spectral region. Relatively narrow spectral regions can be isolated using an *interference filter*. Such a filter for a wavelength λ is, in its simplest form, a Fabry-Pérot interferometer with a mirror separation of $\lambda/2$. Transmission maxima are also obtained for the wavelengths $\lambda/2$, $\lambda/3$, $\lambda/4$ etc. Filters for the short wavelength region, $\lambda < 240$ nm, are frequently made in this way with partially reflecting metal layers on each side of a $\lambda/2$ dielectric layer. Transmission maxima at shorter wavelengths are effectively absorbed by the substrate material (e.g., quartz; see below). The absorption of the metal layers reduces the transmission of the filter. Because of this, *multiple dielectric layers* are used almost exclusively at longer wavelengths. In order to achieve a high reflectivity and low absorption many layers of thickness $\lambda/4$ with alternating high and low refractive indices (e.g., ZnS n=2.3 and MgF_2 n=1.35) are used. The number of layers used determines the sharpness of the filter. The centre wavelength is determined by the thickness of the spacer layer. By using two or more filters of this kind in series (multi-cavity filter), with coupling layers to reduce reflection losses, sharper filters can be constructed. Suppression of transmission maxima at shorter wavelengths can easily be achieved by combining the interference filter with a high-pass coloured glass filter (Sect.6.4.2). Unfortunately there are no simple efficient low-pass filters for suppressing long wavelengths. If such suppression is required a filter with *induced transmission* can be used. Such a filter, which is a hybrid between a metal-dielectric-metal filter and an all-dielectric filter, is characterized by the absence of long-wavelength transmission. In order to obtain a sharp interference filter with complete suppression outside the pass-band it can be necessary to combine a multi-cavity filter, an induced transmission filter and a high-pass coloured glass filter to form an integrated composite filter. Examples of individual transmission curves are shown in Fig.6.39.

The transmission maximum of an interference filter is shifted towards shorter wavelengths if it is used at an inclined angle. The new transmission wavelength λ_θ is related to the nominal wavelength λ_0 by

$$\lambda_\theta = \lambda_0 \frac{\sqrt{n^2 - \sin^2\theta}}{n}. \tag{6.39}$$

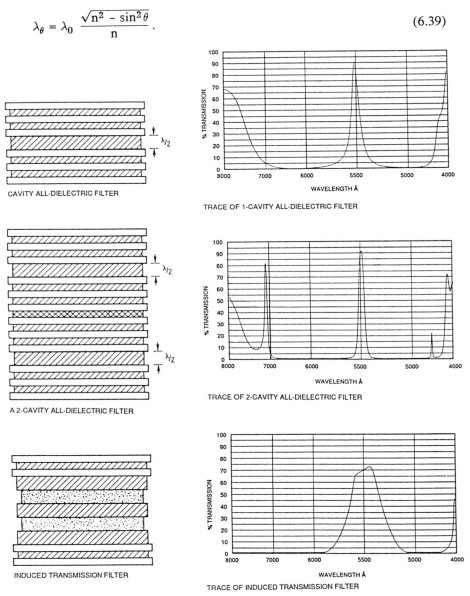

Fig.6.39. Interference filter layers and resulting transmission curves [6.86]

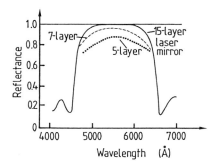

Fig.6.40. Multiple-layer mirror reflectivity [6.87]

Here θ is the angle of incidence and n is the refractive index of the spacer layer. As an example, for $\theta = 10°$, n = 1.45 and a filter wavelength of 600 nm we obtain $\Delta\lambda_\theta = \lambda_0 - \lambda_\theta \simeq 4$ nm. This means, that if a filter is used in a converging or diverging beam of light the transmission profile is somewhat shifted towards shorter wavelengths. For a cone of light with an angle at the apex of 2θ the effective shift is about $\Delta\lambda_\theta/2$.

Mirrors are also made using multiple dielectric layers of thickness $\lambda/4$. The more layers that are added, the better the reflectivity that can be achieved in a certain wavelength region. In Fig.6.40 the effect of an increase in the number of layers is demonstrated. A reflectivity better than 0.999 can be obtained using 30 layers. Clearly, thin-film techniques of this kind are of great importance for the construction of laser cavities and for interferometers with sharp frequency discrimination characteristics.

We will also briefly discuss *anti-reflection layers*. A surface without any special preparation between an external medium (frequently air), with refractive index n_0, and an optical component with refractive index n, has a reflectivity R given by

$$R = \left[\frac{n - n_0}{n + n_0}\right]^2 . \tag{6.40}$$

For a quartz-air boundary we have R = $[(1.5-1)/(1.5+1)]^2 \simeq 4\%$. It can be shown that if a layer of refractive index n_1 and thickness $\lambda/4$ is evaporated onto the optical component the reflectivity will be zero provided that $n_1 = \sqrt{n}$. MgF_2 with n = 1.35 is frequently a suitable material. In order to achieve anti-reflection properties in a larger wavelength region, e.g. in the whole visible region, multiple-layer techniques are used, as illustrated in Fig.6.41. Advanced anti-reflection techniques are clearly a prerequisite for the construction of modern multiple-element lenses, etc. which would otherwise be cluttered by scattered light from the many surfaces.

Mirrors that are intended to be used over a wide range of wavelengths generally consist of *metal coatings* on a substrate. A layer thickness of the order of 150 nm is adequate. Reflectivity curves for some metal coatings are shown in Fig.6.42. Silver has a reflectivity of 96-98% in the visible and IR regions. Since a silver surface tarnishes a protective layer of, e.g., SiO is needed. Gold is a highly suitable material for the IR region. Aluminum can

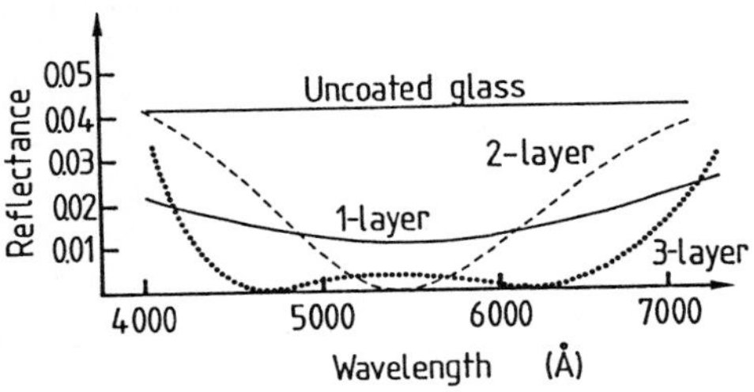

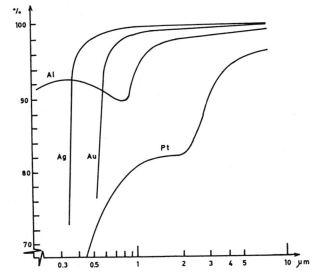

Fig.6.41. Action of anti-reflection coatings [6.87]

Fig.6.42. Reflectivity of metal coatings

be used over a wide range of wavelengths from UV to IR. Aluminum coatings are especially useful far down in the UV region, i.e. 200 nm and below. The reflective properties in the UV region can be improved by using a surface layer of SiO_2 or MgF_2, which also allows the otherwise very vulnerable mirror to be cleaned. In techniques involving high-power pulsed laser beams aluminum mirrors have a limited applicability because of the possibility of evaporation of the coating. For such applications dielectric mirrors or totally reflecting 90° prisms are used.

6.4.2 Absorption Filters

If the narrow region of transmission typical for an interference filter is not required a simpler and much cheaper absorption filter can frequently be used. In *coloured glass filters* or plastic filters the absorption can be caused by simple or complex ions (e.g., nickel, cobolt, neodymium, praseodymium, uranium). Semiconductor-doped glasses are used as sharp cut-off filters. Using CdS_xSe_{1-x} doping the absorption edge moves downwards for increasing x-values. Such filters are very efficient for suppressing short wavelengths. In Fig.6.43 some transmission curves for coloured glass filters are given.

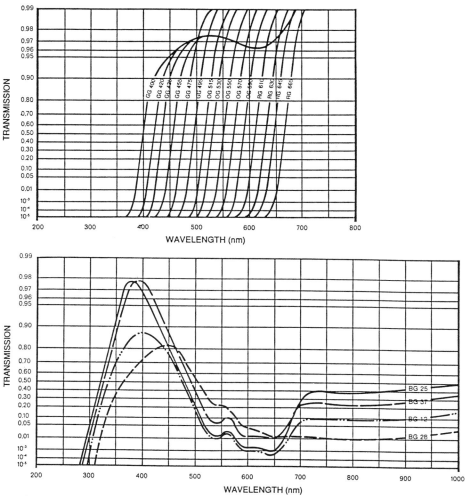

Fig.6.43. Examples of transmission curves for coloured-glass filters (Courtesy: Schott Glaswerke)

Sometimes *liquid filters* consisting of a quartz cuvette filled with a particular liquid can be quite useful. Organic liquids can be used as high-pass filters with cut-offs in the UV region, where coloured glass filters cannot be used. The cut-off wavelengths are schematically given in Fig. 6.44 for a number of liquids. Water solutions of inorganic salts are also of interest. For example, a mixture of nickel sulphate and cobalt sulphate is transparent in the UV region from 230-330 nm but completely blocks the visible region. Liquid filters have been discussed in [6.88, 89].

Absorption filters, in contrast to interference filters, can be used at any angle of incidence and can thus be used in strongly converging or diverging beams. It should be noted that certain coloured glass filters can exhibit rather strong broad-band fluorescence when used to block UV light, which

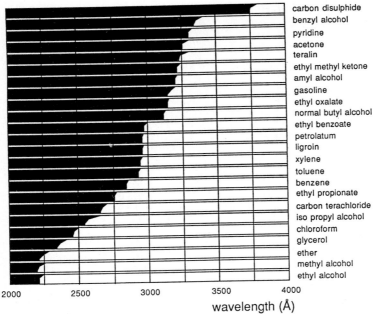

Fig.6.44. Spectral characteristics for some liquid filters. Black indicates blocked transmission [6.88]

can sometimes cause problems. When a combination of an interference and a coloured glass filter is used, placing the interference filter in front of the fluorescing filter can improve the performance.

Neutral-density filters are used to reduce the intensity of a light beam in a well-defined way, e.g. for testing the linearity of an optical detector. Filters of this kind should have a constant attenuation over large spectral ranges to facilitate their use. Normally, semitransparent metal films of chromium or nickel on quartz substrates are used. A filter is characterized by its transmittance T, or optical density D. D and T are related according to

$$D = \log_{10}(1/T) . \tag{6.41}$$

Thus, a filter of optical density $D = 1.0$ has a transmittance T of 0.1 (10%). When two neutral density filters are combined the transmittances are multiplied while the optical densities are added. In Table 6.2 some pairs of D and T values are given. For high-power laser beams, bulk absorbing glasses or combinations of inclined quartz plates (Fresnel reflection) must be used [6.90].

Table 6.2. Optical densities D and transmissions T

D	T [%]	D	T [%]
0.0	100	0.8	16
0.1	79	1.0	10
0.2	63	1.5	3.1
0.3	50	2.0	1.0
0.4	40	3.0	0.1
0.5	32	4.0	0.01

6.4.3 Polarizers

Polarized light can be obtained or analysed using certain prisms or polarizing films. Reflection at a non-normal angle of incidence at the flat surface of an optical material also results in a certain degree of polarization according to the Fresnel formulae (Fig.6.45). In particular, at Brewster's angle ($\tan\theta = n$) the reflected light is fully polarized perpendicularly to the plane of incidence (the plane containing the normal to the surface) (S-polarized light). Light components polarized in the plane of incidence (P-polarized light) penetrate into the medium without loss. A stack of glass or quartz plates at Brewster's angle will successively reflect almost all components perpendicular to the plane of incidence leaving a highly polarized transmitted beam.

A convenient way to attain a very high degree of linear polarization is to use prisms made of the birefringent material calcite ($CaCO_3$). The arrangements in *Glan-Taylor* and *Glan-Thompson* polarizers are shown in Fig.6.46. Both these polarizers consist of a combination of two prisms, in the first type air-spaced, in the second case cemented. The prism angle has been chosen such that the ordinary beam is totally internally reflected and absorbed laterally in the prism while the extra-ordinary beam is transmitted into the next prism and leaves it undeviated. Glan-Thompson prisms have a larger acceptance angle but cannot be used in the UV region because of ab-

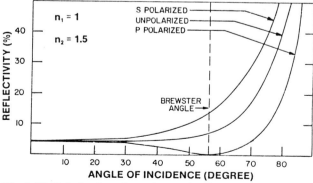

Fig.6.45. Polarization-dependent reflectivity at an optical surface (Fresnel formulae) [6.87]

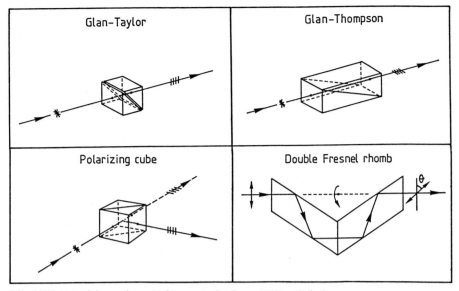

Fig.6.46. Polarizing prisms (After material from Melles Griot)

sorption in the optical cement. Both types of Glan prisms exhibit an extinction ratio of better than 10^5 for crossed polarizers.

A *polarizing beam-splitter* is formed by combining two right-angle prisms into a cube. A special multi-layer film between the prisms acts as a stack polarizer as described above. An unpolarized beam is thus divided into two perpendicular highly polarized components. Conversely, two polarized beams can be combined into a single beam using this type of component.

Optical retarder plates, of appropriate thicknesses, made of birefringent material, normally crystalline quartz or mica, can be used to rotate the plane of polarization for linearly polarized light or to produce circularly polarized light. To rotate linearly polarized light by an arbitrary angle, a *double Fresnel rhomb* is very useful (Fig.6.46).

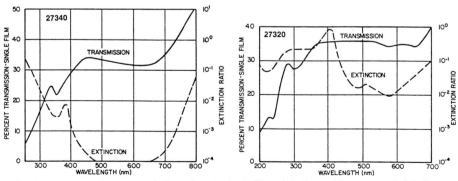

Fig.6.47. Spectral characteristics of polyvinyl film polarizers (Courtesy: Oriel Corp., Stratford, CT, USA)

Linear polarizers can also be made of *polyvinyl film* that has been stretched so as to align long chains of molecules. Different kinds of polarizers of this type are available for wavelengths from 200 nm up to the near IR. In Fig.6.47 examples of transmission and extinction curves are shown.

6.4.4 Optical Materials

Optical components are manufactured from transparent materials of certain refractive indices. We have already discussed the refractive properties of glass and quartz in Sect.6.1.1. Here we will consider the transmission properties of optical materials. Optical glass is transparent from about 350 nm to 2.6 µm. This region of transmission can be extended by using quartz. In Fig.6.48 transmission curves for different qualities of quartz are shown. As can be seen the best quartz has a transmission down to 170 nm. However, strong absorption bands can occur in the near IR region, particularly at 2.7 µm. These are due to the presence of water (O-H bonds) in the quartz. Water-free quartz can be used up to about 3.5 µm.

The region of UV transmission can be extended down to 125 nm using CaF_2 and to 110 nm using LiF. These materials are also transparent in the

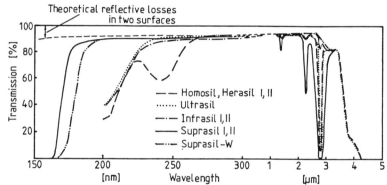

Fig.6.48. Transmission curves for different qualities of quartz [Courtesy: Heraeus Quarzglas GmbH)

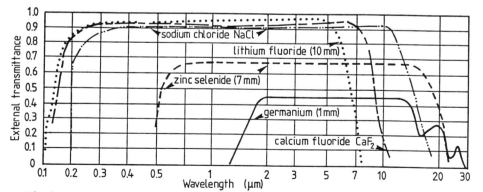

Fig.6.49. Transmission curves for far-UV and IR-transmitting materials (After material from Oriel Corp., Stratford, CT, USA)

IR region out to 9 and 6 μm, respectively. At still longer wavelengths NaCl can be used. The transmission curves for these materials are shown in Fig.6.49.

6.4.5 Influence of the Transmission Medium

In connection with a discussion of the transmission of different optical materials we should also consider the properties of air. For laboratory spectroscopy it is important to note that oxygen, water vapour and carbon dioxide in the air strongly absorb wavelengths below 200 nm. Thus the spectroscopic equipment must be evacuated when working in short-wavelength regions (*vacuum spectroscopy*). Alternatively, for wavelengths above 145 nm the equipment can be flushed with nitrogen. Wavelengths measured in air must be corrected for the refractive index of air to yield *vacuum wavelengths*, which are those used to calculate the positions of energy levels. The wavelength corrections, $\Delta\lambda$, to be *added to an air wavelength* (15°C, 760 torr) are given by the *Edlén formula* [6.91]

$$\Delta\lambda = (n-1)\lambda_{air} > 0 \tag{6.42}$$

$$(n-1) = 10^{-8}\left[8342.13 + \frac{2406030}{130-\sigma^2} + \frac{15997}{38.9-\sigma^2}\right], \quad (\sigma=1/\lambda, \sigma \text{ in } \mu m^{-1}).$$

The correction is graphically displayed in Fig.6.50.

Air has strong absorption bands in the IR region, which are mainly due to H_2O and CO_2. It has been suggested that the slowly increasing CO_2 content of the atmosphere might lead to a gradual atmospheric heating [6.92]. In astronomical observations very long absorption paths through the atmosphere are obtained. The stratospheric ozone layer, at a height of about 25 km, absorbs strongly in a band around 9.5 μm. More interestingly, it ab-

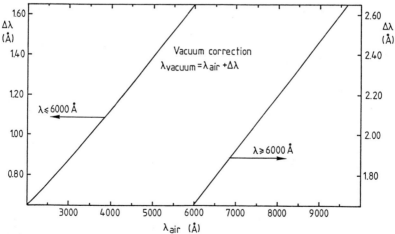

Fig.6.50. Wavelength correction curve (Courtesy: P. Grafström)

sorbs all radiation below 300 nm, which is essential for shielding life from mutagenic short UV radiation. During recent years there has been a lively debate on the possibility of a depletion of the protecting ozone layer due to human activities. One area of concern is the injection of nitrogen oxides (NO_x) into the stratosphere in the form of exhaust from high-flying, SuperSonic Transport (SST) aeroplanes. Tropospheric NO_x is normally washed out as acid rain. More important is the influence of freons (chlorofluorocarbons), from spray cans and cooling equipment, which are inert with respect to the normal atmospheric gases. Freons diffuse into the stratosphere where they can react with the ozone molecules. In many countries strong regulations against freons are being enforced. As the natural variation in the thickness of the ozone layer is quite large, it is difficult to establish trends. However, since the time constants for processes of the kind discussed here are tens of years it is important to identify trends as early as possible. Recently, clear evidence for an Antarctic "ozone hole" during the (Antarctic) spring months has been given [6.93,94]. Spectroscopic ozone layer monitoring is being performed through absorption measurements from the ground (Dobson instruments) or from satellites and balloons. Ground-based vertical lidar sounding of ozone is also a powerful technique (Sect.10.2.3). Air pollution and atmospheric chemistry have been discussed in [6.95-98].

The spectral distribution of the sun's radiation is shown in Fig. 6.51, both outside the atmosphere and at sea-level with different absorption bands indicated. The vertical transmission of the earth's atmosphere, in-

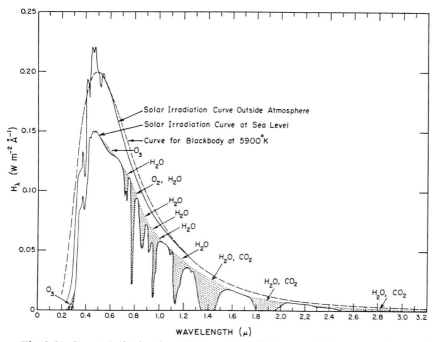

Fig.6.51. Spectral distribution of the sun's radiation outside the atmosphere and at sea level with absorption bands indicated [6.99]

cluding longer IR wavelengths, is shown in Fig.6.52. The curve for the whole atmosphere is depicted at the bottom, and individual constituent curves are included, too. These data will determine the spectral range of ground-based astronomical observations. A corresponding curve for horizontal atmospheric absorption at sea-level over a distance of one nautical mile (1.8 km) is given in Fig.6.53. This curve is of interest when considering free-laser-beam optical communications or passive and active forms of atmospheric remote sensing (Sects.6.6.1 and 10.2). The absorption properties of the atmosphere have been discussed in [6.102-104].

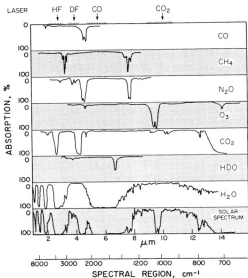

Fig.6.52. Vertical absorption curve for the earth's atmosphere (*bottom*), and individual constituent absorptions [6.100]

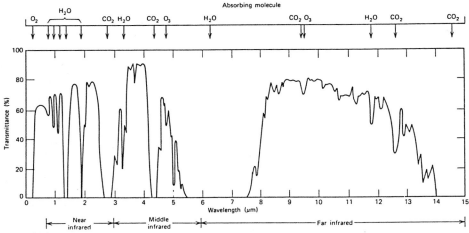

Fig.6.53. Horizontal atmospheric absorption curve over a distance of 1.8 km at sea level [6.101]

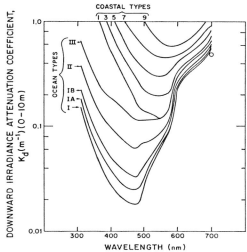

Fig.6.54. Spectral dependence of the absorption coefficient for different types of sea water (From [6.106], based on data from [6.105 and 107])

We will conclude this section by giving the transmission properties of sea-water [6.105]. In Fig.6.54 the absorption coefficients for different types of water are given in the visible region [6.106]. The "blue-green window" of water is clearly seen. The curves determine vertical visibility and the possibility of laser bathymetry (Sect.10.3.1) etc. Communication with submarines using satellite-based blue-green lasers is being considered. The only alternative for such communication purposes is the use of radiowaves of extremely long wavelength (VLW).

6.5 Optical Methods of Chemical Analysis

Several highly applicable methods of analysis rely on the interaction between light (UV-visible-IR) and matter. The wavelengths (energies) at which the interactions occur are characteristic for the individual substances (spectral "fingerprint"), and this forms the basis of qualitative analysis using light. Further, the intensities that are measured in such interactions are closely related to the concentration of the particular substance. Therefore quantitative analysis, which is frequently very accurate, can be performed.

The optical methods of analysis, as well as those based on X-ray transitions (Chap.5) are of great interest for determinations of the presence and concentration of a very large number of substances. The presence of chemical elements can be determined and molecular identification can also be made. Molecular analysis can be performed by IR absorption spectroscopy and also by XPS (Chap.5) and NMR (Chap.7). We will briefly describe some applications of optical analysis methods.

In the steel industry, as well as in the chemical and pharmaceutical industries there are clearly many fields of application for analysis techniques.

In the field of medicine, determinations of haemoglobin and metal concentrations in blood and of glucose, albumin and heavy metals in urine are often performed. In the field of agriculture, it is of great interest to determine the concentration of certain trace elements in soil samples to assess the need for the addition of fertilizers. In environmental conservation it is necessary to measure the concentrations of a large number of substances in the air and water in low concentrations. Low concentrations are frequently given in ppm (parts per million: $1:10^6$), ppb (parts per billion, $1:10^9$) or ppt (parts per trillion, $1:10^{12}$). Frequently, it is necessary to be able to perform the analysis on quantities as small as 10^{-15} to 10^{-20} g. Using laser-spectroscopic techniques single atoms can be detected (Sect.9.2.7).

The development of analytical methods has been in progress since the beginning of the 19th century. J.J. Berzelius (1779-1848) determined the constituents of about 2000 chemical compounds. The periodic table of the elements was put forward by D. Mendeleyev in 1872. At that time only 67 elements were known. By 1900 the number had increased to 83.

Analysis by optical techniques is frequently performed by measuring absorption. It is important to be able to correctly relate the absorption to the concentration. The relation which is called the *Beer-Lambert law* will now be considered.

6.5.1 The Beer-Lambert Law

Consider monochromatic light of intensity P_0 impinging on a sample of thickness b as illustrated in Fig.6.55. The sample can be a solution in a cuvette or atoms in a flame from a specially designed burner. An intensity P_t is transmitted through the sample. (We disregard possible effects from the sample confinement). We now consider the conditions over a small interval Δx in the sample. Before the considered space interval, the intensity has been reduced to P, and it will be further reduced by ΔP in the interval Δx. The fractional attenuation $\Delta P/P$ is proportional to the number of absorbers, Δn, in the small interval Δx

$$\frac{\Delta P}{P} = - k \Delta n = - k_1 c \Delta x \:. \qquad (6.43)$$

Here k and k_1 are constants and the last equality is valid for a uniform concentration c throughout the sample. When light passes through the sample, P changes from P_0 to P_t, n from 0 to N and x from 0 to b. By integration we obtain

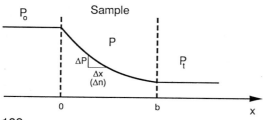

Fig.6.55. Illustration of the Beer-Lambert law

$$\int_{P_0}^{P_t} \frac{dP}{P} = -\int_0^N k\,dn = -\int_0^b k_1 c\,dx, \quad \ln\frac{P_0}{P_t} = kN = k_1 bc. \tag{6.44}$$

We now define the *absorbance* A

$$A = \log_{10}\frac{P_0}{P_t} = 0.434\ln\frac{P_0}{P_t}. \tag{6.45}$$

Thus

$$A = k_2 c \quad (k_2 \text{ proportionality constant}). \tag{6.46}$$

This is the *Beer-Lambert law*, stating that the absorbance is proportional to the concentration of absorbers in the sample.

The ratio P_t/P_0 is defined as the transmittance T. Thus we have

$$A = \log_{10} T^{-1} = 0.434\ln T^{-1}. \tag{6.47}$$

We note that the absorbance A, and the optical density D, previously introduced (Sect.6.4.2) are synonymous concepts.

We will now consider the accuracy of a concentration determined by a measurement of the transmission through the sample. In analytical instruments based on absorption measurements the transmittance or absorbance value is read off directly on a scale or is given digitally. In Fig.6.56 a linear scale for transmittance and a corresponding logarithmic scale for absorbance are given (compare with Table 6.2). From (6.46,47) we find

$$\frac{\Delta c}{c} = \frac{\Delta A}{A} = -\frac{0.43}{A}\frac{\Delta T}{T}.$$

The relative accuracy in the concentration determination $\Delta c/c$ depends on the error in reading the scale (or the error in digitizing). Clearly, a small error ΔT in the T reading results in a large uncertainty in the absorbance if T is small (then also $A \cdot T$ is small). In the same way it is important that the fullscale deflection (T=100%) is suitably set in order to be able to determine small concentrations. If we assume that the uncertainty in the full-scale setting and in measurements is the same we obtain the resulting relative concentration error by quadratic addition of the errors:

$$\frac{\Delta c}{c} = \frac{\Delta A}{A} = -\frac{0.43}{A}\sqrt{\left(\frac{\Delta T}{T}\right)^2 + \left(\frac{\Delta T}{1}\right)^2} = -\frac{0.43\sqrt{1+T^2}}{AT}\Delta T. \tag{6.48}$$

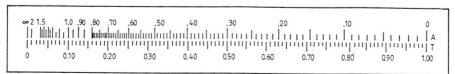

Fig.6.56. Linear scale for transmittance and corresponding logarithmic scale for absorbance

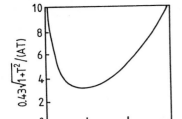

Fig.6.57. Illustration of error distribution curve

Hence, the error in the relative concentration depends on ΔT and on the factor $0.43(1+T^2)^{1/2}/AT$. The variation in this factor as a function of A is given in Fig.6.57. The factor is large for small and large values of A and has a minimum for A = 0.48. The analysis performed above reflects the almost self-evident fact, that it is much more difficult to measure accurately a concentration corresponding to very little absorption or very little transmission than to measure in a situation for which about half of the light is absorbed.

In absorption measurements the atoms or molecules in the ground state are monitored. In contrast, the excited-state population is monitored in *emission* studies. The excitation can be performed in many ways, thermally or in a spark discharge. The intensity of the emitted light clearly depends on the concentration of the substance. If the excitation has been brought about through irradiation with light from an external light source *fluorescence* light is obtained. The fluorescence light intensity can be directly related to the ground-state population.

6.5.2 Atomic Absorption/Emission Spectrophotometry

The colouring of flames when a metal salt is introduced was studied as early as the 18th century. In 1860 R. Bunsen and G.R. Kirchhoff established the connections between spectral emission lines and certain elements, i.e. *flame emission spectroscopy* was first used for qualitative analysis. Since then flame emission spectroscopy techniques have been further developed, including the construction of special burners into which the sample can be injected, and electronic light-measuring equipment. The light emitted by thermally excited atoms is measured. However, only a small fraction of the atoms are excited in the flame, while most of them remain in the ground state. The Boltzmann law describes the relation between the number of excited and ground-state atoms N_{exc} and N_{gr}, respectively:

$$\frac{N_{exc}}{N_{gr}} = \frac{g_{exc}}{g_{gr}} e^{-\Delta E/kT} \ . \tag{6.49}$$

Here g_{exc} and g_{gr} are the statistical weights of the states (g = 2J+1 for atomic states characterized by the total electronic angular momentum quantum number J). As an example, at T = 2500 K we find for Na with the ground state 3s $^2S_{1/2}$ and the first excited state 3p $^2P_{1/2}$ ($\Delta E = 2.1$ eV, $g_{exc} = g_{gr} = 2$) $N_{exc}/N_{gr} = 5 \cdot 10^{-5}$.

The accuracy and sensitivity in a measurement should be improved if the determination were instead based on the large number of ground-state atoms. Relatively speaking, the number of ground-state atoms does not change significantly for a small change in temperature, while, on the other hand, the number of excited-state atoms depends strongly on the temperature. Observations of this kind suggest the introduction of the *atomic absorption method*, which has become a very common standard method for accurate measurements of low concentrations of elements.

Atomic absorption spectroscopy differs from normal spectrophotometry (Sect.6.5.6) in one important aspect. As we have described before (Sect.6.1.1), the linewidth of an atomic line is a few hundredths of an Å. It is therefore necessary to use a line light source, since a monochromator of reasonable dimensions cannot resolve such a line. With a continuum light source filtered by a monochromator to a pass-band with a width of a few nm the absorption due to the atoms in the flame would be totally negligible (Fig.6.58). On the other hand, if a *hollow-cathode lamp* is used the linewidths for the corresponding element match and strong absorption can be obtained. The monochromator is still necessary to suppress other lines that do not originate in the ground state. The presence of such lines would clearly dilute the absorption by the atoms. Generally, a special hollow-cathode lamp is used for each element to be measured. In certain cases several elements can be combined in one hollow cathode. However, since the burning time of a lamp is limited to about 1000 hours, the advantage is limited. Instead, provisions for rapidly changing the lamps can be utilized. Recommended spectral lines for atomic absorption spectrophotometry are given in Table 6.3. The lines are generally strong resonance lines from the ground state and in some cases lines originating in metastable states.

A basic set-up for atomic emission, as well as absorption, spectrophotometry is shown in Fig.6.59. In the absorption mode it is necessary to modulate the output of the lamp, either by modulating the discharge current or by using a mechanical chopper. The signal at the PMT will then consist of a dc contribution due to the thermally excited atoms emitting the line studied and an ac contribution due to the transmitted hollow-cathode

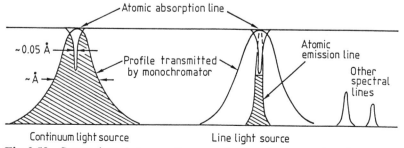

Fig.6.58. Comparison between the use of a continuum and a line light source in atomic absorption measurements

Table 6.3. Recommended lines for atomic absorption spectrophotometry

Element	Wavelength [nm]	Element	Wavelength [nm]
Aluminium	309.3	Neodymium	492.5
Antimony	206.6	Nickel	232.0/341.5
Arsenic	193.7	Niobium	405.9
Barium	553.5	Osmium	290.9
Beryllium	234.9	Palladium	244.8/247.6
Bismuth	223.1/306.8	Platinum	265.9
Boron	249.8	Potassium	766.5
Cadmium	228.8	Praseodymium	495.1
Calcium	422.7	Rhenium	346.0
Caesium	852.1	Rhodium	343.5
Cerium	520.0	Rubidium	780.0
Chromium	357.9	Ruthenium	349.9
Cobalt	240.7	Samarium	429.7
Copper	324.8	Scandium	391.2
Dysprosium	421.2	Selenium	196.0
Erbium	400.8	Silicon	251.6
Europium	459.4	Silver	328.1
Gadolinium	368.4	Sodium	589.0
Gallium	287.4	Strontium	460.7
Germanium	265.2	Tantalum	271.5
Gold	242.8	Tellurium	214.3
Hafnium	307.3	Terbium	432.7
Holmium	410.4	Thallium	276.8
Indium	303.9	Thorium	371.9
Iridium	208.9	Thulium	371.8
Iron	248.3	Tin	224.4/233.4
Lanthanum	550.1	Titanium	364.3
Lead	217.0/283.3	Tungsten	255.1
Lithium	670.8	Uranium	358.5
Lutetium	336.0	Vanadium	318.4
Magnesium	285.2	Ytterbium	398.8
Manganese	279.5	Yttrium	410.2
Mercury	253.6	Zinc	213.9
Molybdenum	313.3	Zirconium	360.1

light. By detecting the ac component with a phase-sensitive ac amplifier (lock-in amplifier) it is possible to study the absorption without background problems. Frequently a dual-beam system is used (Fig.6.60). The light is then intermittently sent through the flame (sample beam) or beside it (reference beam). The two beams are balanced on the lock-in amplifier (zero ac component) for the case of no sample injection into the flame. The presence of atoms to be measured results in an imbalance between the two beams (an ac component), the amplitude of which is proportional to the absorption. Using this method drifts in the lamp intensity, the sensitivity of the detector and the electronic amplification are compensated for.

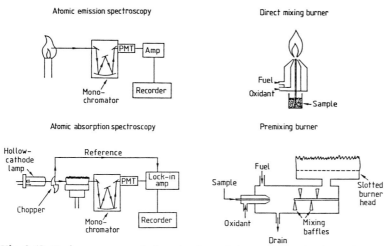

Fig.6.59. Basic arrangements for atomic emission and atomic absorption spectrophotometry. Burner arrangements are also shown

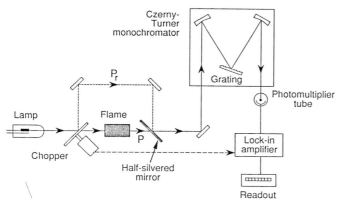

Fig.6.60. A dual-beam spectroscopic system

6.5.3 Burners, Flames, Sample Preparation and Measurements

Two types of burners are used for flame spectrometry - direct mixing burners and premixing burners. In the former type the gas and the oxidant are mixed immediately before ignition at the burner mouth. The sample, which is kept as a water solution is sucked into the flame through a thin tube. In premixing burners the gases are mixed in a chamber into which the sample solution is also sprayed. The latter type of burner normally has a slit-shaped head of about 10 cm in length, yielding a long absorption path through the laminar flame obtained with this type of burner. Any gas mixtures can be used in direct mixing burners while only gas combinations with a burning velocity slower than the gas flow velocity out of the burner can be used in premixing burners to avoid flash-back. This eliminates the hot acetylene/oxygen flame (3100° C) while acetylene/air (2200° C) and

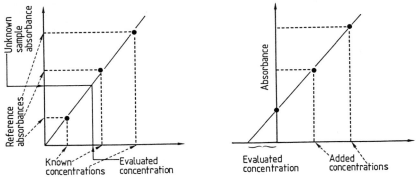

Fig.6.61. Illustration of concentration evaluations using standard solutions and the standard-addition method

acetylene/N_2O (2800° C) can readily be used. In the flame the small sample droplets are dried and then the salt particles are evaporated forming free atoms. The flame temperature must be chosen in a trade-off between the desired high degree of molecular dissociation and undesired atomic ionization, leading to a reduction in the signal. Practical measurements are performed with water solutions containing the sample which has been dissolved, sometimes using an acid. The measurements are performed by comparing the sample solution with standard solutions of known concentrations. Using a number of such standards a calibration curve, which should exhibit linearity in absorbance versus concentration, see (6.46), is obtained. Another possibility is to add known amounts of an element to the sample solution and observe the increase in absorbance ("standard addition method"). Sensitivities obtained are frequently in the ppb region. The use of these techniques is illustrated in Fig.6.61.

6.5.4 Modified Methods of Atomization

In burners of the types discussed above only about 20% of the solution is atomized while the rest is lost due to the formation of large droplets that are drained away. Further, the concentration in the flame at any one time is low because of the continuous aspiration of the sample solution. A number of techniques have been developed in which the whole sample is vaporized at the same time yielding a stronger, transient absorption peak. In the *graphite oven* technique the sample is injected into an electrically heated graphite tube through which inert gas is flowing as indicated in Fig.6.62. The heating is performed in three steps. First the sample is dried at a low temperature. Then it is ashed at a higher temperature and is finally evaporated as the heating current is further increased. The measurement is performed during the third stage while the vaporized sample remains in the oven.

As compromises between flame and flameless atomization techniques the Delves cup and sample boat methods have been developed (Fig.6.63). In the first of these two techniques the sample is kept in a small cup made of nickel. It is introduced into the flame of an ordinary atomic absorption slit burner. The sample is vaporized and passes through a hole into a nickel

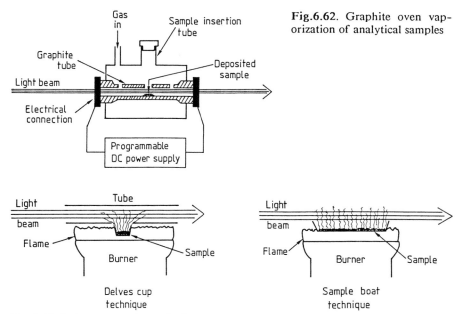

Fig.6.62. Graphite oven vaporization of analytical samples

Fig.6.63. Delves cup and sample-boat arrangements

tube that is mounted horizontally in the flame above the cup. The whole sample is used for the measurement and is present at the same time in the light path. In the sampling boat technique the sample is placed in an elongated "boat" made of tantalum. The sample is brought into the flame just below the light path and an absorption peak is recorded for a short time. The modified methods of atomization discussed above allow the detection limit for many elements to be lowered by 1-4 orders of magnitude.

Atomic absorption and emission techniques have been discussed in [6.108-110].

6.5.5 Multi-Element Analysis

It is frequently desirable to be able to determine the concentration of several elements simultaneously. Atomic absorption is then impractical, since different hollow-cathode lamps are required. Clearly, the atomic emission method is more suitable. However, a drawback with this technique is that thermal excitation in a flame is relatively poor, but alternative methods of excitation can be used. For a long time arcs or sparks have been used in metallurgical analysis. In this case the sample, which must be solid, forms one of the electrodes.

In order to eliminate the drawback of being limited to solid samples a new type of excitation source, the *Inductively Coupled Plasma* (ICP) source, has been developed. The working principle for this source is illustrated in Fig.6.64. Three concentric quartz tubes carry flowing gases. An inductance coil, fed by an RF generator is placed around the upper end of the outer

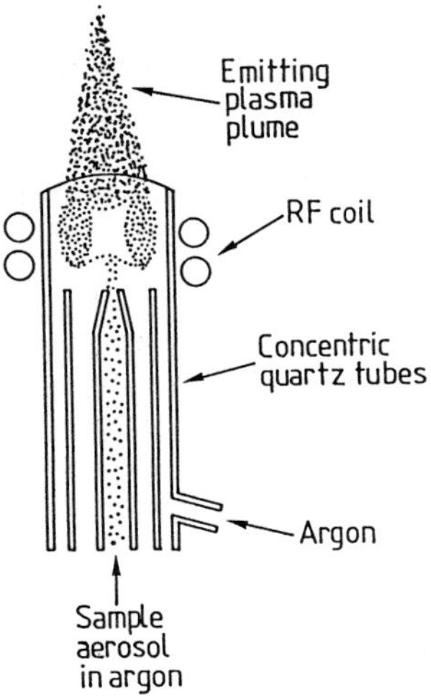

Fig.6.64. Principle of the inductively coupled plasma source

tube. In the oscillating field argon gas flowing out between the outer and middle tube will be ionized and forms a plasma at a temperature in the 5,000–10,000 K range. Some argon gas also flows between the inner and middle tubes to stabilize the plasma and prevent it from coming into contact with the inner tubes. When the sample is introduced, as fine droplets in the argon stream through the inner tube, complete atomization and efficient thermal excitation are obtained in the hot plasma.

The light emitted from the excitation source is dispersed and recorded by multiple photodetectors placed behind suitably located exit slits (Fig. 6.65). Fibre-optical techniques can also be used to collect light at a suitable location in the spectrometer focal plane for transmission to a battery of PMT:s. Calibration can be provided using standard solutions or standard electrodes. Using a computer-controlled data collection system the concentrations of the selected elements can be printed out shortly after the introduction of the sample. A complete system frequently incorporates means for automatic sample exchange.

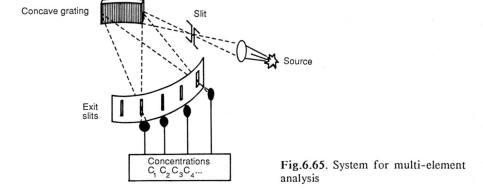

Fig.6.65. System for multi-element analysis

6.5.6 Molecular Spectrophotometry

In a spectrophotometer the absorption of a molecular sample (gas, liquid or solid) can be measured as a function of the wavelength. The absorption normally occurs in not too narrow bands and therefore a continuum light source can be used in connection with a spectral resolving apparatus, which is normally a grating monochromator but can also be a Fourier transform spectrometer [6.111, 112]. Spectrophotometers based on grating monochromators are frequently of the dual-beam type, with one beam passing the sample cell and the other one a reference (empty) cell of the same type before the transmitted intensities are compared at the detector. The lay-out for such an instrument is shown in Fig.6.66.

Instruments are normally designed for the UV/visible region or the IR region. Tungsten or deuterium lamps (Sect.6.1.2) are used in the former instruments while e.g. a Nernst glower is employed in the IR region. Correspondingly, PMT:s or PbS cells are used as detectors.

The sample to be studied in the spectrophotometer can be solid, liquid or gaseous. Liquids are investigated in cells or cuvettes made of quartz ($\lambda < 3\mu$m) or NaCl (3μm $< \lambda < 15\mu$m). The cell walls and the solvent will affect the measurements, and to correct for these effects a reference cell of the same kind filled with the same solvent but without the sample molecules is used. The transmittance can then be written

$$T = \frac{P_{solution}}{P_{solvent}} \simeq \frac{P}{P_0} . \qquad (6.50)$$

Commonly used solvents are carbon disulphide CS_2 (7.5-16μm) and carbon tetrachloride CCl_4 (2.5-7.5μm). If highly absorbing solvents are used the

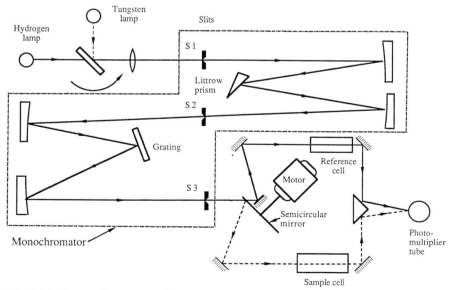

Fig.6.66. Spectrophotometer of the dual-beam type [6.110]

length of the cell must be kept short, e.g. 0.01 mm. Solids that cannot be dissolved are ground into fine particles (<2 μm) that are suspended in an oil.

Most inorganic and organic compounds in liquid or solid phase give rise to broad absorption peaks. Exceptions are ions of the lanthanides and actinides that have sharp, well-defined peaks in the UV and visible regions. The peaks are due to transitions of the optically active 4f or 5f electrons that are shielded by external electrons. Gaseous samples investigated in multi-pass cells give very sharp absorption features with corresponding demands on spectral resolution. A specially valuable device for gas studies is the *White cell* [6.113], for which multiple reflections can be used to achieve an effective absorption path of more than 100 m even if the physical length of the cell is only 1 to 2 m. The optical arrangement in a White cell is illustrated in Fig.6.67.

Spectrophotometry in the UV/visible/IR regions can be used for qualitative as well as quantitative analysis. Qualitative analysis is performed by empirically comparing measured spectra with reference spectra that have been catalogued for a large number of compounds [6.115-119]. For quantitative analysis the same principles as those used in atomic absorption spectrometry are utilized. The structures observed in the IR region are due to vibrational transitions. These give rise to rather sharp spectral features, even for liquids and solids, in comparison with those obtained for normal electronic transitions observed in the UV/visible regions. The rotational motion is quenched and thus only broadened vibrational transitions are observed. In a complicated molecule many fundamental frequencies, corresponding to different atomic groups, are obtained (Sect.3.4). This facilitates the identification of substances. In Table 6.4 examples of characteristic frequencies are given. Two examples of IR absorption spectra are given in Fig.6.68.

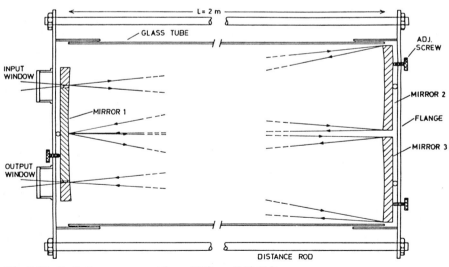

Fig.6.67. Optical arrangement for a White cell [6.114]

Table 6.4. Characteristic bond stretching vibration frequencies of certain molecular groups

Group	Wave number [cm^{-1}]	Group	Wave number [cm^{-1}]	Group	Wave number [cm^{-1}]
$\equiv$C−H	3300	$>$N−H	3350	$\geq$C−C$\leq$	900
$>$C−H	3020	$>$C=O	1700	$\geq$C−F	1100
$\geq$C−H	2960	−C$\equiv$N	2100	$\geq$C−Cl	650
		−C$\equiv$C−	2050	$\geq$C−Br	560
−O−H	3680 (gas) 3400 (liquid)	$>$C=C$<$	1650	$\geq$C−I	500

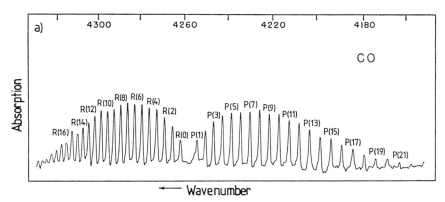

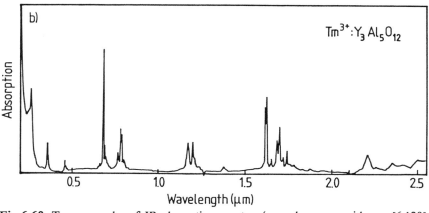

Fig.6.68. Two examples of IR absorption spectra. (*a*: carbon monoxide gas [6.120], *b*: thulium ions in a crystalline matrix [6.121])

143

6.5.7 Raman Spectroscopy

As we have previously noted, Raman spectroscopy is complementary to IR spectroscopy. A special feature of Raman spectroscopy is that molecules without a dipole moment can be investigated, e.g. H_2, N_2, and O_2. In early Raman spectroscopy measurements a strong Hg lamp was used as a light source. In modern commercial instruments an Ar^+ laser (Sect.8.4.5) with an output power of several watts in a single line (normally 514.5 or 488.0 nm) is used. The light scattered from the sample is analysed with a spectrometer. Normally a double or a triple spectrometer is used, since it is of utmost importance to suppress the intense Rayleigh-scattered light. The polarization of the Raman-scattered light can give valuable additional information on the structure of the investigated molecules. In Raman spectroscopy it is important to work with pure substances in order to avoid the masking of weak Raman components by broad-band fluorescence from impurities. (The Stokes components are on the same side of the exciting line as the red-shifted fluorescence light). Raman techniques have been discussed in [6.122-125].

As an example of a Raman spectrum from a liquid a recording for CH_2Cl_2 is shown in Fig.6.69. Raman spectra for gases will be discussed in Sect.10.1.3.

The Raman scattering can be enhanced by a factor of the order of 10^6 if the exciting line coincides with an allowed electronic transition (*resonance Raman effect*). During recent years several types of *coherent Raman spectroscopy* [6.127] have been introduced, e.g. *coherent anti-Stokes Raman spectroscopy* (CARS) (Sects.8.6 and 10.1.4) and *Raman gain spectroscopy*.

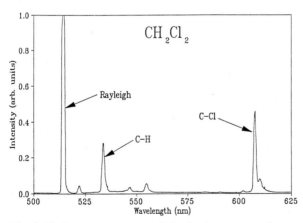

Fig.6.69. Raman spectrum of liquid CH_2Cl_2 obtained using an argon-ion laser emitting at 515 nm [6.126]

6.6 Optical Remote Sensing

In the previous section it has been described how optical spectroscopy can be used to analyse samples in the laboratory. In this section we will describe how spectroscopy can provide information on the environment by performing measurements at a distance. *Remote sensing* is a general term used for techniques, by means of which information on physical or chemical conditions at a spatially remote location can be gained using suitable equipment. In the concept of remote sensing the possibility of quickly changing the object of study is included. The principle of remote sensing is illustrated in Fig.6.70. By means of electromagnetic waves, the remote sensing equipment (generally called the *sensor*) is in contact with the spatially separated measurement volume, which is characterized by parameters P_1, P_2, P_3 etc. Using the sensor the value of a certain parameter, P_i, in the measurement volume can be read off. The measurement process involves an analysis of the radiation reaching the sensor from the object and can be characterized as a type of applied molecular spectroscopy [6.129].

Remote sensing can be performed using *passive* or *active* techniques. While changes in the spectral distribution of the background radiation (e.g., the sunlight) are analysed using passive techniques, changes in the radiation transmitted by the equipment due to interactions with the measurement volume are studied in active techniques. These concepts can be illustrated by a simple example.

In colour photography of a sunlit object the absorption and reflection properties (the colour) of the object can be assessed knowing the spectral composition of the sun's radiation and the colour sensitivity of the film. The photograph constitutes a remote sensing recording using passive techniques. If instead the object had been photographed in a dark room using a flashlight with a certain spectral distribution (frequently simulating that of sunlight) an active recording would have been obtained. This example illustrates that remote sensing is a very general concept. The information obtained can be of different kinds: geometrical information (maps), natural resources, meteorological information or concentrations of atmospheric pollutants.

Remote-sensing systems can be mounted on different platforms depending on the type of measuring task. A sensor mounted on a satellite can have global coverage. Many of the numerous satellites that orbit above the

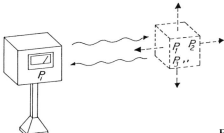

Fig.6.70. Principle of remote sensing [6.128]

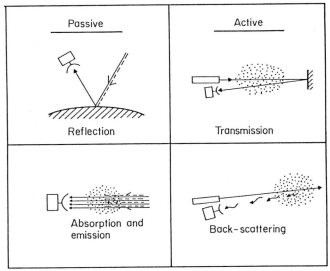

Fig.6.71. Passive and active remote sensing techniques [6.128]

earth are equipped with sensors for monitoring meteorological conditions, earth resources or for military surveillance. Examples of series of satellites transmitting their information to ground-receiving stations via microwave links are TIROS, LANDSAT, and SPOT. Airborne sensors or mobile land systems can be used for regional remote sensing. Finally, fixed systems are suitable for monitoring urban or industrial areas.

In Fig.6.71 different passive and active remote-sensing techniques are illustrated. The selective reflection of solar radiation from the earth's surface, the absorption or the thermal emission of the atmosphere can be studied with passive techniques. Active techniques frequently make use of laser or microwave radiation. The transmission or scattering of such radiation can yield information on the atmosphere or the land and sea surface.

6.6.1 Atmospheric Monitoring with Passive Techniques

Some passive atmospheric monitoring techniques will be discussed here while active techniques for such purposes will be described in Chap. 10. The general principle of passive atmospheric monitoring is illustrated in Fig. 6.72. The absorption of the sun's or sky's radiation, caused by atmospheric constituents, can be measured using a suitable spectrometer. The observed frequencies of absorption are used for gas identification, and the strengths of the absorption lines determine the concentration of absorbing molecules according to the Beer-Lambert law (Sect.6.5.1). If the absorption coefficient $\sigma(\nu)$ of the gas is known from laboratory measurements the total number of absorbing molecules can be determined from the ratio between the received radiation intensity $P_t(\nu)$ and the original intensity $P_0(\nu)$. If $N(r)$ is the molecular concentration at the range r we have

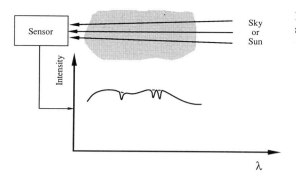

Fig.6.72. Principle of passive atmospheric monitoring [6.130]

$$\frac{P_t(\nu)}{P_0(\nu)} = \exp\left[-\sigma(\nu)\int_0^\infty N(r)\,dr\right]. \qquad (6.51)$$

Thus, range-resolved concentrations cannot normally be obtained using passive techniques since the path of absorption is undefined. If there is a temperature inversion in the atmosphere no mixing occurs through the inversion layer and a value for the average concentration up to the inversion layer can be obtained in vertical measurements. Using satellites, absorption measurements can be made at sunrise and sunset through varying thicknesses of the atmosphere ("limb" absorption). A vertical concentration profile can then be calculated from the absorption data. By carefully measuring the lineshape of absorption lines it is possible to determine a vertical concentration profile through a deconvolution procedure taking the pressure broadening (Sect.6.1.1) at different heights into account [6.131, 132] (Sect.10.2.1). Grating, Fourier transform or heterodyne spectrometers can be used to analyse the spectral distribution.

A special problem in optical atmospheric monitoring is the ever-present turbulence of the air that makes the recorded light intensity vary substantially (hence the "twinkling" of the stars). However, the amplitude of such fluctuations is negligible for frequencies above 100 Hz. This observation calls for the registration of certain small wavelength regions with peaks and valleys in a time of the order of 10 ms if small net absorptions are to be detected. This can be performed by exchanging the fixed exit slit of the spectrometer for a rotating disc with radial slits scanning the spectrum in front of a sufficiently large detector surface (e.g., a 5cm diam. PMT). To ensure a good signal-to-noise ratio thousands of individual scans can be added in a computer allowing absorption features as small as 0.1 per mille to be detected. With such a sensitivity not only common air pollutants can be detected but also the presence of various low-concentration atmospheric radicals such as HONO and NO_3. This technique is normally combined with the use of a distant artificial continuum light source such as a high-pressure xenon lamp (Sect.6.1.2). The technique is then normally referred to as *doas* (*d*ifferential *o*ptical *a*bsorption *s*pectroscopy) [6.133, 134, 114]. The same effect can be obtained by simultaneously recording a high-resolution spectrum and a low-resolution (normalizing) spectrum using a spectrometer

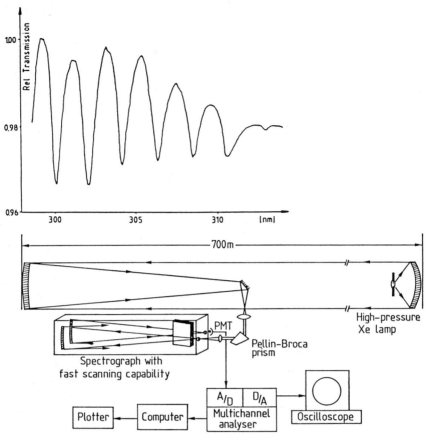

Fig.6.73. Set-up for doas spectroscopy and a measurement example with signals due to an average concentration of 22 ppb of SO_2 over 600 m [6.114]

with dual output slits and then dividing one signal by the other [6.135]. A doas set-up and a measurement example are shown in Fig.6.73.

Instruments called *correlation spectrometers* can also be used for passive gas analysis. In these systems a spectrum stored in the instrument is compared (correlated) with the spectrum of the incoming light and a signal that is proportional to the number of absorbing molecules of a particular kind is generated. The correlation can be performed in different ways. In a dispersive, photoelectrically recording spectrometer a metal mask with slits made for the lines of a particular gas can be vibrated back and forth in the image plane and the ac component of the transmitted intensity can be recorded (Fig.6.74). If the light does not exhibit the particular lines there will be no correlation, but if the lines are present systematic increases and decreases in the transmitted intensity are obtained when the mask is moved. The mask pattern can be made as concentric rings vaporized on a round quartz disc, that is rotated in the focal plane. For half the circumference the pattern is slightly displaced out from the expected spectral lines [6.136].

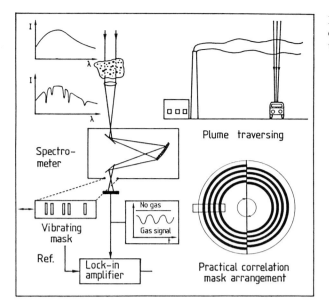

Fig.6.74. Principle of a dispersive mask correlation spectrometer

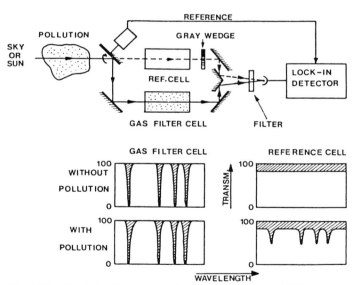

Fig.6.75. Principle of a gas correlation spectrometer [6.130]

In a different type of instrument a gas cell is instead used for the correlation. A *gas correlation spectrometer* is shown in Fig.6.75. The incoming light is alternately sent through a cell containing the gas to be studied at such a high concentration that little or no light can pass at the absorption wavelengths, and through an empty reference cell. For the case of no external gas a gray wedge (continuously graded neutral density filter) is adjusted so that equal intensities are obtained through both spectrometer channels in a particular pass-band isolated by a filter. Now, if an external pollution cloud is present the signal through the gas correlation cell is not affected since no higher absorption than full absorption can be obtained. On the other hand, the light through the reference channel is reduced due to the absorption lines of the cloud. Thus an imbalance is recorded at the initially balanced lock-in detector. For calibration purposes a cell with a known (concentration × length)-value is inserted in the light path in front of the correlation spectrometer [6.137].

6.6.2 Land and Water Measurements with Passive Techniques

In passive monitoring of land or water surfaces the reflective properties or the IR thermal emission are normally used. In earth resource monitoring the selective reflection properties of different types of materials, exposed to sunlight, are used. In Fig.6.76 some reflectance curves are shown.

Multi-spectral aerial photography represents the simplest type of reflective remote sensing. A number of aligned identical cameras equipped with different filters, are activated at the same time. This technique produces images directly and can be used to monitor land vegetation, and oil, algae and turbidity in water. The sensitivity of photographical materials only extends into the near IR region which limits the possibility of discriminating between different materials.

Many substances and materials have characteristic spectral signatures in that part of the IR region which is not covered by photographic emulsions. As a general-purpose technique in wide wavelength regions *Multi-Spectral Scanning* (MSS) performed from an aeroplane, or better still a satellite, can be used. The principle of this technique is illustrated in Fig.6.77.

Different points on the earth's surface are sensed sequentially using a scanner mirror placed in front of a light-collecting telescope. As the sensor platform moves forward, a certain strip ("swath") on the surface is covered in a manner resembling the scanning on a TV screen. The radiation is divided up into different wavelength bands that are recorded in parallel detectors. After digitizing, the information is stored on a recorder or trans-

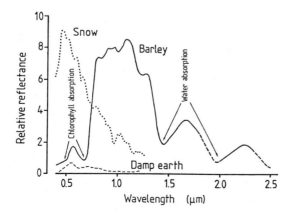

Fig.6.76. Reflectance curves for different ground surfaces

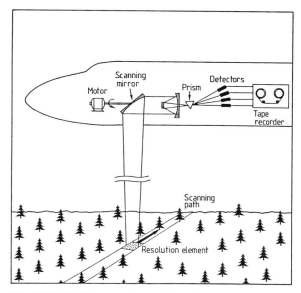

Fig.6.77. Multi-spectral scanning of the earth's surface

mitted to a receiving station on the ground. Presently, American LANDSAT satellites operate from a height of about 900 km and have a swath of about 180 km. The surface reflectivity is sensed in 7 bands centred at 0.49, 0.56, 0.66, 0.83, 1.65, 2.25 and 11.45 μm. The spatial resolution is 30×30 m^2. New generations of satellites, e.g. the French SPOT satellites operate with an improved spatial resolution (10×10 m^2). Receiving stations are distributed around the globe. The digital data must first be rectified to a normal map grid eliminating the influence of the observation angle. Packages of thus preprocessed data, called "*scenes*", covering a surface of, e.g., 180×180 km^2 are made available to users in the form of *Computer-Compatible Tapes* (CCTs). Image processing can then be performed with advanced interactive computer systems. A classification of the information in a scene can be made using stored spectral signatures. Thus maps can be produced showing, for example, different kinds of crops identified by their specific spectral signatures. Present trends in digital image processing include an increasing utilization of the geometrical line and granulation patterns (texture). This field is developing very rapidly. Array detectors can be used to eliminate the need for mirror scanning ("push-broom" sensors). Two-dimensional detectors can combine spatial and spectral information (imaging spectrometer) [6.138]. Clearly, military satellites have performances vastly superior to those of civilian satellites.

IR measurements for assessing surface temperature are of interest in many contexts. One application is to monitor heat plumes in the sea outside nuclear power plants. The measurements can be performed with scanners in suitable wavelength bands or with special heat cameras, which are also used for monitoring heat leaks from houses in cold regions, and for industrial and medical purposes. By monitoring temperature inertia in a diurnal cycle, additional information on, for example, different types of rocks can be obtained.

The techniques and applications of multi-spectral analysis from space have been discussed, for example, in [6.139-142].

6.7 Astrophysical Spectroscopy

In the same way as in laboratory optical analysis and in remote sensing it is possible to use characteristic wavelengths and line intensities in light from astronomical sources for qualitative and quantitative analysis of the chemical constituents of stars and other objects [6.143-146]. The radiation is collected with large telescopes [6.147, 148]. In this section we will give a few examples illustrating how knowledge can be obtained on astronomical objects by the analysis of electromagnetic radiation in the optical region. The "surface temperature" of a star can be determined from its spectral features. Stars are divided up into classes according to their temperature (highest → lowest): O, B, A, F, G, K and M-type stars. In Fig.6.78 some classified star spectra are given. Table 6.5 lists special spectral features pertaining to the different classes.

Stellar spectra normally consist of absorption lines which occur when the intense continuum radiation from the hot interior of the star is filtered on passing through the cooler outer stellar atmosphere. The strength of the absorption line is a measure of the abundance of the element. As a measure in the determination of the number of absorbing atoms the so-called *equivalent width* of the spectral line is used. The equivalent width is defined as the width of a square box that covers the same surface as the actual absorption profile of the line, as illustrated in Fig.6.79. An example of an experimentally recorded stellar line is included in the figure.

The relation between equivalent width and the number of absorbing atoms is called the *curve of growth*. Clearly, the transition probability for

Table 6.5. Spectral features characterizing different classes of stars

O	Lines of neutral and ionized He, and of multiply ionized light elements.
B	Lines of He and of singly ionized light elements. Hydrogen Balmer series appears.
A	Strong hydrogen Balmer series. Neutral and singly ionized metal lines.
F	Hydrogen Balmer series weaker. The Fraunhofer K and H lines due to Ca^+ become very strong.
G	Lines of neutral metals dominate the spectrum. Molecular bands of CN and CH appear.
K	The lines of neutral metals and the molecular bands become even stronger than in class G. Bands of TiO appear.
M	TiO bands become stronger and many other molecular bands are present.

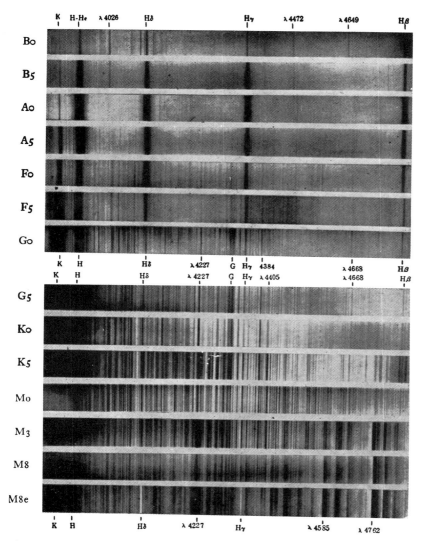

Fig.6.78. Classified star spectra [6.146]

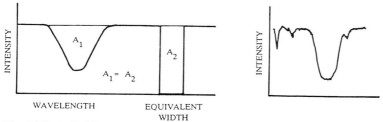

Fig.6.79. Definition of equivalent width and an experimentally recorded spectral line [6.149]

the transition considered is of paramount importance in this context and we will return to this point later (Sect.9.4.4). In calculations it is necessary to assume a certain model for the stellar atmosphere. In Fig.6.80 a small wavelength interval of the sun's spectrum is shown, recorded at a high spectral resolution. The resonance line of Ca II is very prominent. The relative abundances of the elements in the sun, as obtained using spectral analysis, are given in Fig.6.81.

Emission lines also occur in stellar radiation, especially from hot stars, e.g. from Wolf-Rayet stars. In the solar corona, emission lines from very highly ionized atoms occur. Several of these lines that are observed in the visible or UV regions correspond to "forbidden" transitions within the ground configuration [6.152]. Lines from very highly ionized atoms in general fall in the extreme UV region because of the strong excess nuclear charge. Telescopes for X-ray and short UV wavelengths have been placed on astronomical satellites that are controlled from ground observatories. Examples of such astronomical satellites are OSO (Orbiting Solar Observatory), Einstein, Exosat and IUE (International Ultraviolet Explorer) [6.153]. Exploration of the sky in the IR region from space is also of great interest [6.154]. The Hubble Space Telescope, which is the largest space-borne astronomical facility so far, was launched in 1990 [6.155].

The Doppler effect causes displaced spectral lines from objects in motion. The rotation of a star around its axis causes an extra broadening of the spectral lines, as does turbulence in the stellar atmosphere. A detailed analysis of the line profiles can clearly yield interesting information on the physical condition of the object.

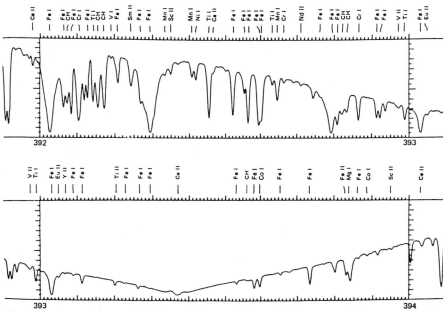

Fig.6.80. A small interval of the sun's spectrum around the strong CaII resonance line [6.150]

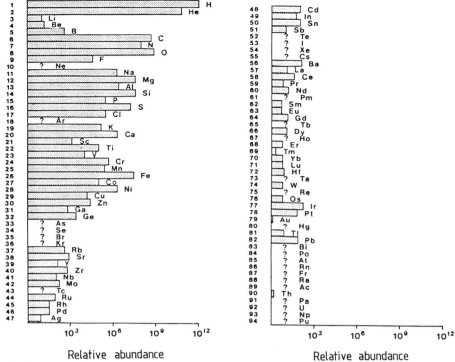

Fig.6.81. Relative abundances of the elements in the sun, as obtained using spectral analysis [6.151]

Distant galaxies exhibit large red shifts. Extremely large red shifts are observed for quasars [6.156]. The red shifts are normally interpreted in terms of radial motion. The largest equivalent velocity observed exceeds 0.9c. In Fig.6.82 a spectrum for a quasar with a velocity of 0.85c is shown. It can be noted how, for example, the Lyman α line, normally at 121.6 nm, has been shifted up to 445.1 nm in the blue spectral region! The study of other extraordinary celestial objects has been further stimulated by the sudden occurrence of the 1987 supernova in the Large Magellanic Cloud [6.158]. It was the first supernova in 383 years that could be seen with the naked eye. A spectrum of the nova recorded 7 days after the explosion is shown in Fig.6.83. The spectrum exhibits a continuum background and superimposed Balmer α and β lines. The hydrogen lines are strongly broadened due to different velocities in the expanding gas shell. Note that both emission and absorption line components occur. By studying Doppler-shifted narrow absorption lines from Na and Ca II with the bright nova continuum as background it has been possible to identify more than 10 cold gas clouds moving at different speeds (see figure insert).

The Zeeman effect of spectral lines can be used to measure magnetic field strengths in the region of line formation. Clear Zeeman patterns occur in lines originating from sun-spots, which are characterized by strong magnetic fields.

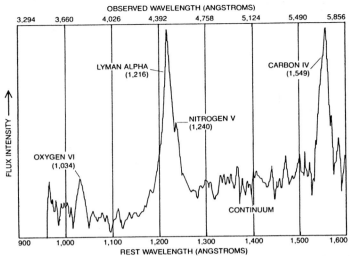

Fig.6.82. Spectrum of a quasar exhibiting huge Doppler shifts [6.157] (Copyright 1982 by Scientific American, Inc.. All rights reserved)

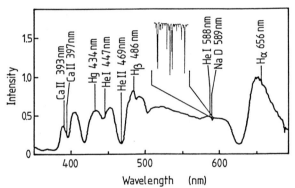

Fig.6.83. Spectrum of the supernova 1987 A in the Large Magellanic Cloud, recorded 7 days after the explosion. The prominent lines are due to the hydrogen Balmer series, Doppler-shifted and broadened due to very high expansion velocities (10.000km/s). The H_α line is partly seen also in emission. An insert shows narrow Na absorption lines due to cold gas clouds moving with different velocities along the line of sight [6.159]

The study of the planetary atmospheres has become a field of increasing interest during recent years. Terrestrial observations have been complemented by recordings from spacecraft. In Fig.6.84 terrestrial IR recordings of CO_2 in the atmosphere of Venus are shown, illustrating the increase in information content with increasing spectral resolution. As an example of a spacecraft recording, a spectrum of the atmosphere of one of the Saturn moons, Titan, is shown in Fig.6.85. The IR recording exhibits lines from many hydrocarbons. Our knowledge of Jupiter, Saturn, Uranus, and Neptune has been vastly extendend through the Voyager missions [6.161-163].

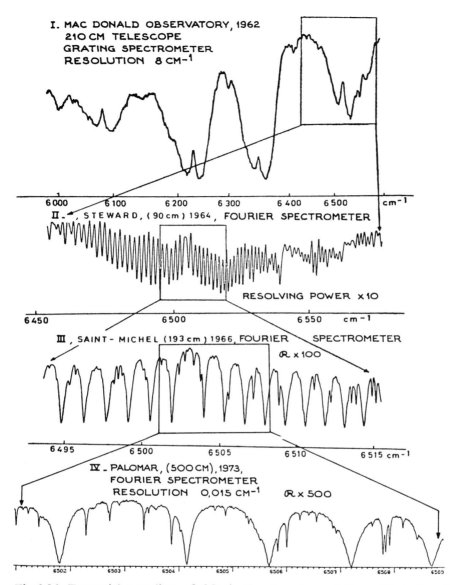

Fig.6.84. Terrestrial recordings of CO_2 in the atmosphere of Venus illustrating the continuing refinement of the available spectral resolution [6.160]

At the 1986 return of Halley's comet mankind had also become ready to launch a whole fleet of well equipped scientific space vehicles towards it including the European Giotto and the Soviet Vega sondes [6.164]. A 16×8×8 km^3 sized comet nucleus could be photographed using Giotto's multi-colour camera. Comet spectra are rich in molecular emission bands. A terrestrial photographic recording of a Halley's comet spectrum after the

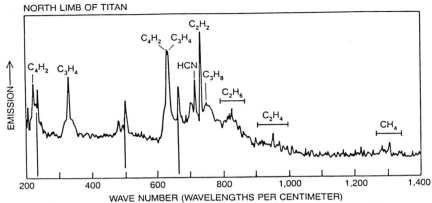

Fig.6.85. Recording of emission bands in the atmosphere of the Saturn moon Titan, obtained by the spectrometer of the Voyager I space-craft entering the Saturn-moon system [6.162] (Copyright 1982 by Scientific American, Inc.. All rights reserved)

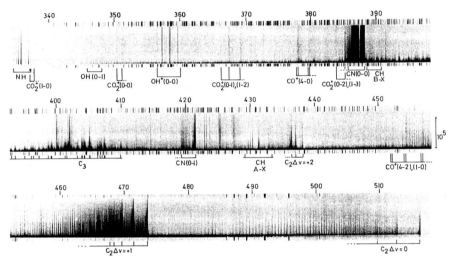

Fig.6.86. Molecular emission spectrum from Halley's comet on its 1986 return to its perihelion. Along the vertical scale the whole length of the comet tail is imaged using a long spectrometer slit. C_2, C_3 and CN bands are very prominent. Note that the C_3 emission (the "Comet" band) occurs only from a region close to the comet nucleus [6.165]

passage of the perihelium is shown in Fig.6.86. By using a long spectrometer slit, imaging a distance of 10^5 km from the nucleus (bottom) through the comet tail, the strength of the emission through the comet structure could be recorded.

7. Radio-Frequency Spectroscopy

Whereas the resolution in optical investigations of free atoms is limited by different broadening mechanisms in the light source and the spectral equipment, resonance methods yield a linewidth which is limited essentially only by the Heisenberg uncertainty relation. For investigations of ground- and meta-stable states two methods, *Optical Pumping* (OP) and *Atomic-Beam Magnetic Resonance* (ABMR) can be utilized. In the first method, a spatial deflection of free atoms is used while the second method is an optical resonance method. For studies of short-lived excited states two additional optical precision methods are available: *Optical Double Resonance* (ODR) and *Level Crossing* (LC) spectroscopy. Resonance techniques can also be used for investigating liquids and solids. *Nuclear Magnetic Resonance* (NMR), *Electron Spin Resonance* (ESR) and *Electron-Nuclear Double Resonance* (ENDOR) will be discussed. As the radio-frequency techniques make use of *magnetic resonance*, a general description of this phenomenon will be given.

Radio- and microwaves also have several other fields of application in spectroscopy. Molecular rotational transitions correspond to this wavelength region. Radiometers can be used for passive remote sensing, of e.g. temperature and air humidity, and radar systems can be utilized for active measurements of e.g. oil slicks at sea. Finally, radio astronomy is a fascinating field, yielding information on the most remote parts of the universe.

7.1 Resonance Methods

7.1.1 Magnetic Resonance

We will now study the influence of a rotating magnetic field $\mathbf{B}_1$ on a magnetic moment $\boldsymbol{\mu}_L$, which precesses with an angular frequency ω_0 in an external magnetic field $\mathbf{B}_0$ as illustrated in Fig. 7.1 (Sect.2.3.1)

$$\boldsymbol{\mu}_L = - g\mu_B \mathbf{L} \tag{7.1}$$

and

$$\omega_0 = g\mu_B \mathbf{B}_0 , \tag{7.2}$$

where $\mathbf{L}$ is the angular momentum vector associated with $\boldsymbol{\mu}_L$. The field $\mathbf{B}_1$ rotates with an angular frequency ω around $\mathbf{B}_0$. If ω and ω_0 are very different the relative positions of $\boldsymbol{\mu}_L$ and $\mathbf{B}_1$ vary quickly and only a minor influence on $\boldsymbol{\mu}_L$ can be expected. However, if $\omega = \omega_0$, the relative positions of $\boldsymbol{\mu}_L$ and $\mathbf{B}_1$ are fixed and it can be seen intuitively that $\mathbf{B}_1$ will have a large

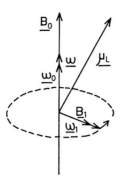

Fig. 7.1. Illustration of magnetic resonance

effect on μ_L. We will now show this more rigorously. We have, see (2.10),

$$\frac{d\mathbf{L}}{dt} = \mathbf{M} = \mu_L \times (\mathbf{B}_0 + \mathbf{B}_1) , \qquad (7.3)$$

and

$$\frac{d\mu_L}{dt} = - \mu_L \times (g\mu_B \mathbf{B}_0 + g\mu_B \mathbf{B}_1) = (\omega_0 + \omega_1) \times \mu_L , \qquad (7.4)$$

where we have defined ω_1 by the relation $\omega_1 = g\mu_B \mathbf{B}_1$ in analogy with (7.2). In order to evaluate the influence of the rotating field we now introduce a coordinate system, which rotates with $\mathbf{B}_1$ around $\mathbf{B}_0$. We then have

$$\left.\frac{d\mu_L}{dt}\right|_{\text{lab}} = \left.\frac{d\mu_L}{dt}\right|_{\substack{\text{rel} \\ \text{rot} \\ \text{syst}}} + \left.\frac{d\mu_L}{dt}\right|_{\substack{\text{rot} \\ \text{syst}}} , \qquad (7.5)$$

$$(\omega_0 + \omega_1) \times \mu_L = \frac{d\mu_L}{dt} + \omega \times \mu_L , \qquad (7.6)$$

$$\left.\frac{d\mu_L}{dt}\right|_{\substack{\text{rel} \\ \text{rot} \\ \text{syst}}} = (\omega_1 + \omega_0 - \omega) \times \mu_L = (\omega_1 - \delta\omega) \times \mu_L . \qquad (7.7)$$

When $\delta\omega = \omega_0 - \omega = 0$, μ_L rotates around ω_1. The projection of μ_L on $\mathbf{B}_0$ then changes sign periodically and the energy increases and decreases regularly. This is the semi-classical description of *magnetic resonance*. A quantum-mechanical description corresponding to the one given for *electrical* dipole transitions can be given for these *magnetic* dipole transitions.

As we have seen, a resonance will be induced by a rotating field. However, it is easier to produce a linearly oscillating field and such a field can always be resolved into two counter-rotating components, out of which one can be made resonant. The other component is then very far from resonance and will normally have a negligible effect.

7.1.2 Atomic-Beam Magnetic Resonance

In many spectroscopic techniques, e.g. in ABMR, atomic beams are employed [7.1]. Such beams are generated in a vacuum system by evaporation of atoms. The energy of the atoms will be of the order of kT, normally corresponding to thermal velocities of a few hundred m/s. The temperature needed to produce an atomic beam is determined by the vapour pressure of the element [7.2,3] (typically a value of 10^{-3} torr is used in the evaporating oven). In Fig.7.2 vapour pressure data for different elements are given.

The first atomic beam experiments were performed between 1910 and 1920. In 1922 O. Stern and W. Gerlach performed their famous experiment, in which they showed that a beam of silver atoms was divided up into two components in an inhomogeneous magnetic field [7.4]. This was the first direct experimental demonstration of space quantization. In order to explain the results of this experiment we will first consider an *electric* dipole placed in an inhomogeneous *electric* field, as illustrated in Fig.7.3. The dipole is influenced by the force

$$F = edE, \qquad (7.8)$$

where dE has the components dE_x, dE_y and dE_z.

$$dE_x = \frac{\partial E_x}{\partial x}dx + \frac{\partial E_x}{\partial y}dy + \frac{\partial E_x}{\partial z}dz = (d\mathbf{r}\cdot\nabla)E_x. \qquad (7.9)$$

Corresponding expressions are obtained for the y and z components. The electric dipole moment is defined as $\mathbf{P} = e\mathbf{dr}$ and thus

$$F_x = (\mathbf{P}\cdot\nabla)E_x, \quad \text{etc.} \qquad (7.10)$$

An analogous expression is valid for a magnetic dipole μ in an inhomogeneous magnetic field **B**

$$F_x = (\boldsymbol{\mu}\cdot\nabla)B_x, \quad \text{etc.} \qquad (7.11)$$

If μ is the dipole moment μ_J of an atom, only the force component in the direction of the field will yield a net effect due to the precessional motion.

If we choose the field direction at the atom to be the z direction the time-averaged value of (7.11) is reduced to

$$\langle F_x \rangle = \langle F_y \rangle = 0; \quad \langle F_z \rangle = \langle \mu_{J_z} \rangle \frac{\partial B}{\partial z}. \qquad (7.12)$$

In the Zeeman region for the fine structure, in which M_J is well-defined, we can write

$$\langle \mu_{J_z} \rangle = -\mu_B g_J M_J, \qquad (7.13)$$

i.e., the atom is influenced by a force which is proportional to M_J. Silver atoms have $J = 1/2$ in the ground state and thus the beam is separated into two components, as shown in Fig.7.4.

If we take the coupling between I and J into account, the total magnetic moment (which is essentially due only to the electronic shell), will

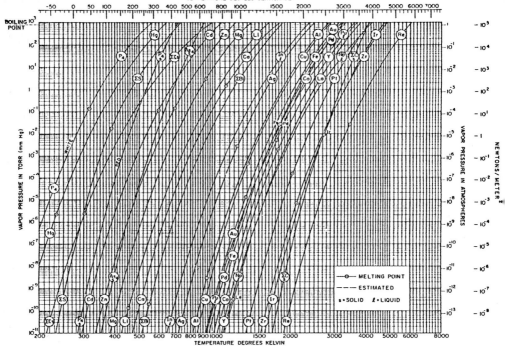
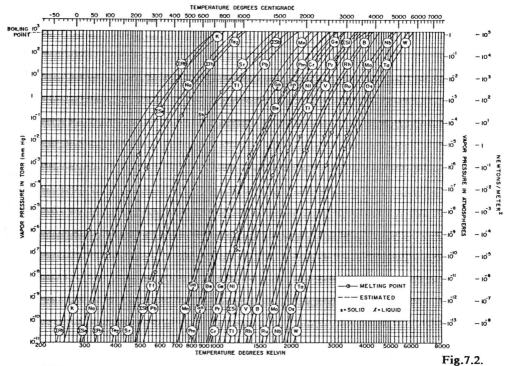

Fig. 7.2.

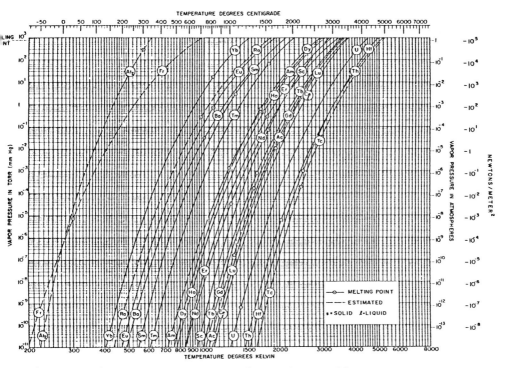

Fig.7.2. (Continued) Vapour pressure data for the elements [7.3]

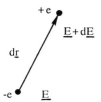

Fig.7.3. An electric dipole in an inhomogeneous electric field

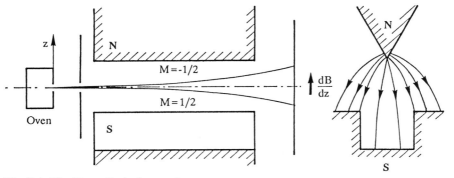

Fig.7.4. The Stern-Gerlach experiment

vary with the field strength due to the successive decoupling of I and J which occurs for increasing magnetic fields. The general expression for the force acting on an atom is

$$F = -\nabla W = -\frac{dW}{dB}\nabla B \tag{7.14}$$

where W is the atomic energy in the magnetic field, which depends on the absolute value of the magnetic field strength $B = |\mathbf{B}|$. If we compare this equation with (7.12), we can define an effective magnetic moment

$$\mu_{eff} = -\frac{dW}{dB} \tag{7.15}$$

and the force is still given by (7.12). Figure 2.20 shows an energy-level diagram for an atom with hyperfine structure, subject to an external magnetic field. As can be seen, μ_{eff} varies strongly with the field for certain sublevels. For special values of the field $\mu_{eff} = 0$ for one or several sublevels ($M_F = -1$ in Fig.2.20). This was used by Rabi and co-workers for the determination of several nuclear spins and hfs constants (zero moment method).

A great improvement in the atomic beam technique was made with the introduction of the *ABMR technique* by Rabi and co-workers in 1938 [7.5]. This was the first application of the magnetic resonance principle, which has later been successfully used in many other cases (see below). The ABMR technique is illustrated in Fig.7.5. An atomic beam, produced by an oven, passes through two inhomogeneous magnetic fields A and B, which are of the same type as those used in the Stern-Gerlach experiment. If the field gradient is directed upwards in both magnets, an atom with $\mu_{eff} < 0$ will, according to (7.12), be deflected downwards as indicated by the full line in the figure. A magnet C is placed between the A and B magnets, generating a homogeneous field, in which an oscillating magnetic field can be applied through an RF loop. If the applied frequency corresponds to the energy difference in the C field between the originally selected state and another higher or lower state of the atom, a transition to such a state can occur through absorption or stimulated emission of the corresponding radiation. This is the resonance phenomenon discussed in Sects.4.1 and 7.1.1. If the transition occurs to a state with $\mu_{eff} > 0$ in the B field, the atom will be deflected according to the dashed line in the figure and will hit a particle detector. This so-called *flop-in* arrangement gives a strong increase in signal at resonance. If the A and B fields have oppositely directed field gradients, the detector normally detects a large signal, which is decreased at resonance (*flop-out* arrangement). Note that the atoms normally experience a strong magnetic field in the A and B magnetic fields facilitating the sign determination of μ_{eff} for the Paschen-Back region atoms.

The theoretical linewidth in resonance experiments is determined by the Heisenberg uncertainty relation. In ABMR the width is determined by the time spent by the atoms in the oscillating field and corresponds to 10-50

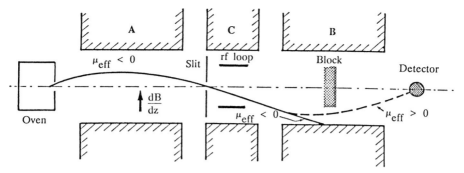

Fig.7.5. Atomic-beam magnetic-resonance apparatus (flop-in arrangement)

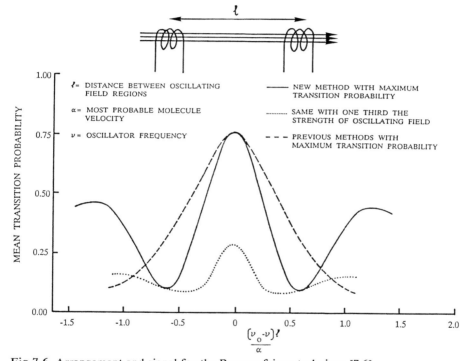

Fig.7.6. Arrangement and signal for the Ramsey fringe technique [7.6]

kHz. The resonance frequency can therefore normally be determined to within a few kHz, whereas the accuracy in conventional optical spectroscopy is of the order of 100 MHz. In a modified method introduced by *Ramsey* (Nobel Prize 1989) [7.6,1], two synchronously oscillating RF fields, widely separated in a long C field are used to effectively increase the interaction time and the linewidth can be reduced to a few hundred Hz, enabling extremely accurate hfs determinations. Due to the interference in the two separated interactions, fringe patterns are obtained, as illustrated in Fig.7.6 (*Ramsey fringe method*) (see also Sect.9.5.3).

The selection rules and transition probabilities for RF transitions are determined by the magnetic dipole operator for the whole system

$$\mu = \mu_I + \mu_J . \quad (7.16)$$

In a weak field (Zeeman effect of the hfs), the selection rules for RF transitions are

$$\Delta M_F = 0, \pm 1 , \quad (7.17)$$

$$\Delta F = 0, \pm 1 , \quad F = 0 \leftrightarrow F = 0 \quad \text{forbidden}. \quad (7.18)$$

$\Delta M_F = 0$ transitions are induced by a magnetic RF field parallel to the static field, while $\Delta M_F = \pm 1$ transitions occur for an oscillating field perpendicular to the static field. The transitions are denoted σ and π transitions, respectively. (The magnetic field is perpendicular to the electric field; compare Sect.4.2.1). In a strong magnetic field (Paschen-Back effect of hfs), we have the selection rules

$$\Delta M_I = 0 , \quad \Delta M_J = 0, \pm 1 . \quad (7.19)$$

This means that the RF field can flip the decoupled electronic angular momentum but not the nuclear spin.

According to (2.45) the resonance frequency for a $\Delta F = 0$, $\Delta M_F = \pm 1$ transition in a weak field is given by

$$\nu = \frac{|\Delta E_m|}{h} = \left| \frac{\mu_B B g_J}{h} \frac{F(F+1) - I(I+1) + J(J+1)}{2F(F+1)} \right| , \quad (7.20)$$

where the small contribution from the nuclear magnetic moment has been neglected. (If ν is expressed in MHz and B in Gauss (1 Gauss = 10^{-4} T), the value of μ_B/h becomes 1.400 MHz/Gauss). The above expression is used to determine nuclear spins, I, with the ABMR method. Since I is limited to integers or half-integers the g_J value can be determined by the same relation if an approximate value is known. A prerequisite for the detection of a resonance is that the value of μ_{eff} in strong fields is changed through the transition. (For the flop-in arrangement the sign must also be changed.) Therefore, for J=1/2 only one $\Delta F=0$ transition can be studied (between M_F = -1 and M_F = -2 in the example given in Fig.2.20).

In stronger fields, F is no longer a well-defined quantum number. The influence of the other F states can be calculated using higher-order perturbation theory or matrix diagonalization. The deviation from the linear expression given in (7.20) in this field region thus yields approximate information on the hfs splitting. Using the approximate value as a starting point, a search for $\Delta F = 1$ transitions in low fields can be performed. From the hfs constants a and b, the magnetic dipole moment μ_I and the electric quadrupole moment Q of the nucleus can be determined, provided that reliable theoretical calculations of the electronic magnetic field and the electric field gradient at the nucleus can be performed; see (2.39,44). If the magnetic moment has been determined by a direct method (like NMR, Sect.7.1.6) for one isotope it is possible to determine the corresponding moment for another isotope using the ratio between their interaction constants. According to (2.36) and (2.39) we have $a = -kg_I\mu_N$ and $\mu_I = g_I\mu_N I$. The maximum component $\mu_I = g_I\mu_N I$ is defined as the scalar dipole moment of the nucleus. For two isotopes 1 and 2 we then have

$$\frac{\mu_{I_1}}{\mu_{I_2}} = \frac{a_1}{a_2}\frac{I_1}{I_2}. \tag{7.21}$$

This relation is based on the assumption that the k value is the same for both isotopes. Since k, in a first approximation, is only related to the electronic shell, such an assumption is very reasonable. However, small differences in k can occur due to electron penetration into the nucleus (s-electrons). This results in a deviation from the relation in (7.21), or a so-called *hyperfine anomaly*. However, this anomaly is generally less than 1%. The dipole moment of the nucleus can also be determined directly by high-precision ABMR measurement in high magnetic fields, by observing the small direct contribution from the nucleus.

The first ABMR measurements were performed on alkali atoms, which can easily be detected using a hot wire of W, Pt or some other material with a high work function (Langmuir-Taylor detector). An atom with a low ionization energy is ionized with a high probability when it hits the wire and the resulting ion current is measured. For elements with a higher ionization potential the ionization can be accomplished through electron bombardment, but the yield is then many orders of magnitude lower. Atoms which have passed the apparatus at resonance, can also be detected by observing the fluorescence light which is released upon irradiation with a laser, tuned to a strong transition (Sect.9.5.1). Radioactive isotopes can be detected with a very high sensitivity by using the radioactive decay. The atoms are then collected on a foil which, after exposure, is transferred to a scintillation or Geiger-Müller detector. For a further increase in sensitivity, focusing sextupole magnets can be used instead of the conventional A and B magnets. In a sextupole magnet, such as the one shown in Fig.7.7, the field strength is proportional to r^2 (r is the distance from the symmetry axis), which, according to (7.14,15), means that a focusing force is obtained for $\mu_{eff} < 0$, while defocusing occurs for $\mu_{eff} > 0$. A large number of short-lived radioactive nuclei have been studied with regard to spin and moment using the ABMR method [7.7]. Such measurements are made possible by placing the ABMR apparatus on-line with an isotope-producing target at an accelerator facility. Nuclear spins for isotopes with a half-life shorter than 1 minute can be determined.

Apart from being applied to atomic ground states, the ABMR method can also be used for investigating metastable states with a lifetime of at least a few milliseconds; a time allowing passage through the apparatus.

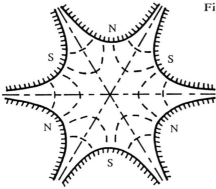

Fig. 7.7. A sextupole magnet

Metastable levels not higher than 0.5 eV above the ground state are often sufficiently populated thermally [7.8]. For refractory elements, special techniques have been developed for producing atomic beams of elements such as Mo, Ta and W [7.9]. Metastable levels can be efficiently populated with a plasma discharge in the oven [7.10], by light irradiation etc. It is valuable to study several states belonging to the ground configuration, allowing for a more complete analysis of the hyperfine interactions. ABMR techniques and results are discussed in [7.11]. Atoms with J = 0 (diamagnetic atoms) cannot be studied by the ABMR method since they are not deflected in inhomogeneous fields. Spin and moment determinations can then be made only after populating a state with J > 0.

The ABMR method has an important application as time and frequency standard. Presently, the hfs splitting between the F = 4 and F = 3 levels of the 6 $^2S_{1/2}$ ground state in ^{133}Cs in zero magnetic field is used as a frequency/time standard. This splitting has been determined to (and then been *defined* as)

$$\Delta\nu(^{133}\text{Cs}) = 9,192,631,770 \text{ Hz} .$$

This value agrees with the astronomical definition based on the tropical year of 1900 within the accuracy of the latter definition. The stability and reproducibility are presently about $1:10^{13}$. The Ramsey-fringe technique is used to attain this high degree of accuracy.

7.1.3 Optical Pumping

In magnetic resonance experiments, differences in population between substates are detected using different techniques. In the ABMR method the differences are created by spatially deflecting the atoms. In 1950 A. Kastler showed that large differences in population between the sublevels of the ground state can be obtained using polarized light [7.12]. This results in a high sensitivity compared with the case of Nuclear Magnetic Resonance (NMR) and Electron Spin Resonance (ESR), for which methods the normal thermal population differences are used (Sect. 7.1.6). The basic principle for the Optical Pumping (OP) method is illustrated in Fig. 7.8.

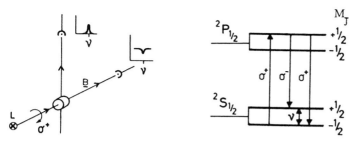

Fig.7.8. Principles of optical pumping

Here alkali atoms are considered with a $^2S_{1/2}$ ground state and a first excited $^2P_{1/2}$ state. First we disregard the influence of the nuclear spin. In a weak magnetic field both states are split into two magnetic sublevels with $M_J = \pm 1/2$. If the field is 1 Gauss, the lower of the two levels of the ground state has a thermal population excess of a few parts in 10^7. The optical pumping is performed on a low-density gas with a vapour pressure of about 10^{-6} torr corresponding to about 10^{11} atoms/cm^3. At this low pressure the atoms can be considered to be free from the influence of collisions. For example, if we want to study sodium, a resonance cell containing sodium vapour is irradiated with a sodium lamp, preferably of the RF-discharge type (Sect.6.1.1). Using an interference filter, the transition 3p $^2P_{1/2}$ - 3s $^2S_{1/2}$ at 5896 Å (D$_1$ line) is selected. If the light is circularly polarized with right-hand helicity, and if it is radiated along the magnetic field direction the atoms must take up the angular momentum of the photon upon absorption. The angular momentum is directed in the light propagation direction; σ^+ excitation (Fig.4.9). This means that the ground-state atoms with $M_J = -1/2$ are transferred to $M_J = +1/2$ ($\Delta M_J=+1$) in the excited state, from which they decay after about 10^{-8} s. The relative probabilities for decays to the $M_J = -1/2$ and $M_J = +1/2$ levels of the ground state are 2/3 and 1/3, respectively. Atoms in the $M_J = +1/2$ state cannot be excited since the angular moment of the photon cannot be accommodated. Thus atoms which were originally in the $M_J = +1/2$ state remain uninfluenced, whereas the $M_J = -1/2$ atoms are pumped over into the $M_J = +1/2$ ground level after a sufficient number of excitation processes. This is equivalent to an atomic orientation. After the pumping, the gas is totally transparent to the pumping light and a detector placed in the field direction will register a high light intensity. If an RF field, fulfilling the resonance condition for the ground state, is now applied perpendicularly to the static field direction, magnetic dipole transitions $\Delta M_J = \pm 1$ will be induced, transferring atoms into the $M_J = -1/2$ state, where they immediately start to absorb light. At the detector, a decrease in the transmitted intensity is observed. After the excitation, fluorescence light from σ and π transitions is obtained (Fig.4.9), which can be observed by a detector placed perpendicularly to the magnetic field. The RF resonance, which often occurs at an energy of about 10^{-8} eV, thus results in a release of energetic optical photons (2-3 eV in the visible region) and a strong internal atomic amplification is obtained.

If we take the influence of the nuclear spin into account, we have analogous conditions. With a nuclear spin I = 3/2 (e.g., ^{23}Na, ^{39}K or ^{87}Rb) the ground state has an energy-level scheme of the type illustrated in Fig.2.20. The excited $^2P_{1/2}$ state has a Breit-Rabi diagram of essentially the same appearance as that of the ground state. When σ^+ light is radiated, the magnetic quantum number is increased by one unit in each absorption process whereas the magnetic quantum number can remain the same or be decreased or increased by one unit in the emission process. The systematic increase in the quantum number in the absorption process results in a final accumulation of atoms in the state with the highest M_F quantum number, after a sufficient number of excitations. (Since $\Delta F = \pm 1$ transitions are also allowed, the hyperfine level groups communicate with each other). When the F vector has been oriented in this way there will also be a nuclear orientation due to the coupling between I and J. This nuclear orientation can be used for radioactive detection of unstable atoms [7.13]. Oriented β-emitting nuclei will give rise to an anisotropic β-particle intensity distribution. At resonance the orientation is lost and an isotropic angular distribution is obtained. Several short-lived isotopes have been studied with this method, particularly at on-line production facilities [7.14].

If the electronic shell has no angular momentum (J=0), direct nuclear orientation is obtained by pumping. The nuclear moments can then be determined directly, as in NMR experiments (Sect.7.1.6), but with a much higher sensitivity.

In practice, the optical pumping process is made more difficult by the possibility to leave a pumped state without being affected by an applied resonant field. This phenomenon is called *relaxation*, and is primarily caused by collisions with the cell walls. By filling the cell with 10-50 torr of inert gas (buffer gas) the time between wall collisions can be much prolonged due to diffusion. After collisions with inert gas atoms with spherically symmetric electron shells, the orientation of the alkali atom is maintained since the inert gas atoms cannot take up angular momentum. Alternatively, the cell walls can be covered with some suitable organic compound, such as paraffin or teflon. In this way relaxation times longer than one second can be attained. According to the Heisenberg uncertainty principle, this corresponds to a resonance linewidth of less than 1 Hz. A large number of precision determinations of hyperfine structure have been made using optical pumping. Optical pumping experiments are discussed in more detail in [7.15-7.17].

Since extremely narrow resonance lines can be obtained in optical pumping experiments, frequency standards of comparatively simple design can be achieved. The hyperfine transitions used in the atomic-beam clock are also used in the optically pumped frequency standards. However, the resonance frequency is comparatively strongly dependent on the pressure of the buffer gas [7.18]. It is also dependent on the intensity of the pumping light ("light shifts") [7.16]. Thus, it would seem that an *absolute* frequency standard of maximal precision cannot be achieved. On the other hand, optically pumped systems have proven to be very suitable for relative measurements and as *secondary* standards. By observing sharp $\Delta F = 0$, $\Delta M_F = \pm 1$

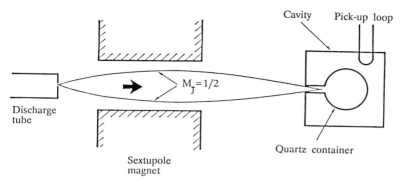

Fig.7.9. The hydrogen maser

transitions, which are strongly magnetic-field-dependent, fields can be measured with a very high precision. Optically pumped magnetometers can also be used at low magnetic fields in contrast to NMR devices, which require a field of a few hundred Gauss to produce a measurable signal.

The *hydrogen maser* provides another way of obtaining an atomic clock [7.19]. The principle of this device is shown in Fig.7.9. Atomic hydrogen is formed through dissociation of hydrogen molecules in an electric discharge. A beam of hydrogen atoms is passed through a sextupole magnet. Atoms in the $M_J = +1/2$ state are then focused while atoms with $M_J = -1/2$ are defocused (ABMR, Sect.7.1.2). The former atoms arrive in a quartz container inside a cavity which is tuned to the hfs separation for the ground state of hydrogen, about 1420 MHz. The cavity is subjected to a weak magnetic field, and the $M_J = +1/2$ state is then transformed into the F=1 state, since the nuclear spin is I = 1/2. Transitions to the lower state, F = 0, can then occur through stimulated emission (c.f. the NH_3 maser, Sect.8.1). If there is a sufficient number of F=1 atoms in the cavity, self-oscillation (maser action) occurs at a frequency corresponding to the hfs separation of hydrogen. The linewidth is influenced by relaxation processes. By coating the walls with teflon, long relaxation times can be achieved. Since buffer gas is not needed in the cavity there are no pressure shifts. On the other hand, the resonance frequency is slightly dependent on the shape of the cavity and the wall surface etc. However, these effects are several orders of magnitude lower than the pressure shift in optical pumping. The most accurate determination of the hfs separation in hydrogen is

$$\Delta\nu(H) = 1,420,405,751.7667 \pm 0.0009 \text{ Hz}$$

using the ABMR caesium clock as the reference. The reproducibility of the hydrogen maser does not exceed that of the ABMR Cs clock which still constitutes the official standard. Atomic clocks have been discussed in [7.20-21].

7.1.4 Optical Double Resonance

In 1949, A. Kastler and J. Brossel proposed a resonance method for excited atomic states [7.22]. The principles of this method are best described by discussing an explicit example. A suitable case is the very first experiment of this kind, performed by *Brossel* and *Bitter* for spin-zero mercury isotopes, as illustrated in Fig.7.10 [7.23]. A resonance cell containing mercury vapour is illuminated with light from a mercury lamp and the released fluorescence light is observed. In an external magnetic field, the first excited state, 6s6p $^3P_1$, is split up into three magnetic sublevels while the ground state, $6s^2$ $^1S_0$, is not affected. If the electric field vector of the exciting light is parallel to the external field only π transitions, $\Delta M=0$, are induced (Fig.4.9). The fluorescence light, which is emitted after about 10^{-7} s, will also be due to π transitions, with an angular distribution $I_\pi \propto \sin^2\theta$. Thus, no light is emitted in the field direction. By subjecting the cell to a resonant RF field, transitions to the $M = \pm 1$ levels can be induced and σ components ($\Delta M=\pm 1$) are now obtained in the fluorescent light. Since the σ light has an angular distibution $I_\sigma \propto (1+\cos^2\theta)$ with maximum intensity in the field direction, the resonance can be observed as an increase in the light signal at a detector in this direction. Of course, the π emission is correspondingly decreased, and this can be observed at right angles to the field direction through a linear polarizer, selecting the π components. The width of the resonance curve will be

$$\Delta \nu = \frac{1}{\pi \tau}, \tag{7.22}$$

since the two sublevels involved both have a natural radiation width of $\Delta \nu_N = (2\pi\tau)^{-1}$. As in optical pumping, the influence of low-energetic RF photons is detected using high-energetic optical photons. The name double-resonance indicates that both optical and RF resonances are required in the technique. In the chosen example, experimental information on the g_J factor and the natural radiative lifetime is obtained for the excited state. In Fig.7.11 examples of RF resonance in excited S states of ^{39}K are shown together with the corresponding Breit-Rabi diagram. In this case, of hyperfine structure, the signals in the Paschen-Back region yield information on the nuclear spin (2I+1 resonances), the magnetic dipole interaction constant

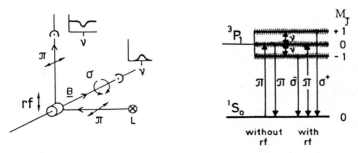

Fig.7.10. Principles of optical double resonance

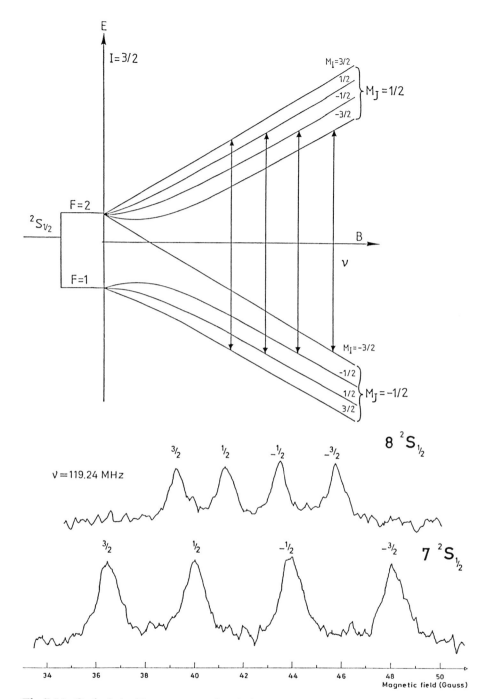

Fig.7.11. Optical double-resonance signals in the Paschen-Back region for excited S states of ^{39}K (I=3/2). Stepwise excitations from the ground 4 $^2S_{1/2}$ via the 4 $^2P_{1/2}$ state with circularly polarized light were used [7.24]

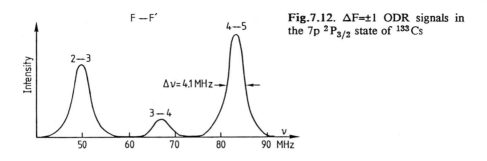

Fig.7.12. $\Delta F=\pm 1$ ODR signals in the 7p $^2P_{3/2}$ state of ^{133}Cs

(the signal separations are $a/\mu_B g_J$) and the Landé g_J factor (the centre of gravity of the 2I+1 signals corresponds to the pure electronic Zeeman-effect magnetic-field position). Hyperfine structure can also be measured by optical double-resonance even if the magnetic field is zero. (The direction of the electric field vector of the light is then chosen as the quantization axis). A schematic example of $\Delta F=1$ transitions induced in zero field for the 7 $^2P_{3/2}$ state of ^{133}Cs (I=7/2) is given in Fig. 7.12. The corresponding energy-level diagram can be found in Fig.2.18 (left; b(^{133}Cs) is almost zero).

Elements with such a low vapour pressure that a sufficient atomic density cannot be obtained in a quartz resonance cell before the softening point of quartz is reached, can be investigated in an atomic beam which is produced in a vacuum system (Sect.7.1.2). A section of the beam is illuminated with light and is subject to an RF field. The intensity, or polarization distribution, of the fluorescence light is observed. The optical double-resonance technique is a very general one allowing important information to be extracted for excited states. (See also Sect.9.3). Similar information is obtained in experiments using the level-crossing technique which will be described next. The experimental arrangements are quite similar, too.

7.1.5 Level-Crossing Spectroscopy

In the level-crossing (LC) method interference effects in the fluorescence radiation are studied instead of RF resonances. Since the technique is very closely related to the resonance methods we will treat it in this context. Level-crossing effects were first observed by W. Hanle in 1924, who studied the magnetic-field dependence of mercury fluorescence light close to zero magnetic field [7.25]. A classical description of the effect was given. The phenomenon was investigated again in 1959 in a study of fine-structure level crossings for the 1s2p ^3P multiplet of ^{4}He, performed by *Colgrove* et al. [7.26].

Let us consider two magnetic sublevels belonging to an excited state of lifetime τ. In a particular field B_0 the two levels cross. We also consider two sublevels belonging to the ground state. In Fig.7.13 we assume that the magnetic quantum numbers M are 1 and 3 in the excited state and 2 and 3 in the ground state.

The atoms are irradiated with light that is polarized at right angles to the external magnetic field. Such light will induce σ^+ and σ^- transitions

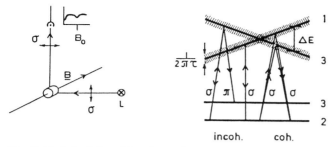

Fig.7.13. Principles of level-crossing spectroscopy

($\Delta M = \pm 1$) as described in Fig.4.9. First we consider the case where the magnetic field is set far from B_0. The M=3 level in the excited state is excited from the M=2 level in the ground state and in the decay, σ as well as π light is obtained. The M=1 level can be excited from the M=2 level, and σ light is emitted in the decay. The scattering processes, in which the two excited levels participate, are completely independent of each other. This is called *incoherent* scattering. In the magnetic field region in which the two excited levels overlap within their natural radiation widths $(2\pi\tau)^{-1}$, a different scattering process can occur. *Coherent* excitation of the two levels is possible from the M=2 level, since every incoming photon has σ^+ ($\Delta M=+1$) as well as σ^- ($\Delta M=-1$) character. Because of the uncertainty in energy the process is allowed in the overlap region from an energy point of view. The atom is excited into a linear superposition of the two excited substates. The difference in phase in the factors $\exp(-iEt/\hbar)$ for the wavefunctions of the two states is

$$\Delta\phi = \frac{\Delta E t}{\hbar} \tag{7.23}$$

and within the overlap region

$$\frac{\Delta E}{h} \leq \frac{1}{2\pi\tau} \tag{7.24}$$

the phase difference will not exceed $\Delta\phi = 1$ during the lifetime τ. Thus, a "phase memory" is retained from the excitation to the decay and we get interference between the amplitudes from the scattering processes involving the two sublevels. The total intensity of the scattered light is independent of the strength of the magnetic field, but in the overlap region the angular distribution of the fluorescence light is changed. The detected intensity change can be used to localize the level-crossing position. Conceptually, there are many parallels between the level-crossing phenomenon and Young's double-slit experiment. The levels then correspond to the slits. There is interference only when the slits are close to each other, not when they are far apart. The total intensity is constant, and only the distribution changes. The interference effect occurs only if the two slits are illuminated coherently.

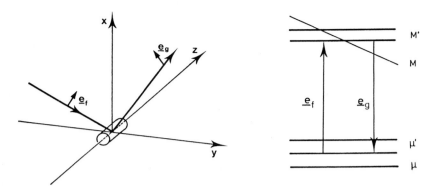

Fig.7.14. Resonance scattering

The complete theory of resonance scattering is given by *Breit* [7.27] and *Franken* [7.28]. The *Breit formula* from 1933 gives the intensity S of emitted photons with polarization vector e_g obtained after absorption of photons with polarization e_f (e_g and e_f are unity vectors) in the interaction with an atomic system (Fig.7.14).

If the excited state has magnetic sublevels with indices M and M' and the ground state has sublevels with indices μ and μ' the Breit formula can be written

$$S = C \sum_{\substack{\mu\mu' \\ MM'}} \frac{f_{M\mu} f_{\mu M'} g_{M'\mu'} g_{\mu'M}}{1 + 2\pi i \tau (\nu_M - \nu_{M'})} . \tag{7.25}$$

Here $f_{M\mu}$, $g_{M'\mu'}$ etc. are matrix elements of the type

$$f_{M\mu} = \langle M | e_f \cdot r | \mu \rangle ,$$
$$g_{M'\mu'} = \langle M' | e_g \cdot r | \mu' \rangle , \tag{7.26}$$

where er is the electric dipole operator. The transition probability per time interval between sublevels with quantum numbers M and μ is proportional to $|f_{M\mu}|^2$. τ is the natural radiative lifetime for the excited state and $\nu_M - \nu_{M'}$ is the energy separation in frequency units between the sublevels with quantum numbers M and M'. In (7.25) the summation should be performed for all pairs of levels in the excited state and in the ground state. To differentiate between summations over the ground-state levels and over excited-state levels we define

$$K_{MM'} = \sum_{\mu\mu'} f_{M\mu} f_{\mu M'} g_{M'\mu'} g_{\mu'M} . \tag{7.27}$$

We then have

$$S = C \sum_{MM'} \frac{K_{MM'}}{1 + 2\pi i\tau(\nu_M - \nu_{M'})}. \qquad (7.28)$$

Since the order of summation is immaterial we can write

$$S = C \sum_{MM'} \frac{1}{2}\left[\frac{K_{MM'}}{1 + 2\pi i\tau(\nu_M - \nu_{M'})} + \frac{K_{M'M}}{1 + 2\pi i\tau(\nu_{M'} - \nu_M)}\right]$$

$$= C \sum_{MM'} \frac{1}{2}\left[\frac{K_{MM'}}{1 + 2\pi i\tau(\nu_M - \nu_{M'})} + \frac{K^*_{MM'}}{1 - 2\pi i\tau(\nu_M - \nu_{M'})}\right], \qquad (7.29)$$

since $f^*_{M\mu} = f_{\mu M}$ etc. We then have

$$S = C \sum_{MM'} \left[\frac{\mathrm{Re}\{K_{MM'}\}}{1 + 4\pi^2\tau^2(\nu_M-\nu_{M'})^2} + \frac{\mathrm{Im}\{K_{MM'}\}2\pi\tau(\nu_M-\nu_{M'})}{1 + 4\pi^2\tau^2(\nu_M-\nu_{M'})^2}\right]. \qquad (7.30)$$

Because of the selection rule $\Delta M = 0, \pm 1$ we find that $K_{MM'} \neq 0$ only for $|M-M'| = 0, 1$ or 2. For $|M-M'| = 0$ a level crossing cannot occur because of the non-crossing rule. Instead a related phenomenon, "anti-crossing" can be observed under certain circumstances [7.29]. For the arrangement shown in Fig. 7.13 it can be shown that $K_{MM'} = 0$ for $|M-M'| = 1$. If we now consider the special case where the sublevels of the excited state are linear Zeeman levels we have $|\nu_M - \nu_{M'}| = 2\mu_B gB/h$ for $|M-M'| = 2$ yielding

$$S = C\left[\sum_M \mathrm{Re}\{K_{MM}\}\right. \qquad (7.31)$$

$$\left. + \frac{\sum_{\substack{MM'\\|M-M'|=2}} \mathrm{Re}\{K_{MM'}\}}{1 + \left(\frac{2g\mu_B\tau B}{\hbar}\right)^2} + \frac{\sum_{\substack{MM'\\|M-M'|=2}} \mathrm{Im}\{K_{MM'}\}\frac{2g\mu_B\tau B}{\hbar}}{1 + \left(\frac{2g\mu_B\tau B}{\hbar}\right)^2}\right].$$

The first term represents the field-independent background, the second denotes a Lorentzian with FWHM $\Delta B = \hbar/g\mu_B\tau$, and third one describes a dispersion curve with the extrema separation $\Delta B = \hbar/g\mu_B\tau$

This formula describes the signal around zero magnetic field, the so-called *Hanle effect*. The signal is generally a superposition of a Lorentzian and a dispersion curve on a constant background. The background corresponds to the incoherent excitation processes, in which each level scatters individually. We note that

$$K_{MM'} = \sum_{\mu\mu'} |f_{M\mu}|^2 |g_{M\mu'}|^2 . \tag{7.32}$$

Depending on the chosen polarization directions, $K_{MM'}$ can be made purely real or purely imaginary. It can be shown that when the polarization of the detected light is chosen to be parallel or perpendicular to the polarization of the exciting light $K_{MM'}$ will be real, yielding a pure Lorentzian, whereas at 45° $K_{MM'}$ is imaginary. By changing the polarization it is thus possible to choose the signal shape. From (7.31) it can be seen that the signal half-width depends, not only on τ, but also on g. The g value describes the rate of crossing and it is obvious from Fig.7.13 that the region of level overlap depends on the crossing angle. For a general level-crossing signal in a field B_0, where B_0 is non-zero the signal full width at half maximum (FWHM) of a Lorentz-shaped curve is given by

$$\Delta B = \frac{1}{\pi\tau(d\nu/dB)_{B_0}} . \tag{7.33}$$

The Hanle effect is, in general, much stronger than the high-field signals since many pairs of $|\Delta M|=2$ levels contribute to the zero-field signal.

While the level-crossing phenomenon in general requires a quantum-mechanical treatment, a simple, semi-classical model can be given for the Hanle effect. We then consider the absorbing atoms as electrical dipoles which oscillate along the horizontal axis, as shown in Fig.7.15. Such a dipole will radiate in a $\sin^2\theta$ pattern, where θ is measured relative to the horizontal axis. Thus, no intensity will be observed at the detector shown in the figure. If a magnetic field is applied, the atom seen as a magnetic dipole will precess around the magnetic field with the Larmor frequency ω_L

$$\omega_L = \frac{g_L \mu_B B}{\hbar} . \tag{7.34}$$

The dipole radiation pattern will then revolve and the direction θ can be replaced by $\omega_L t$, where t is the time after the excitation. Since the dipole radiates energy, a factor $\exp(-t/\tau)$ must be added. τ is the mean lifetime of the state. In level-crossing spectroscopy continuous excitation of dipoles occurs and the reradiated light is observed continuously. The observed intensity I(B) will then be the result of an integration from 0 to infinity in time. We thus have

$$I(B) = C \int_0^\infty e^{-t/\tau} \sin^2(\omega_L t) dt , \tag{7.35}$$

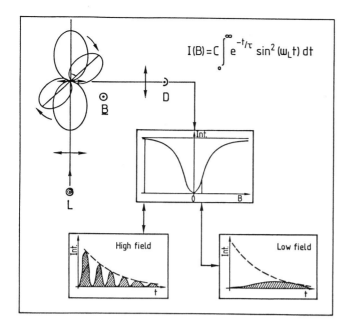

Fig.7.15. Semiclassical picture of the Hanle effect

where C is a constant. This is a standard integral yielding

$$I(B) = C'\left[1 - \frac{1}{1 + (2g_J \mu_B \tau B/\hbar)^2}\right] \qquad (7.36)$$

which is an inverted Lorentzian.

In Fig.7.16 an example illustrating the use of the level crossing method for the determination of hyperfine structure is given. The state 6p $^2P_{3/2}$ in ^{87}Rb is studied. By employing magnetic field modulation in conjunction with lock-in detection the derivative (approximately) of the $\Delta M=2$ level-crossing signals is recorded. Clearly, the positions of the crossings depend not only on the factors a and b, but also on the g_J factor. As a matter of fact, only the ratios a/g_J and b/g_J can be determined by level-crossing spectroscopy.

Using optical double-resonance and level-crossing spectroscopy much experimental information, particularly on hyperfine structure, has been obtained. To perform a complete hyperfine structure analysis, it is necessary to have data for many states belonging to the same configuration. Using excitation from metastable states and utilizing cascade population of lower states from high-lying states opens up a wide field of application for these methods [7.31]. By using tunable lasers as excitation light sources still further possibilities are created (Sect.9.3). The consistency of theoretical calculations for evaluating nuclear moments can be tested by studying a large number of levels for the same atom. (Clearly, the nucleus can only

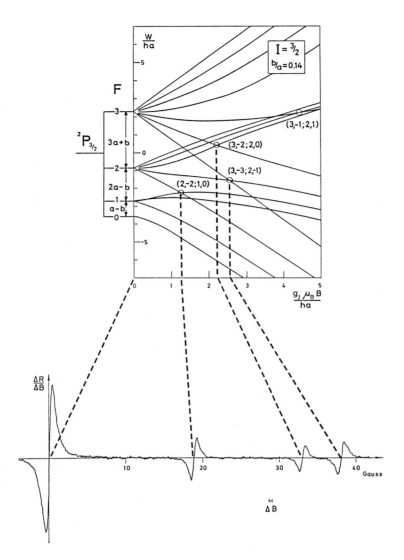

Fig.7.16. Hfs level diagram for the 6p $^2P_{3/2}$ state in ^{87}Rb and corresponding level-crossing signals [7.30]

have well-defined nuclear moments.) The influence of polarization, correlation and relativistic effects can be investigated in this way.

As we have seen, lifetimes for excited states can be determined using optical double-resonance and level-crossing spectroscopy. The results are very reliable since these methods are subject to only small systematic errors.

Resonance scattering by atoms has further been discussed in [7.32]. ODR and LC techniques were reviewed in [7.33-35] and extensive results for alkali atoms have been presented in [7.36].

7.1.6 Resonance Methods for Liquids and Solids

We will now discuss resonance techniques that are useful for studying liquid and solid samples. In these methods the differences in population between magnetically separated sublevels, due to the thermal Boltzmann distribution, are utilized. Magnetic field splittings are always small in comparison to kT, which is about 1/40 eV at room temperature. Thus population differences will always be small, and the number of atoms required is much larger than for the optical resonance techniques. However, very sensitive resonance detection techniques based on RF signals have been developed. The field of resonance spectroscopy of non-gaseous media is covered in [7.37, 38].

When sample atoms are not in a gaseous phase, coupling to the surrounding medium occurs. This has the effect that a new equilibrium value p_L of the macroscopic magnetization, μ_L, is reached in the sample following a perturbation, e.g. an abrupt magnetic field change, only after a certain time, the relaxation time characterized by the time constant τ. The effect of the relaxation can be taken into account by adding a term $(p_L - \mu_L)/\tau$ to (7.7). The new mathematical relations, expressed in component form, are known as the *Bloch equations*. The solution reveals that the magnetic resonance curve is a Lorentzian with a half-width inversely proportional to τ. From the equations it also follows that apart from fast changes of sign for $(\mu_L)_z$ at resonance, there will also be a transverse macroscopic component $(\mu_L)_T$, which rotates at the resonance frequency. Here the relaxation time τ is introduced in a vector relation and thus all components of μ_L will have the same value of τ. However, it is generally necessary to separate the *longitudinal relaxation time* τ_1 for $(\mu_L)_z$ and the *transverse relaxation time* τ_2 for $(\mu_L)_{x,y}$.

Magnetic resonance is normally detected utilizing the transverse component $(\mu_L)_T$, which is formed when the resonance condition is fulfilled. In the detection method devised by F.B. Bloch, a detection coil, placed at right angles to the inducing RF coil, is used. A sinusoidal voltage component is induced in the detection coil by the rotating magnetization. Because of the perpendicular arrangement the driving field is not picked up. A simple measurement system based on this principle is shown in Fig.7.17. In a dif-

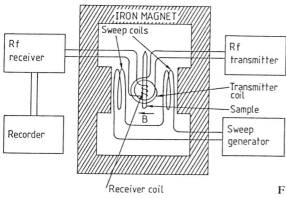

Fig.7.17. An NMR spectrometer

ferent detection technique, devised by E.M. Purcell, a single coil is employed and a bridge circuit is used to measure the induction changes in the coil that occur at resonance.

a) Nuclear Magnetic Resonance

The NMR technique utilizes the macroscopic magnetic moment resulting from the nuclei in the sample. In an external magnetic field B the nuclear spin I can have 2I+1 different orientations. The energy E_m of a magnetic sublevel is given by

$$E_m = - g_I' \mu_B B m_I \ . \tag{7.37}$$

A macroscopic longitudinal magnetization $(\mu_I)_z$ due to the slightly different thermal level populations is obtained. Using an RF field, magnetic dipole transitions between the levels are induced and g_I can be measured. Since the magnetic moments of nuclei are small ($\mu_N = \mu_B/1836$), the resonance frequencies for commonly used field strengths are low. At a field of 1 T, typical for an iron magnet, resonance frequencies of a few tens of MHz are obtained. With superconducting magnetic field coils fields of 10 T corresponding to ~500 MHz (for protons) have been attained with a corresponding increase in signal strength.

The nucleus most commonly observed is the hydrogen nucleus (proton). The proton resonance frequency is slightly dependent on the position of the hydrogen atom in the molecule. This is because a varying degree of shielding of the magnetic field occurs depending on the properties of the neighbouring atoms. Thus, the NMR method can be used for structural determinations and chemical analysis (see ESCA, Sect.5.2.2). As an example, an NMR spectrum for ethanol C_2H_5OH is shown in Fig.7.18. The proton resonance frequency is characteristic for each of the structural groups. At higher resolution it is observed that the resonances are broken up into a number of separate signals. This is due to *spin-spin interactions* between the different nuclei in the molecule. Further information on the molecule can be obtained from the number of resonances and the size of splittings.

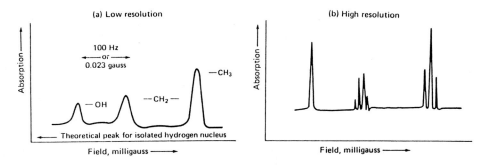

Fig.7.18. NMR spectra of the protons in C_2H_5OH taken at low and high resolution [7.39]

The technique we have discussed so far is the continuous-wave (stationary) method. For a constant magnetic field, the frequency of the oscillator, that generates a field with an amplitude of about 10^{-7} T, is slowly swept and the signal is recorded as a function of frequency. An NMR spectrum in the *frequency domain* is thus obtained.

Modern NMR spectroscopy frequently uses pulsed RF radiation and the observation is performed instead in the *time domain*. Pulsed NMR can be utilized in different ways. If a short pulse of high field strength B_1, ($\sim 10^{-3}$ T) and duration t_p is applied, the magnetic moments will be rotated by an angle α around B_1

$$\alpha = g_I' \mu_B B_1 t_p \tag{7.38}$$

and a macroscopic transverse magnetization is created for nuclei with a resonance frequency covered by the Fourier-broadened frequency distribution, $\Delta\nu$, centred around the nominal oscillator frequency ν_0 ($\Delta\nu \propto 1/t_p$). For each kind of nuclei (e.g., the protons in the ethanol molecule) a rotating macroscopic transverse magnetization will occur that will induce an oscillating voltage in the pick-up coils. The oscillation will die away with the relaxation time constant. The recorded signal is called the *Free Induction Decay* (FID) signal. Every type of nucleus covered by the pulse will give rise to separate FID signals which will interfere with each other. Using a Fourier transform, the individual Larmor precession frequencies $F(\omega)$ (resonance frequencies) can be calculated from the time domain spectrum (the interferogram) $f(t)$

$$F(\omega) = 2\pi \int_{-\infty}^{+\infty} f(t) \exp(-i\omega t) dt . \tag{7.39}$$

Clearly, this *Fourier Transform NMR* (FT-NMR) has much in common with optical Fourier transform spectroscopy (Sect. 6.2.4) and quantum-beat spectroscopy (Sect. 9.4.6), including the multiplex advantage, allowing simultaneous detection of the signals from different nuclei. In a mechanical analogy, the sharp pulse can be seen as a stroke that excites several tuning forks, which for some time ring out their eigenfrequencies. A very good signal-to-noise ratio is obtained in FT-NMR, allowing the detection of biologically less abundant nuclei with non-zero spin, e.g., ^{13}C and ^{31}P.

For liquids high-resolution NMR spectra, such as the one shown in Fig. 7.18 for ethanol, can be obtained. Sharp signals, separated due to spin-spin or quadrupole interactions, are obtained. The signals are not broadened since the molecular motion in the liquid averages out the inhomogeneities that would otherwise make the resonance frequency unsharp. This is not the case for solids, in which the constituents are fixed and yield permanent different field distributions broadening the resonance lines. A remedy for the broadening is to simultaneously apply a strong radio-frequency field that is resonant with the nuclei that are not under (resonance) investigation. These nuclei will then rotate and the spin-spin coupling to the studied nuclei will be averaged out. The technique is called *high-power decoupling*. Broadening due to coupling between nuclei of the same kind can obviously not be eliminated in this way, but this can be accomplished instead by choosing a *magic angle* θ ($3\cos^2\theta - 1 = 0$, $\theta = 54.7°$) between an axis of rapid mechanical sample rotation and the B_0 direction (see also Sect.9.4.6).

Many NMR techniques that utilize pairs of RF pulses or sequences of pulses have been devised for measurements of τ_1 and τ_2. The angle of rotation due to the $\mathbf{B}_1$ field can be chosen by varying the length of the pulses, see (7.38). In one technique a "90° pulse" brings the magnetization down into the x,y plane from the initial z direction. Because of slight inhomogeneities in the magnetic field, the individual nuclear spins will have slightly different Larmor precession frequencies, and thus the spins, which were initially aligned by the 90° pulse, will dephase leading to a decay of the ac signal picked up from the rotating transverse magnetization. However, if a "180° pulse" is applied after some time T, all the individual spins will be inverted and start precessing in the opposite direction, again with slightly different Larmor frequencies. After a time T the individual spins will again be in phase creating a macroscopic transverse magnetization, that will result in a strong pick-up signal, a *spin echo*, which will begin to die away when the spins begin to be out of phase again. The situation is similar to that in a race between fast and slow runners, where the runners change direction at a second signal and run back to the starting line, which they will all reach at the same time. The height of the spin echo will reflect the true "dephasing time", τ_2, due to the spin-spin relaxation, that can now be measured in spite of the unavoidable field inhomogeneities. However, if a molecule moves in a liquid subject to an inhomogeneous field the amplitude of the echo will be further reduced and diffusion coefficients or even flow velocities can be measured. The spin-echo phenomenon has its counterpart in the optical region, where photon echoes can be induced by laser pulses. Echo techniques have been discussed in [7.40].

NMR techniques are used for studying the chemical composition as well as the detailed structure and bond character of molecules. Kinetic processes, such as rotations and inversions of molecules, can also be studied. NMR has found many industrial applications, especially in measurements on foodstuffs. The strength of the proton signal yields the concentration of a hydrogen-containing substance. The signal of protons in a solid matrix decays faster than that for a liquid sample. This property can be used for fast measurements of moisture content or oil assessment in seeds, etc. NMR spectroscopy has been discussed in [7.41-43].

Very recently, a new diagnostic application of NMR in the medical field has been made possible through the development of the *NMR-imaging (tomographic) technique* [7.44-46]. Here the object, e.g. the head of a patient, is placed in a magnetic field that has a well-defined field gradient produced by gradient coils in the x, y and z directions. For a given oscillator frequency, the resonance condition is fulfilled only in a particular layer which can be moved in the gradient direction through the object by varying the frequency. By sequentially measuring the signals from layers at different orientations in the object it is possible to reconstruct the proton

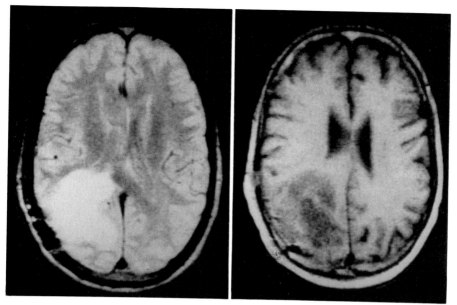

Fig.7.19. NMR tomographic images of a human brain revealing a malignant tumor in the lower left region. The image to the left is τ_2 weighted and the image to the right is τ_1 weighted. (Courtesy: O. Jarlman)

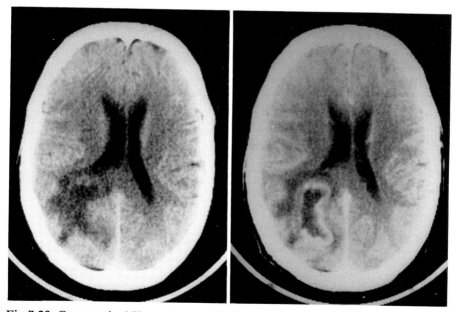

Fig.7.20. Computerized X-ray tomographic images of the same human head, as shown in Fig.7.19. The image to the right is enhanced with an iodine contrast medium. (Courtesy: O. Jarlman)

density of the object in three dimensions by a mathematical algorithm (back-projection), similar to the one used in X-ray tomographic imaging [Computer-Aided Tomography (CAT) scanners]. In this way high-resolution images can be obtained which discriminate between different types of tissues. The diagnostic capability is much increased by using the different relaxation times in different tissues, e.g. in tumours compared with normal tissue. Using pulse techniques it is also possible to produce dynamic images of blood flow, etc. Current developments are aimed at increasing the sensitivity, e.g. by using higher fields, so that imaging of other nuclei of high medical interest but low biological abundance (^{43}Ca, ^{31}P, ^{23}Na, ^{19}F, ^{14}N and ^{13}C) may be realized. In Fig. 7.19 an example of an NMR tomographic image and in Fig. 7.20 a corresponding X-ray tomographic image are shown illustrating the detection of a brain tumour with the two techniques.

At the end of this discussion of NMR techniques it should also be pointed out that a well-defined, small probe sample, e.g. water in an RF coil, can be used to measure magnetic fields with high accuracy. The field is calculated by inserting the measured frequency in (7.37). Because the population differences are due to thermal effects, fields above 0.02 T are normally necessary. This technique is also frequently used for stabilizing a magnetic field by regulating the magnetizing-coil power supply using a servo-loop locked into the NMR signal.

b) Electron Spin Resonance

ESR is the counterpart of NMR, dealing with magnetic splitting due to electron spin rather than nuclear spin. In a crystal field the projection of the orbital angular momentum **L** on a given direction is frequently averaged to zero. This is called *quenching* and has the effect that orbital angular momentum does not need to be considered when discussing the energy-level splittings in an external field. Thus the splitting will be given only by the magnetic moment due to the total electronic spin **S**, as shown in Fig. 7.21. In the simplest case we have

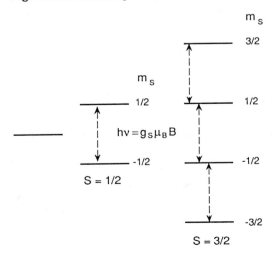

Fig. 7.21. ESR transitions

$$E_m = g_s \mu_B B m_s \ . \qquad (7.40)$$

Since the magnetic moment of the electron is large ($g_s \simeq 2$) in comparison with that of the nucleus, much higher resonance frequencies are obtained than for NMR. In commercial instruments it is customary to work at 0.34 T, corresponding to a resonance frequency of about 9.5 GHz. Thus, the experimental technique involves resonant cavities and waveguides for the microwaves used. In the presence of crystal fields the position of the resonance depends on the orientation of the crystal with respect to the magnetic field. The interaction must thus be described by vectors or tensors rather than by a scalar relation such as (7.40).

Because of the larger energy separations in ESR the signals are much stronger than in NMR. ESR can be applied in much the same way as NMR for analytical purposes. A prerequisite is that the net spin is non-zero, which is the case for *paramagnetic substances*. In particular, free radicals and molecular ions can frequently be investigated. ESR techniques have been discussed in [7.47, 48].

c) Electron-Nuclear Double-Resonance

In the ENDOR method NMR and ESR are combined. The technique is illustrated in Fig. 7.22, in which a system with $S = 1/2$ and $I = 1/2$ is considered. In the Paschen-Back region we have two m_I values for each m_S value. If a microwave field saturates the ESR transition between the states with $m_I = +1/2$ only a weak absorption signal is obtained when an equilibrium situation is reached. The upper level then has a high excess population compared with the Boltzmann distribution. If an NMR transition is then induced in the upper group of levels, the state with $m_I = +1/2$ is depopulated. This results in a strong increase in the ESR resonance. In an ENDOR experiment the ESR resonance in the microwave region is thus constantly driven and a search is made in the radio-frequency region until the NMR resonance occurs, resulting in a strong increase in the ESR amplitude. Using the ENDOR technique, NMR signals with an intensity typical for ESR signals are thus obtained.

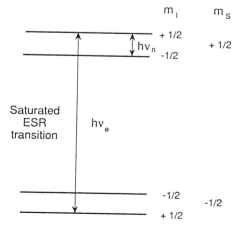

Fig. 7.22. The principle of ENDOR

We have earlier discussed the high sensitivity in optical pumping, for which a much more complete orientation is achieved than in the thermal case. Certain solids can also be pumped optically and ESR and NMR signals can be detected optically [7.49]. However, the efficiency is reduced due to competing radiationless transitions.

7.2 Microwave Radiometry

Spectroscopy in the microwave region yields valuable information on the rotational structure of molecules [7.50-52]. Apart from this type of fundamental experiment, important applications of microwave radiometry include atmospheric and ocean monitoring [7.53-56]. Passive, as well as active techniques are used for different types of remote sensing. Active microwave techniques are pursued with radar systems.

In Fig.7.23 the vertical absorption of the atmosphere in the 1-300 GHz region is shown. The absorption bands are due to H_2O and O_2. The 60 MHz O_2 band consists of about 25 lines which can be resolved in the stratosphere but which overlap in a single band at low altitudes due to pressure broadening. The atmospheric pressure can be measured by studying pressure broadening effects. At high altitudes the broadening is given by the Doppler effect, which yields information on the temperature. The water vapour content can be determined using the strong absorption at 22 and 190 GHz. Vertical measurements can be performed from the ground (towards zenith) or from satellites (towards nadir). So-called *limb emission* or *limb absorption* measurements can be made from satellites, as illustrated in Fig.7.24. In absorption measurements radiation through the atmosphere at

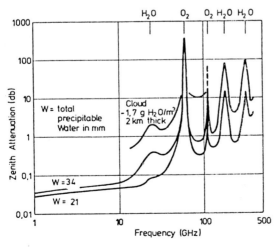

Fig.7.23. The integrated vertical microwave absorption of the atmosphere with varying water content in the air [7.57]

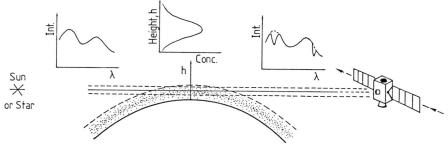

Fig.7.24. Limb absorption measurements from a satellite

sunrise and sunset is employed. A deconvolution procedure must be used to obtain the vertical profiles of pressure, temperature and humidity. Satellites equipped with microwave radiometers can be used to determine temperatures to an accuracy of better than 1°C.

Active microwave systems (radar) have many applications apart from the normal ranging applications that are used, for example, for navigation. A weather radar station is capable of locating and quantifying rainfall over considerable distances. To detect the small particles a short operating wavelength is necessary to achieve adequate scattering. Radar systems for environmental applications are air- or satellite-borne. In SLAR (Side Looking Airborne Radar) systems the imaging sweep is obtained through the forward movement of the propelling craft. The principle of SLAR is shown in Fig.7.25. A disc-shaped radiation lobe is transmitted from an elongated antenna. The ground surface is hit at successively later times and the echo from the radar pulse will return to the antenna at different times and a picture line is thus generated. Then a new pulse is transmitted and a new picture line, corresponding to the new position of the aeroplane is obtained. The strength of the return signal depends not only on the distance, which can easily be compensated for, but also on the structure of the surface, water content, etc. Ocean waves especially, yield echoes whose strength depend on the wavelength of the water waves (sea clutter). A 10 GHz system ($\lambda = 3$cm) is especially sensitive to the fine capillary waves, which occur at sufficiently large wind speeds (>8m/s). If the sea is covered by oil the

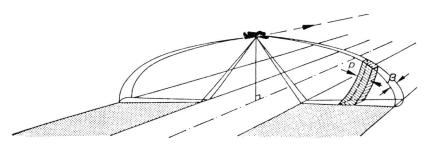

Fig.7.25. The principle of SLAR [7.58]

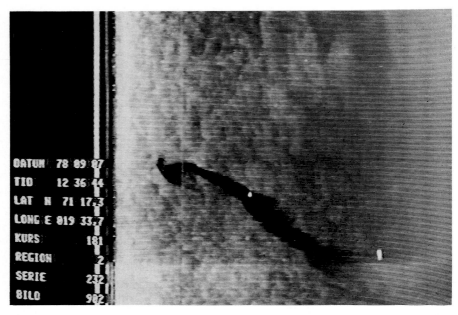

Fig.7.26. 20×20 km² SLAR image of a marine oil slick and ships. 24 hours before recording 25 tons of crude oil had been dumped. (Courtesy: The Swedish Space Corporation)

capillary waves disappear and the oil slick can be detected as a decrease in the sea clutter. In Fig. 7.26 oil detection with a SLAR system is illustrated. Such a system is also useful for sea ice monitoring.

Because of the long wavelength of the radar waves a high spatial resolution cannot be obtained for antennae of reasonable size. This is a particular problem especially for satellite applications, which can however, be circumvented using so-called SAR (Synthetic Aperture Radar) techniques. A high lateral resolution is obtained by using short pulses in the same way as in a SLAR system. In order to obtain a high resolution in the flight line direction, the slightly different Doppler shifts in the back-scattered radiation from the diverging antennae lobe are used to sort the return signals into different spatial channels along the flight line [7.59]. Remote sensing from space using SAR techniques will be a rapidly expanding field with the launching of, e.g., the ERS series of satellites.

7.3 Radio Astronomy

In the science of radio astronomy continuum radiation and absorption or emission lines due to celestial phenomena are observed in the frequency region of 1-300 GHz (300mm < λ < 1mm). The absorption bands due to the terrestial atmosphere, as illustrated in Fig. 7.23, will clearly place limitations

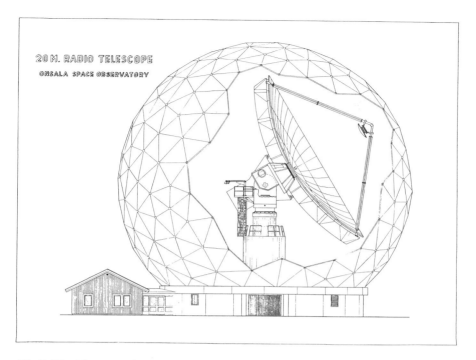

Fig.7.27. Diagram of the 20 m diameter mm-wave telescope at Onsala, Sweden [Courtesy: Onsala Space Observatory]

on such measurements. The diffraction-limited resolution for a telescope is determined by the ratio λ/d, where d is the diameter of the telescope. In a comparison with optical telescopes it can be noted that a radio telescope has 10^4 to 10^5 times worse angular resolution for a given value of d. For practical reasons a telescope diameter of the order of 100 m is an upper limit, especially since a surface precision of about $\lambda/20$ must be maintained. Examples of radio telescopes are the 100 m telescope near Bonn in FRG which is used in the frequency region 1-45 GHz and the 20 m Onsala telescope in Sweden, which can be used for frequencies up to 120 GHz ($\lambda = 2.5$ mm). A diagram of the latter telescope is shown in Fig. 7.27.

To amplify the weak radio signals picked up by the telescopes, travelling-wave masers and superconducting mixers are sometimes used. An example of a maser medium is iron-doped rutile (TiO_2) which can be magnetically tuned in a certain frequency range. Signal analysis is often performed using auto-correlation techniques, in which the incoming signal is compared with a replica of itself that has been displaced in time by a variable interval. Other signal analysis techniques employ analogue filter banks and acousto-optical spectrometers.

In order to increase the angular resolution in radio astronomy the technique of *aperture synthesis* is used, in which signals from spatially separated antennae are combined, monitoring the correct relative phases. For example, a system of telescopes that can be displaced up to 5 km relative to each other, due to the fact that they are mounted on rails, is used in Cambridge, Great Britain. The lengths of the connecting cables are known with an uncertainty of 1 mm. The largest such interferometer is the Very Large Array in New Mexico, a Y with 3 arms of length up to 27 km! The matematical processing of data from different telescope positions to form a high-resolution radio image includes, as its most critical component, a fast Fourier transform, and the fast development of computers has been a prerequisite for high-resolution radio astronomy. As in optical Fourier transform spectroscopy (Sect.6.2.4), a multiplex advantage is obtained in radio interferometry: information from a solid angle interval, which is determined by the (wide) diffraction lobe of each individual small telescope, is obtained simultaneously. An extreme angular resolution (10^{-4} arcseconds) can be obtained in so-called VLBI (Very Long Base line Interferometry) measurements. The first trans-oceanic radio-interferometric measurements were performed in 1968 between telescopes in Sweden, Massachusetts, West Virginia and California. The largest telescope separation was 8000 km. Perfect time synchronization using atomic clocks (Sect.7.1.3) is needed in order to utilize the detailed phase difference information. VLBI can also be used to obtain precise information on the continuing small displacements of continents. From such measurements it is known that Europe and North America drift apart at a rate of 2 cm/year!

Radio signals are obtained primarily from interstellar gas clouds but also from distinct, localized sources. The gas clouds are observed against the general cosmic background corresponding to black-body radiation of T_b = 3 K. Depending on whether the local radiation temperature T of the cloud is lower or higher than T_b, molecular lines from the cloud are observed as absorption lines in the background radiation or as emission lines, as illustrated in Fig.7.28. If a population inversion between the molecular levels

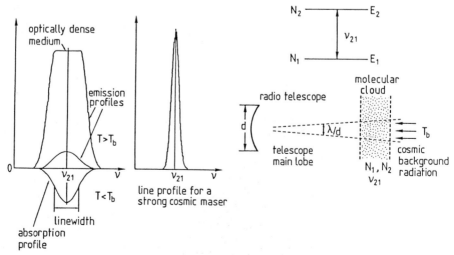

Fig.7.28. Absorption and emission lines in the microwave region. The narrowing of a maser transition is also illustrated [7.60]

exists (at least 3 levels must then be involved) the background radiation can be amplified by maser action. Such action can be recognized through the characteristic line narrowing for stimulated emission.

Molecular lines observed in the microwave region are due to pure rotational or hyperfine transitions. Hydrogen is the most abundant element in the universe. The H_2 molecule does not possess a dipole moment and thus no allowed electric dipole transitions are observed. This molecule can, therefore, not be observed.

However, atomic hydrogen can be observed at the magnetic dipole hyperfine transition (ν = 1.42 GHz) between the hyperfine levels F=1 and F=0 of the ground state (Sect.7.1.3). For example, using the corresponding 21 cm line the spiral structure of our galaxy has been investigated.

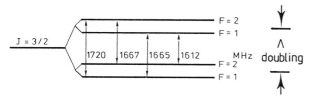

Fig.7.29. OH molecule substructure and microwave transitions [7.60]

More than 70 different interstellar molecules have been detected. The hydroxyl radical OH has a large dipole moment and was observed in 1963 after necessary laboratory measurements had been made. The OH levels and transitions are depicted in Fig.7.29. The rotational level is split into two levels (Λ-doubling, Sect.3.2), each with two hyperfine levels. The observations made were absorption measurements with the strong continuum radiosource Cassiopeia A providing the background. Maser action in OH has been observed in many cases and the pumping is normally due to hot star radiation. The CH radical, also featuring Λ-doubling, was observed at Onsala in 1973. The water molecule H_2O, which has an angle of 105° between the H atoms, has a complex rotational spectrum. Intense maser action has been observed on the 22 GHz transition, which is also seen in atmospheric absorption (Fig.7.23). High-frequency radio astronomy observations have to be performed on dry days. A large number of other molecules, some of them quite complex, have been observed in space. The (very) remote sensing of the physical conditions in the interstellar clouds has been discussed in [7.61].

As with any other spectral lines, radio lines exhibit Doppler shifts if the molecular gas is in motion. In Fig.7.30 velocity distributions are shown based on Doppler observations on OH, CH and HCHO lines observed in the direction of Cassiopeia A. The components around 45 km/s are due to clouds in the Perseus arm of our galaxy, while the components around 0 km/s are due to the Orion arm. While OH and HCHO appear in absorption, CH is seen in (weak maser) emission against the background source.

Apart from studying interstellar clouds a large number of "point" sources have also been investigated using radio astronomy. The observation

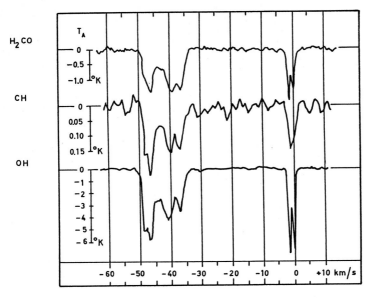

Fig.7.30. Spiral structure of our galaxy mapped out with Doppler-shifted microwave signals [7.60]

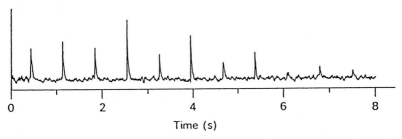

Fig.7.31. Periodic signals due to a pulsar source (Courtesy: Onsala Space Observatory)

of pulsars (1967) is among the more spectacular ones. Pulsars have a pulse frequency of the order of seconds. They are considered to be rapidly rotating neutron stars, which are very small, dense objects. An example of pulsar signals is given in Fig.7.31. The enormous energy generation in quasars, probably associated with black holes at cosmological distances is another fascinating discovery due to radio astronomy. Radio astronomy has been covered in [7.62-66]. The somewhat controversial field of the search for extraterrestial intelligence has been treated in [7.67].

8. Lasers

In this chapter we will discuss the general principles of lasers and study the most important *tunable* lasers that are of primary spectroscopic interest. Since many tunable lasers are optically pumped by fixed-frequency lasers we will also describe the most useful types of such lasers. For a more thorough account of laser physics we refer the reader to standard textbooks [8.1-10].

8.1 Basic Principles

The first operating laser, the ruby laser, was constructed in 1960 by *Maiman* [8.11]. An important step in the development chain resulting in the laser was a theoretical paper from 1958 by *Schawlow* and *Townes* [8.12], who analysed the prerequisites for laser action. The ammonia maser had been introduced as early as 1954 by *Gordon* et al. [8.13]. This device operated in the microwave region, and the laser is the counterpart in the visible wavelength region. As we will see, it is not trivial to bridge the large frequency gap between the microwave and optical regions. The 1964 Nobel Prize in Physics was awarded to N.G. Basov, A.M. Prokhorov and Ch.H. Townes for the development of the maser and the laser [8.14-16]. Maser and laser are acronyms for "microwave" and "light", respectively, "amplification by stimulated emission of radiation". The stimulated emission is essential in both processes, and we will first discuss this phenomenon. Let us, in the same way as in Sect.4.1 consider a system of atoms with two energy levels E_1 and E_2 (Fig.8.1). Photons of energy E_2-E_1 are allowed to impinge on the system and we consider which of the two processes, absorption or stimulated emission, is the most probable. In Chap.4, we showed that $B_{21} = B_{12}$ and therefore the relative probability of the two events is only dependent on the populations of the individual levels. In a system in thermodynamical equilibrium (left), the lower level is strongly overpopulated and therefore the photon is normally absorbed. In order to make the stimulated emission process more probable, we must, in one way or another, induce a greater population in the upper level than in the lower level. Then a greater number of photons are obtained than the number impinging on the medium. These stimulated photons have the same frequency, phase and direction of propagation as the stimulating photons and are described as *coherent*. We have thus shown that *population inversion* is a prerequisite for amplification by stimulated emission of radiation. Note the complete symmetry pertaining

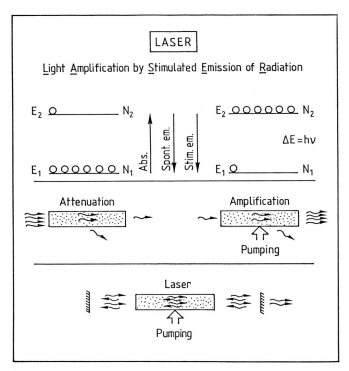

Fig.8.1. Basic phenomena in laser action

to the two processes of absorption and stimulated emission. By arranging a feedback cavity around the amplifying medium a self-oscillating laser can be obtained, as will be discussed in more detail later. A complete discussion of the conditions for reaching the threshold for laser oscillation must also include considerations of linewidths.

The construction of a maser or a laser involves the fundamental task of creating a population inversion. Spontaneous emission is not desired in this case, since it reduces the population difference without leading to the emission of coherent photons. Such decays constitute noise in the amplifier that the inverted system constitutes. A good introduction to laser physics is obtained by considering the mechanism of the ammonia maser. The ammonia molecule is schematically drawn in Fig.8.2. The nitrogen atom oscillates back and forth through the plane defined by the hydrogen atoms. Two eigenstates are possible for this inversion oscillation, corresponding to a symmetric and an anti-symmetric wavefunction. The energy separation between these eigenstates corresponds to 23.87 GHz or a wavelength of 1.25 cm for absorbed/emitted photons. In Fig.8.2 Townes' set-up is also shown (see also, Sect.7.1.3). From the gas reservoir molecules emerge in both energy states. In passing an inhomogeneous electric field, molecules in the upper energy level will be focused into a cavity, while lower-level molecules are defocused and filtered away. In this process the molecular induced

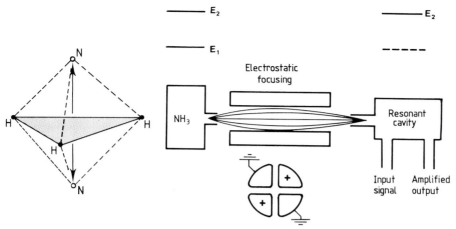

Fig.8.2. The ammonia molecule and schematics of an ammonia maser

dipole moment is active (Sect.7.1.2). The cavity is resonant at the transition frequency between the eigenstates. An inverted population is sustained in the cavity. If a microwave signal of the correct frequency is then fed into the cavity it will give rise to stimulated emission and the signal will be amplified. If sufficiently many molecules in the upper energy state are fed into the cavity, self-oscillation can be initiated by noise photons also present in the cavity. The ammonia maser was used as a clock before more accurate atomic clocks were developed.

Whereas amplification (oscillation) is obtained comparatively readily at these frequencies, which are of the order of 10^{10} Hz, it is considerably more difficult to achieve a corresponding situation for optical waves with a frequency of 10^{15} Hz. The reason is clear from (4.27).

$$\frac{A_{21}}{B_{12}\rho(\nu)} = 16\pi^2 \frac{\hbar\nu^3}{c^3\rho(\nu)} \,, \tag{8.1}$$

i.e., the ratio between the number of spontaneous and stimulated decays increases as ν^3. Thus, the upper level will be comparatively rapidly depopulated by spontaneous decays and a very efficient pumping mechanism for the upper level is necessary to achieve laser action. From this argument it follows that it is very difficult to construct an X-ray laser. However, a great deal of research is being devoted to this problem and significant progress has been made in this field. In particular, amplification has been demonstrated in neon-like selenium (Se XXV) at 206 and 210 Å (3s-2p transitions) following the impact of a high-energy laser pulse on a selenium foil, which then virtually explodes [8.17]. In another system amplification on the Balmer-α line in C VI (182 Å) has been obtained [8.18]. Further successful demonstrations include a 106 Å line in neon-like Mo. It is particularly attractive to try and reach 44 Å where the Kα absorption edge of carbon occurs. Research activities aiming at short-wavelength lasers have been described in [8.19].

8.2 Coherence

Lasers are characterized by light that is highly *coherent* in comparison with light from conventional light sources. There are two types of coherence: *temporal coherence* and *spatial coherence*. The degree of *temporal coherence* of a light source is a measure of the possibility of predicting phase and amplitude for the light at a given location and at a certain time, provided that these quantities were known at an earlier well-defined time, at the same point. As discussed in Chap.4, there is a time uncertainty τ associated with the spontaneous decay of an excited atom. If the classical wave approach is applied to the light, then the wave will be emitted during $\simeq \tau$ seconds, and therefore the length ℓ_c, the *coherence length* of the connected wave train, will be

$$\ell_c = \tau c .\tag{8.2}$$

Because of the finite emission time in this picture, the wave train is cut off and a Fourier analysis of the train yields a frequency spread of $\Delta \nu = (2\pi\tau)^{-1}$, as previously discussed. From a quantum-mechanical point of view the argument should be given the other way around. The time uncertainty yields a frequency spread and a corresponding associated length of the wave train. Wave trains emitted in different decays are completely unrelated. For $\tau = 10^{-8}$ s a coherence length of 3 m is obtained. Practical gaseous light sources have a Doppler broadening which is considerably larger than the natural radiation width. Because of the increased frequency spread a prediction of the phase is made more difficult. Line light sources have a typical coherence length of $\simeq 10$ cm. Sources, emitting a Planck distribution clearly have a very short coherence length.

A light wave is considered to be *spatially coherent* if there is a constant phase difference between *different* points of observation. Light that is emitted from different parts of a conventional light source is not phase-related. Therefore, the light from an extended light source will not be spatially coherent at close distances.

A good measure of the coherence of a light source is its ability to produce stable interference fringes. (An incoherent light source may produce momentary interference but the fringes continuously move swiftly and unpredictably.) The two types of coherence can be illustrated by two examples. If the path difference in a Michelson interferometer (Sect. 6.2.4) is 1 m no fringes are obtained with a conventional light source because of its poor temporal coherence. If the two slits in Young's experiment are illuminated by light from different parts of a fluorescence tube no fringes are obtained because of the poor spatial coherence of the light source. Lasers have very high temporal and spatial coherence since the photons are produced in chain reactions of stimulated emissions, in which all the generated photons represent waves with a common phase.

8.3 Resonators and Mode Structure

A population-inverted medium will amplify an incoming wave of the correct frequency through stimulated emission. Whereas the maser in the microwave region is an amplifier with a low noise level, finding applications, e.g. in radio astronomy (Sect.7.3), a laser in the visible region is very noisy because of the increasing importance of the spontaneous emission. According to (8.1) the signal-to-background ratio will be reduced by a factor of about 10^{15} for a frequency change of a factor 10^5, from microwaves to visible light. Therefore, lasers are rarely used for light amplification, except for the case of amplifying laser pulses. Lasers are normally used as *light generators (oscillators)*. An amplifier becomes an oscillator if feedback is introduced. This can be achieved by placing the laser medium in a *resonator* consisting of two mirrors (Figs.8.1,3). A spontaneously generated noise photon of the correct frequency and direction of propagation (perpendicular to the end mirror) is amplified and starts the process. The light intensity is substantially increased by reflections back and forth through the laser medium. In order to sustain the laser action, it is necessary to continuously pump in order to maintain the population inversion. In order to obtain a useful external beam, one of the mirrors is made partially transparent. Clearly, the emerging light represents a loss factor and the degree of *output coupling* possible without terminating the laser action depends on the gain of the medium. Ideally, the extracted energy would be the only source of loss. By using multiple dielectric layers the absorption losses due to the mirrors can be kept well below 1%. All optical surfaces inside the cavity must be kept very clean since any absorption is amplified by the multiple passes. In the figure the laser mirrors are indicated as plane. For such a system the alignment of the resonator is critical if losses are to be prevented. A considerably more stable configuration is obtained if at least one of the mirrors is slightly curved. In Fig.8.4 a number of resonator arrangements are illustrated. The relative merits of different configurations have been discussed, e.g. in [8.20].

As for a microwave cavity, several types of transverse electromagnetic oscillations or *modes* are possible for a laser cavity. Generally, one tries to

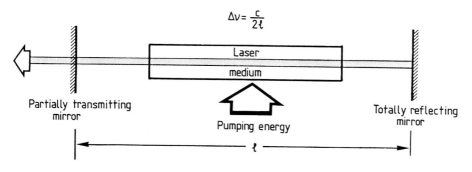

Fig.8.3. Principal arrangements for a laser

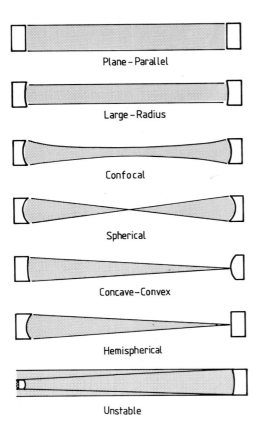

Fig. 8.4. Different resonator arrangements

isolate the mode that has the highest symmetry. This mode is designated TEM$_{00}$. Other modes (TEM$_{pq}$, p, q integers) corresponding to more or less asymmetric radiation fields can be suppressed by making the tube containing the laser medium sufficiently narrow or by introducing an axial aperture. A laser beam due to a laser oscillating in the TEM$_{00}$ mode has a symmetric cross section and the intensity falls off according to a Gaussian distribution. Laser modes were discussed by *Boyd* et al. [8.21].

As a consequence of the coherence, the laser light can be transmitted as an almost parallel bundle. The small divergence is a consequence of the unavoidable influence of diffraction. For radiation in phase and with the same amplitude over a circular aperture of diameter d, the angle θ between the centre of the diffraction pattern and the first dark interference ring is given by

$$\theta = 1.22 \, \lambda/d \, . \tag{8.3}$$

A visible diffraction-limited laser with an emerging beam of diameter 1 mm will have a divergence of about 0.5 mrad. Many practical systems have a divergence that is limited only by diffraction. For such a laser the spot at a distance of 1 km has a diameter of about 0.5 m. By first expanding the beam to a larger value of d it is possible to achieve a correspondingly smaller divergence. By focusing a parallel TEM$_{00}$ beam a very small spot diameter δ can be obtained. For a perfect lens of focal length f we obtain

$$\delta = \frac{4}{\pi} f \frac{\lambda}{d} . \tag{8.4}$$

A very high degree of linear polarization can be obtained for a laser with low losses in one direction of oscillation and high losses in the perpendicular direction. For a laser with a gas as the active medium this is accomplished by placing the windows of the gas container at Brewster's angle to the optical axis of the laser (Fig.8.5). For other lasers some other type of polarizer can be placed inside the cavity.

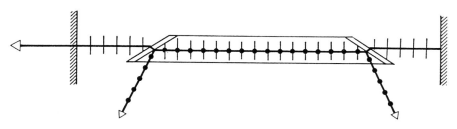

Fig.8.5. Generation of polarized laser light using intracavity Brewster's angle windows

We have now discussed different characteristics of laser light without dealing with its frequency distribution in detail. It is evident that a stable oscillation mode can be achieved in the resonator only if there is constant constructive interference, i.e. standing waves. This occurs when an integer number of half wavelengths fit into the cavity of (optical) pathlength ℓ. By considering two adjacent modes we obtain the mode separation $\Delta\nu$

$$n\frac{\lambda}{2} = n\frac{c}{2\nu} = (n+1)\frac{c}{2(\nu + \Delta\nu)} = \ell ,$$

$$\Delta\nu = c/2\ell . \tag{8.5}$$

This expression has already been derived, see (6.26). Clearly, the laser cavity is a Fabry-Pérot resonator with a free spectral range given by (8.5). With a cavity length of 1 m the mode separation is 150 MHz.

Let us now consider a gas laser, for which the stimulated emission occurs for a certain spectral line, determined mostly by Doppler and pressure broadening (Fig.8.6). Modes not falling within the linewidth are impossible since there is no gain for such modes. As a matter of fact, modes that are not close to the peak of the profile (the gain profile) will have too low a gain to compensate for the losses. Clearly, there is a certain threshold for the pump energy below which no modes at all have a sufficient gain. When lasing, only the most favourable modes will oscillate because of the regen-

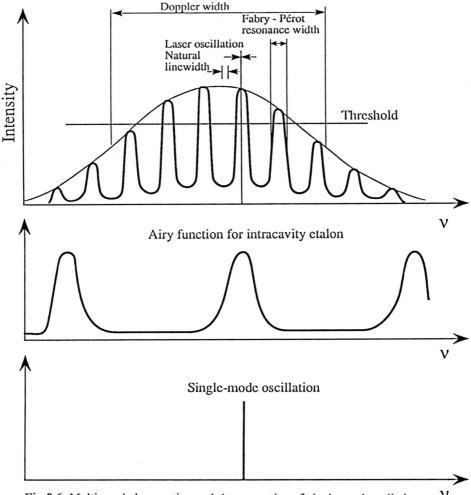

Fig.8.6. Multi-mode laser action and the generation of single-mode radiation

erative nature of the laser action ("strong become stronger, weak become weaker"). When several modes oscillate simultaneously we talk about *multi-mode operation*. By introducing a Fabry-Pérot interferometer into the laser cavity with a free spectral range sufficiently large so that only one of its transmission peaks falls in the region for possible laser action, a single cavity mode can be selected (*single-mode operation*). The interferometer is frequently made as a plane-parallel disc with semitransparent dielectric coatings (an "etalon"). The choice of mode is made by inclining the etalon. The transmission maxima of the etalon are also influenced by the temperature. In the absence of temperature stabilization, "mode hopping" can occur.

When the laser is oscillating in only one mode it produces a beam of very sharp frequency. Because of the generation of light by stimulated emission inside the cavity, the linewidth can become much smaller than the width of the Fabry-Pérot resonance (the Airy function) or the natural radiation width. Using indirect measurement techniques, linewidths of the order of 1 Hz have been established for a He-Ne laser, approaching the Schawlow-Townes limit [8.12]. To obtain narrow lines a careful stabilization of the length of the laser cavity using a servo system is necessary, otherwise vibrations will cause a considerable effective linewidth. Even without such servo stabilization a single-mode laser has very good temporal coherence. In comparison, the coherence length for a multi-mode laser is very small. A high spatial coherence is still maintained due to phase locking in the stimulated emission. Therefore, sharp interference fringes are obtained in most interference experiments.

Now we will study a number of laser types more closely. We distinguish between *fixed-frequency lasers* and *tunable lasers*. In a limited sense, the former type is also tunable since different modes under the gain curve can be selected. In lasers for which the active medium is a solid, the gain curve can be moved slightly by changing the temperature. We will primarily discuss fixed-frequency lasers that are used for the pumping of tunable lasers. By a tunable laser we here mean a laser whose wavelength can be varied over a large region (>100Å). Such lasers, which are of great spectroscopic interest, will be discussed in some detail.

8.4 Fixed-Frequency Lasers

8.4.1 The Ruby Laser

As we have mentioned, laser action was first observed in ruby. Ruby is a red crystal of Al_2O_3 with an addition of about 0.05% Cr_2O_3. Only the Cr^{3+} ions are of interest here since the other ions do not participate in the process. The chromium ion has three d electrons in its unfilled shell and has a ^4F term as the ground-state term. The next higher state is a ^2G term. The ruby crystal has a weakly rhombic structure. Because of the action of the crystal field, the ^4F term will be split into the levels $^4F_1$, $^4F_2$ and $^4A_2$, where the designations are no longer the ordinary ones from atomic spectroscopy but are defined in the group-theory treatment of the crystal field problem. The ^2G term is split into levels designated $^2A_1$, $^2F_1$, $^2F_2$ and ^2E. The $^4F_1$ and $^4F_2$ levels are strongly broadened to energy bands, whereas ^2E has a doublet structure. In Fig.8.7 the energy levels of Cr^{3+} in ruby that are relevant for laser action are indicated. The ruby crystal has strong, broad absorption bands around 550 and 400 nm. Ions that have been pumped to the ^4F bands will fall, within 10^{-7} s, to the ^2E levels, in which a population is quickly built up because the lifetime of this level is very long; about 5 ms. Population inversion with regard to the ground state is most easily achieved for the lower of the two ^2E levels. Stimulated emission is obtained at 6943 Å at room temperature. If the crystal is cooled to 77 K (liquid nitrogen temperature) the wavelength is shifted to 6934 Å. In the ruby laser, which is an example of a *three-level laser*, population inversion must

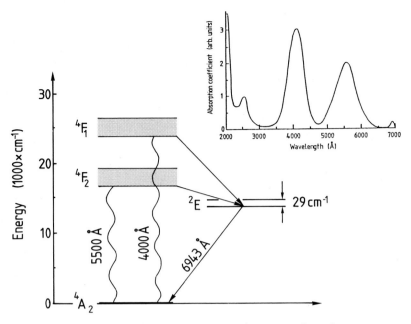

Fig.8.7. Level diagram and schematic absorption curves for ruby

be achieved with regard to the normally well populated ground state. It is thus necessary to pump the ions very efficiently from the ground state to the ^4F levels. In Fig.8.8 one possible pumping arrangement (also used by Th. Maiman) is shown. A ruby rod, about 5 cm long and with flat, parallel silvered end surfaces, is placed inside a helical flash-lamp, filled with e.g. xenon at high pressure (Sect.6.1.2). A capacitor of about 100 μF charged to about 2 kV is discharged through the flash-lamp, which lights up for about 1 ms. Because of the broad absorption bands of ruby a non-negligible part of the light, emitted as a continuum, is absorbed. After about 0.5 ms stim-

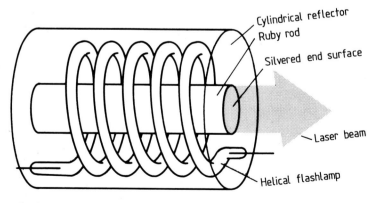

Fig.8.8. Arrangement for a ruby laser

ulated emission is transmitted through the one semi-transparent end surface of the rod. The light is emitted as a sequence of short spikes of about 1 μs duration. This is due to the disappearance of the population inversion when stimulated emission has occurred. The flash-lamp is still on and needs some time to restore the population inversion leading to the emission of a new laser spike. This behaviour, which is frequently not wanted, can be eliminated by *Q-switching*. In this case, laser mirrors, separated from the rod, are used and the light path between the mirrors is initially blocked. The quality factor or *Q value* of the cavity is then low and stimulated emission is not initiated although a strong inversion has occurred. If the blocking is quickly removed the cavity can enhance the light field and the emission occurs in a *giant pulse*. In practice, polarized laser light is used and the Kerr or Pockels effect are employed to rapidly switch the plane of polarization. Giant pulses can have a peak power of 10^8 W and a pulse width of about 10 ns.

8.4.2 Four-Level Lasers

We have already mentioned the difficulties involved with a three-level laser in which the laser transition is terminated in the continuously well populated ground state. The successful operation of the ruby laser relies on the very favourable combination of broad absorption bands and the long upper-state lifetime allowing the storage of energy. In a *4-level laser*, a final level, which is not the ground state, is used. The basic diagram is shown in Fig.8.9a.

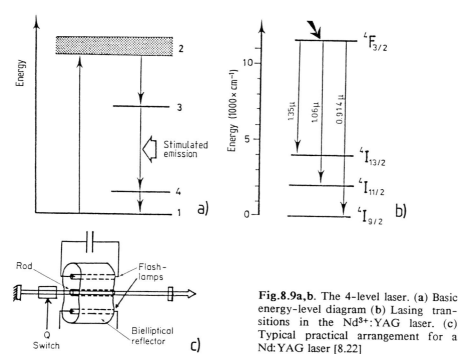

Fig.8.9a,b. The 4-level laser. (a) Basic energy-level diagram (b) Lasing transitions in the Nd^{3+}:YAG laser. (c) Typical practical arrangement for a Nd:YAG laser [8.22]

Suitable energy-level diagrams of this general type can be found, particularly for ions of the rare-earth elements. The ions can be incorporated into certain crystals such as CaF_2 or $CaWO_4$, and also in glass. It is important that level 3 (the storage level) has a comparatively long lifetime so that the ions can be accumulated there. On the other hand, level 4 must be depopulated efficiently. In many elements level 4 is so close to the ground state that cooling is necessary to reduce its thermal population. Laser action has been achieved for most of the rare-earth ions. An especially useful type of 4-level laser is the Nd:YAG laser, for which Nd^{3+} ions have been incorporated into an yttrium-aluminum host crystal of garnet type ("Yttrium-Aluminum-Garnet", $Y_3Al_5O_{12}$). The lasing transitions for this laser are shown in Fig.8.9b; in particular, the most commonly used one at 1.064 μm. In Fig.8.9c a frequently used Nd:YAG laser arrangement with two linear flash-lamps in a bi-elliptic pumping cavity for efficient pumping of the rod is shown. Nd:YAG lasers with a high output power are constructed with an oscillator followed by one or several amplifier stages. This is a very general construction for pulsed lasers. The oscillator is designed for relatively low power. It is then possible to control the mode structure, linewidth and pulse length precisely. A sharp frequency, a short pulse length and a clean TEM_{00} mode can generally only be achieved at the expense of the output power. A laser amplifier, consisting of a larger flash-lamp-pumped rod, will boost the energy of the pulse without changing the beam quality. The diameter of the amplifying medium is successively increased in the amplifying stages and the laser beam is correspondingly expanded. The technology of solid-state lasers has been discussed in [8.23-25]. High-power Nd-glass lasers are used in fusion research based on laser-driven inertial confinement [8.26-8.28].

A typical Nd:YAG laboratory laser system for spectroscopic applications is an oscillator-amplifier arrangement yielding 1.06 μm pulses of 10 ns duration and 1 J energy at a repetition rate of 10 Hz. In such systems, a so-called unstable resonator (Fig.8.4) is sometimes used to allow a more efficient energy extraction from the oscillator rod than that which is possible for a resonator sustaining a TEM_{00} mode. Instead of a semi-transparent output coupler a very small central feedback mirror is used and the output power is extracted in an annulus around this mirror. In rod arrangements, Nd:glass is limited to very low repetition rates because of the poor thermal properties of glass, leading to strong lens effects. However, by using a slab arrangement instead of a rod for the active medium, as illustrated in Fig. 8.10, lensing effects can be shown to self-compensate and it is possible to construct very efficient glass lasers [8.29]. The Nd:YAG and glass materials absorb the pumping radiation in certain spectral regions, as shown in Fig.8.11. New host materials, such as GGG (Gadolinium-Gallium-Garnet, $Gd_3Ga_5O_{12}$) and GSGG (Gadolinium-Scandium-Gallium-Garnet), in which also Cr^{3+} ions have been incorporated as sensitizers, better utilize the

Fig.8.10. Slab arrangement of the gain medium in a solid state laser

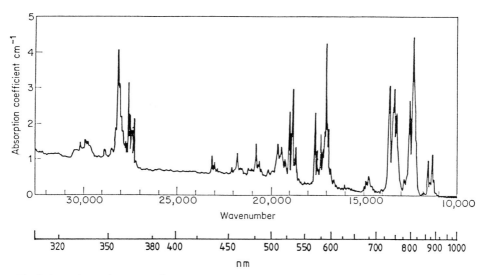

Fig.8.11. Absorption curve for Nd:YAG [8.30]

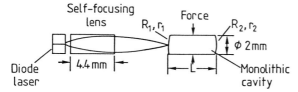

Fig.8.12. Diode laser pumping of a miniature Nd:YAG laser [8.32]

available radiation and show great promise. Different laser crystals have been described in [8.25]. Increased efficiencies (beyond the 2% wall-plug power to laser power-conversion efficiency typical for a Nd:YAG laser) are attainable by pumping a YAG rod with a suitable frequency-matched diode laser (Sect. 8.5.5.) [8.31]. Frequently, the diode laser is chosen to operate at the 808 nm absorption peak. In Fig.8.12 the pumping of a miniature rod is shown yielding single-mode output at a wall plug conversion efficiency of better than 5%. Such a laser is ideal as a high-quality oscillator for subsequent amplification. As high-power diode lasers become available at reduced prices, high-power solid-state lasers not employing flash-lamp technology will emerge.

8.4.3 Pulsed Gas Lasers

The lasers we have considered so far have a solid material as the active medium. Generally, flash-lamp pumping is employed and short pulses are obtained at a repetition rate of typically 10 Hz. High peak powers are obtained in the Q-switched mode (MW-GW). Gaseous media can also be used for the generation of short laser pulses. We shall here consider the nitrogen laser, the excimer laser and the copper vapour laser. The relevant parts of

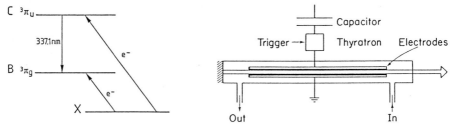

Fig.8.13. Energy-level diagram and basic arrangements for a nitrogen laser [8.22]

the N_2 molecule level diagram and the basic arrangement for the N_2 laser are given in Fig.8.13. Pumping is achieved using an electrical discharge of short duration, transversely, in a tube through which N_2, at a pressure of about 100 torr, is flowing. Recently, sealed N_2 lasers of low power have become available. The transfer probability at electron impact for X→C is much larger than for X→B. Thus, population inversion occurs at level C with regard to B. However, the lifetime of state C is only about 40 ns, whereas the corresponding value for state B is about 10 μs. A very fast discharge, generated using a *Blumlein circuit* is utilized, since the upper level cannot store energy (Q-switching is not possible). The population inversion can obviously be achieved only for a short time (self-terminating transition). The pulse length for the nitrogen laser is normally less than 10 ns and the emission follows at 337.1 nm. Peak powers up to 1 MW can be achieved. Pulse energies are typically a few mJ, and repetition rates exceeding 100 Hz can be obtained. The nitrogen laser has such a high gain that a laser beam can be obtained even without cavity mirrors through amplified spontaneous emission (sometimes the term *super-radiant laser* is used). Normally, a totally reflecting mirror is used at one end of the gain tube while the window at the opposite end has a very high transmission. The divergence of the beam is given by the geometry of the discharge channel and is typically 10 mrad.

The construction of an *excimer laser* [8.33-8.35] is similar to that of an N_2 laser. Excimer molecules ("excited dimers") are characterized by the absence of a stable ground state while short-lived, excited states exist. In excimer lasers, molecules such as KrF and XeCl are used as the active medium. These molecules are formed in the excited state in a fast, electrical discharge in a mixture of the inert gas and F_2 or HCl. The high reactivity of the latter gases makes material selection, gas handling etc. critical for this type of laser. In Fig.8.14 an excimer level diagram and laser lines for different excimer molecules are given. Since no ground state exists excimer molecules constitute the perfect laser medium with automatic population inversion once a molecule has been created. Important laser emission lines are 249 nm (KrF; yields the highest power), 308 nm (XeCl, best suited for dye laser pumping) and 193 nm (ArF; attractive short wavelength for photochemistry). Pulse energies of hundreds of mJ can be achieved and average powers of 100 W can be reached. Excimer laser technology is quickly maturing, and these lasers find many important applications.

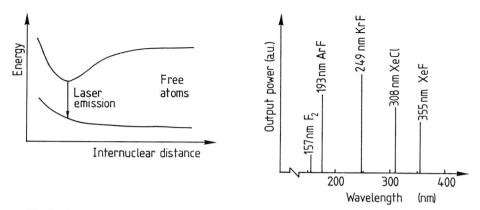

Fig.8.14. Energy-level diagram for an excimer molecule, and laser emission lines from different exciter molecules

Pulsed CO_2 lasers, which were available long before excimer lasers, have a similar construction. These infra-red lasers are of great technological importance. CO_2 lasers will be discussed later.

In the *copper vapour laser* the discharge tube must be heated to high temperatures in order to produce a sufficient Cu vapour pressure. This laser emits at 510 and 578 nm corresponding to the (4p $^2P_{3/2} \to$ 4d $^2D_{5/2}$) and (4p $^2P_{1/2} \to$ 4d $^2D_{3/2}$) transitions, respectively. Since the terminal ^2D states are metastable only pulsed operation is possible. High repetition rates can usually be obtained with this laser, up to 10 kHz. The average power can exceed 10 W. Exchanging copper for gold results in red emission at 628 nm.

Whereas population inversion can often be achieved relatively easily with a pulsed pumping source, it is considerably more difficult, and sometimes impossible, to continuously maintain population inversion while laser action prevails. Clearly, lasing means that excited atoms decay with the emission of stimulated radiation and thus the laser action itself will cause the lasing action to stop. Efficient pumping mechanisms are required. We will now study some important types of continuous fixed-frequency lasers which all work with gaseous laser media.

8.4.4 The He-Ne Laser

The helium-neon laser was the first gas laser and it was designed and built by A. Javan and co-workers [8.36] shortly after the introduction of the ruby laser. The He-Ne laser is one of the most common laser types. The construction can be made comparatively simple and cheap, and continuous laser action is obtained. The active medium is a gas mixture of He and Ne (ratio 5:1) in a glass tube at a total pressure of about 1 torr. Energy is added through an electric discharge through the gas. In order to understand the principle of the He-Ne laser, the energy level diagrams of both He and Ne must be considered. (Fig.8.15). In the discharge (1000V, 10mA) He atoms are excited to the metastable 2s $^3S_1$ and 2s $^1S_0$ states through electron im-

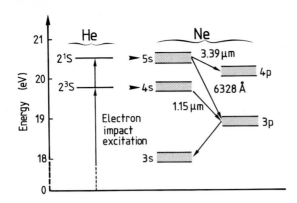

Fig.8.15. Energy-level diagrams relevant to the He-Ne laser

pact. In Ne the 4s and 5s level systems have almost the same energies as the metastable He states. Because of this, the probability of Ne excitation through collisions with metastable He atoms is very large. A population inversion between the s states and the lower-lying 3p and 4p levels is achieved and several laser transitions are possible. Since the lifetimes of the p states are very short ($\simeq 10$ ns) in comparison with the s state lifetimes (~100 ns), the population inversion can be maintained and continuous lasing is achieved. The reflection properties of the laser mirrors determine which line will lase. The most frequently used line is the red line at 6328 Å, but many other visible and IR lines are possible. The Doppler width of this line is about 1700 MHz. With a typical resonator length of 30 cm a mode separation of 500 MHz is obtained. Such a laser will exhibit 2 or 3 modes. With a sufficiently short laser (10-15 cm), single-mode operation can be attained. He-Ne lasers yield low output powers, typically a few mW.

Primary energy storage in He is also used in the *helium-cadmium laser*. Here, a sufficient metal vapour pressure must be achieved by heating. Excited Cd ions are produced by Penning ionization

$$A^* + B \rightarrow A + B^+ + e \; .$$

Two short-wavelength lines, 4416 Å ($4d^9 5s^2\ ^2D_{5/2} \rightarrow 4d^{10} 5p\ ^2P_{3/2}$) and 3250 Å ($4d^9 5s^2\ ^2D_{3/2} \rightarrow 4d^{10} 5p\ ^2P_{1/2}$), are obtained at power levels of 50 and 5 mW, respectively. An alternative way of producing the excited Cd ions is to use soft X-ray photo-ionization of ground-state $4d^{10} 5s^2$ Cd atoms [8.37]. High-power laser pulses impinging on a high-Z material, such as Ta, can be used for soft X-ray production (Sect.6.1.1). Using soft X-ray pumping, gain has also been demonstrated in In III (185 nm) [8.38] and Xe III (109 nm) [8.39].

8.4.5 Gaseous Ion Lasers

In gaseous ion lasers a population inversion between excited states of ionized argon or krypton is achieved. The pumping is accomplished through a strong dc discharge in low-density gas (~0.2 torr). The discharge tube is

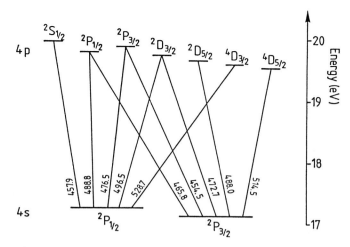

Fig.8.16. Energy level diagram of the Ar$^+$ ion, with argon-ion laser lines indicated

1-2 m long and current densities of about 500 A/cm^2 are used. Thus, a high electrical power is needed (~10 kW) and the discharge tube, which is made of beryllium oxide or graphite, must have an efficient water cooling system. Excited ions are produced in collisions between inert gas atoms and electrons. For ionization of an argon atom an energy of 15.75 eV is needed. The interesting excited ionic states belonging to the configuration 3p^{4}4p are located about 20 eV above the ionic 3p^5 ground state. In Fig.8.16 a partial level scheme for Ar$^+$ is given, with laser lines connecting 3p^{4}4p levels and 3p^{4}4s levels indicated. Several blue and green laser lines are obtained (5145, 5017, 4965, 4880, 4765, 4727, 4658, 4579 and 4545 Å). The strongest lines are

$$5145 \text{ Å} \quad 3p^4 4p \; ^4D_{5/2} \to 3p^4 4s \; ^2P_{3/2}$$
and
$$4880 \text{ Å} \quad 3p^4 4p \; ^2D_{5/2} \to 3p^4 4s \; ^2P_{3/2} \; .$$

When using laser mirrors with a high reflectivity in the blue-green region all the lines are produced simultaneously. Argon-ion lasers with a total output power of up to 30 W are commercially available. In Fig.8.17 the

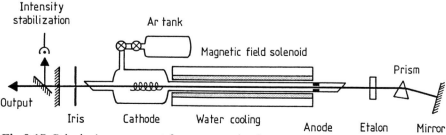

Fig.8.17. Principal arrangement for an argon-ion laser

principle arrangement of an argon-ion laser is shown. By using a prism in front of the totally reflecting mirror the individual spectral lines can be isolated. For a certain position of the prism only one wavelength is reflected back towards the semi-transparent output coupler, while the other wavelengths are refracted out of the cavity. By introducing an intracavity solid Fabry-Pérot interferometer (etalon), single-mode oscillation can be obtained.

At very high discharge currents laser transitions are also obtained in doubly ionized argon. Several UV lines in the wavelength region 300-386 nm are then obtained with a total power up to 5 W.

The *krypton-ion laser* has the same construction as the *argon-ion laser* but the discharge tube is instead filled with krypton. Apart from blue and green lines, several red lines are also obtained with this laser (7931, 7525, 6764, 6471, 5682, 5309, 5208, 4825, 4762, 4680, 4131 and 4067 Å), as well as strong UV lines (3564, 3507 and 3375 Å). Ion lasers are discussed in greater detail in [8.40].

8.5 Tunable Lasers

Whereas fixed-frequency lasers have found important applications in, e.g., measurement techniques, information transmission, holography and material processing, tunable lasers are of greatest interest for atomic and molecular spectroscopy. Using different types of tunable lasers the wavelength range 320 nm to tens of μm can be covered by direct laser action, and the tunability region of an individual laser can be considerably extended using nonlinear optical effects. Tunable lasers comprise dye lasers, F-centre lasers, certain solid-state lasers, spin-flip Raman lasers [8.41], parametric oscillators, high-pressure CO_2 lasers and semiconductor lasers. Here we will discuss the most important of these laser types. Overviews of tunable lasers can be found in [8.42-44].

8.5.1 Dye Lasers

Laser action in organic dyes was discovered independently by *Sorokin* et al. [8.45] and *Schäfer* et al. [8.46] in 1966. Since then, several hundreds of dyes have been shown to have suitable properties, to greater or lesser degrees, for use as laser media. One of the most common laser dyes is Rhodamine 6G dissolved in methanol or ethylene glycol. The complex structure of organic dyes is illustrated in Fig.8.18, where the formula for Rhodamine 6G is shown.

The general energy-level structure of an organic dye is shown in Fig.8.19. The transitions relevant for laser action occur between the two lower electronic singlet states. The ground state, as well as the excited state, is split into a continuous structure of smeared energy sublevels due to the interaction with the solvent. Normally the molecules are Boltzmann-distributed on the lowest sublevels of the ground state. The molecules can be ex-

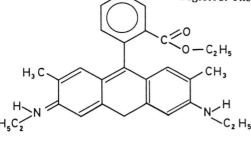

Fig. 8.18. The chemical structure of Rhodamine 6G

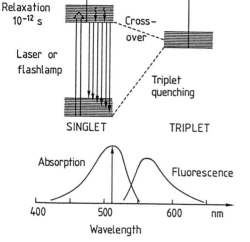

Fig. 8.19. General energy-level structure of an organic dye molecule, and resulting absorption and fluorescence curves

cited to the next singlet band with light. In radiationless transitions the molecules relax very quickly ($\sim 10^{-12}$ s) to the lowest level of the excited state, from where the molecules return to the sublevels of the ground state with a lifetime of about 10^{-9} s, emitting fluorescent light. The resulting absorption and emission curves for Rhodamine 6G are included in Fig. 8.19. Because of the relaxation in the excited singlet band the same fluorescence curve is obtained regardless of the spectral distribution of the absorbed light. For example, the curve given in the figure is obtained regardless of whether a flash-lamp or a fixed-frequency laser is used for the excitation.

The usefulness of an organic dye for laser applications much depends on the position of higher-lying singlet bands, and on the influence of the system of triplet levels that is always present in organic dye molecules. If a higher-lying singlet state has an inappropriate energy, a strong absorption of fluorescence light results. Through radiationless transitions, excited dye

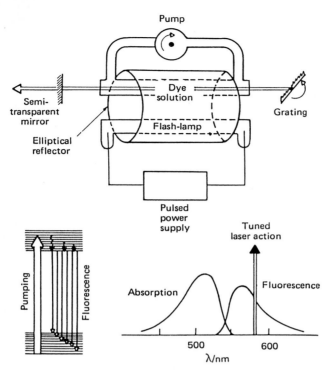

Fig.8.20. Flashlamp-pumped dye laser [8.22]

molecules can be transferred to the triplet system. If a large number of molecules assemble in the lowest triplet state, the fluorescent light from the singlet system can be strongly absorbed by triplet molecules. If the organic dye is pumped sufficiently strongly, a population inversion between the lowest sublevel of the upper singlet state and the sublevels of the ground state is obtained. Amplification by stimulated emission can then be achieved. If a dye solution is contained in a cell, placed in a laser cavity, and sufficient pumping power is supplied, laser action will occur (Fig.8.20). With broad-band laser mirrors stimulated emission is obtained in a wavelength range of a few tens of Å close to the peak of the fluorescence band.

A very important step was taken by *Soffer* and *McFarland* [8.47] in 1967 by replacing the totally reflecting mirror with a grating in a Littrow mount (Fig.6.20). A frequency-selective feedback was obtained and the bandwidth of the stimulated radiation was reduced to about 0.5 Å. By turning the grating the laser could be continuously tuned over the fluorescence band of the dye. In Fig.8.20 an arrangement for flash-lamp-pumping of a tunable dye laser is shown. Here, a tube is used for the dye solution which is pumped with a linear flash-lamp. Laser pulses with a duration of about 1 μs and a peak power of several kW are obtained. The repetition rate is generally about 1 Hz. Pulsed dye lasers are more frequently pumped by a fixed-frequency laser, normally a nitrogen, an excimer, a copper or a fre-

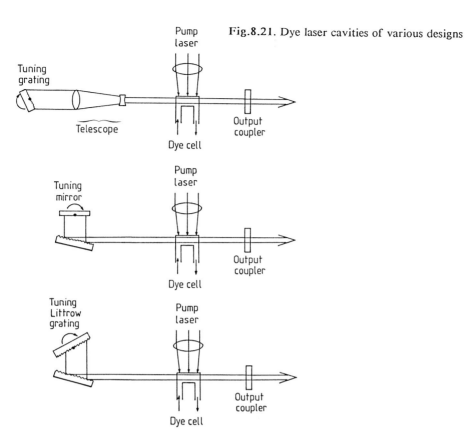

Fig.8.21. Dye laser cavities of various designs

quency-doubled Nd:YAG laser. In Fig.8.21 three different arrangements are shown. In the Hänsch design [8.48], an intracavity beam expander is used to widen the diameter of the beam so that more lines of the Littrow-mounted grating are illuminated. This leads to an increased resolution (Sect.6.2.2) at the same time as the grating is protected from local heating. The linewidth with such an arrangement is typically 0.1 Å. Using an intracavity etalon the linewidth can be reduced by a factor of 10. Instead of using a telescope a multiple-prism arrangement can be used, since expansion *across* the groves is the only consideration from a resolution point of view. Note that because of the short pumping time (5-10ns) there is normally no time for a cavity mode structure to build up in a short-pulse laser. Mode positions are only weakly indicated in the essentially smooth frequency distribution.

In Fig.8.21 an arrangement is shown in which all the grating lines can be utilized without using any form of beam expansion [8.49]. Here, grazing incidence on a fixed grating is employed. The laser is tuned by turning a mirror reflecting the first order of the grating back on itself. The useful laser beam can be taken out as the zeroth order beam from the grating (reflection) or through a partially transmitting end mirror. The latter arrangement reduces the broad-band dye emission. By exchanging the tuning mirror for a grating in a Littrow mount a still higher resolution can be achieved. By eliminating the telescope the cavity can be made very short. This leads to many round trips during the pulse, a large free spectral range and the possibility of single-mode operation [8.50].

If a high pulse energy is needed while retaining a small linewidth and a high beam quality, it is advisable to utilize an oscillator/amplifier arrangement, as mentioned above. In Fig.8.22 two examples are given. In order to achieve a very small linewidth (~100MHz) an external Fabry-Pérot interferometer can be used to filter the oscillation output before amplifica-

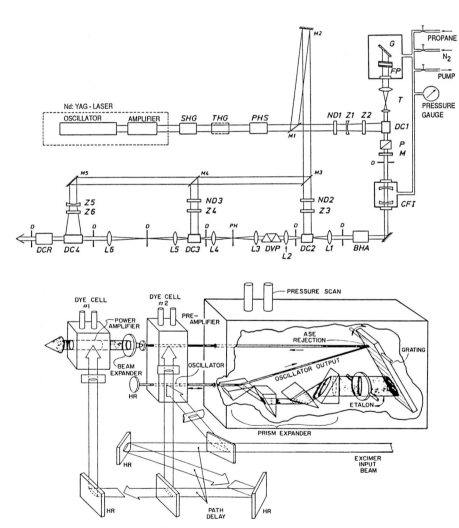

Fig.8.22. Oscillator/amplifier arrangements for tunable dye lasers ([8.51], and Courtesy, Lambda Physik)

tion. In the figure, the possibility of pressure-scanning the laser is also illustrated. Since grating, internal and external Fabry-Pérot interferometer are tuned in the same way when the pressure is changed (the expression $n\ell$ occurs in all the formulae) all the elements will remain synchronized once they have been correctly adjusted.

Pulsed dye lasers can be operated from 320 to 1000 nm using different dyes. In Fig.8.23 tuning curves for a dye laser, pumped by an excimer laser or harmonics of the Nd:YAG laser, are shown. An energy conversion of 10-20% is typical for excimer laser pumping. With a frequency-doubled Nd:YAG laser (532nm) the conversion for Rhodamine dyes can exceed 40%. The frequency range of dye lasers can be extended using nonlinear optics techniques (Sect.8.6).

As for fixed-frequency lasers, it is considerably more difficult to achieve continuous laser action than pulsed operation with a dye laser. The first continuous dye laser was designed in 1970 by *Peterson* et al. [8.52]. A special problem associated with organic dyes is the building up of a population of molecules in the lowest triplet state. For pulsed pumping, laser light can be obtained for a short time before triplet absorption dominates. To achieve continuous laser action, absorbing triplet molecules must be removed. Certain chemicals, such as cyclo-octatetraene (COT), are very active in transferring triplet molecules to the ground state without emitting radiation (triplet quenching). An argon or a krypton-ion laser with a power of several watts is used as a pump source for a continuous dye laser. With Rhodamine 6G, a conversion efficiency of 20% can be achieved in the most favourable wavelength region. However, the efficiency is normally lower. At the present time it is possible to cover the wavelength range 375-950 nm with continuous dye lasers. In order to achieve continuous laser action a certain dye must be pumped at a suitable wavelength, for which good absorption occurs. The green, yellow and red regions can be covered by dyes that can be pumped by the blue-green lines of the argon-ion laser. Certain dyes need to be pumped by the violet or red lines from the krypton-ion laser. Blue and violet dyes, e.g. Stilbene, require pumping with the UV lines from an Ar^+ or Kr^+ laser. In Fig.8.24 an example of the arrangement of a continuous, tunable dye laser is given. Using focused argon-ion laser light a continuous population inversion is maintained at a point on a quickly flowing flat jet of dye that is ejected through a specially designed nozzle. Ethylene glycol is used as the dye solvent, because of its high viscosity. This arrangement ensures an efficient cooling of the dye.

The gain volume is at the focal point of a folded cavity, consisting of two spherical mirrors and a flat output coupler. Tuning is accomplished with a birefringent, so-called Lyot filter [8.53]. This consists of one or several crystalline quartz plates mounted at Brewster's angle. The dominant transmission maximum of this filter can be moved by turning the optical axis of the filter with respect to the plane of polarization of the light in the cavity. A band width of about 0.5 Å is obtained from a cw dye laser when only a primary tuning element is used. The width of the filter peak is much larger but the laser will only oscillate close to the transmission peak because of the regenerative nature of laser action. With a typical cavity length of 30

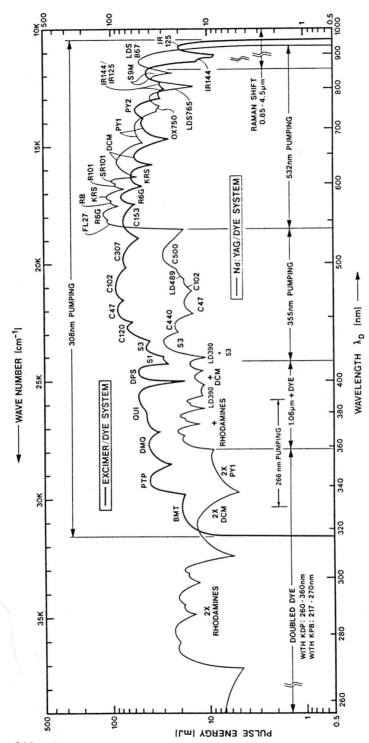

Fig.8.23. Tuning curves for excimer-pumped and Nd:YAG-pumped dye lasers [8.55]

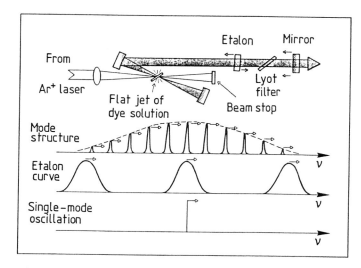

Fig.8.24. Continuous dye laser using a flat dye solution jet as the active medium

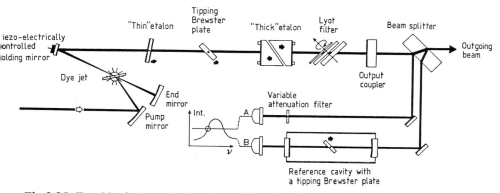

Fig.8.25. Tunable single-mode dye laser with active frequency stabilization (Courtesy: Coherent, Inc.)

cm a mode separation of 500 MHz is obtained and then several hundreds of modes will oscillate under the gain profile, which is given by the frequency selectivity of the cavity (not by any Doppler broadening as is the case in a gas laser). With an intracavity high-finesse etalon (Sect.6.2.3), a single cavity mode can be selected but the resulting frequency width for the outgoing light will still be several 10's of MHz since the frequency stability is limited by vibrations etc. With a frequency stabilization system the linewidth can be reduced to less than 1 MHz. Such a system is shown in Fig.8.25. In order to be able to make a system with etalons work over a wide wavelength region, two low-finesse etalons (broad-band coatings) are used in series. With a 1 mm thick etalon (n=1.5), a free spectral range of 100 GHz is obtained and when operated with a transmission maximum

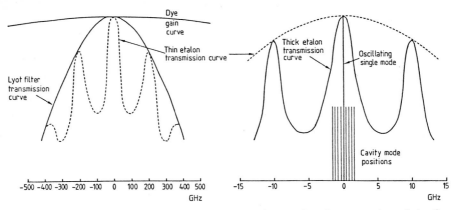

Fig.8.26. Hierarchic sequence of spectral curves illustrating the narrowing of the frequency domain and the isolation of a single cavity mode for a linear cavity

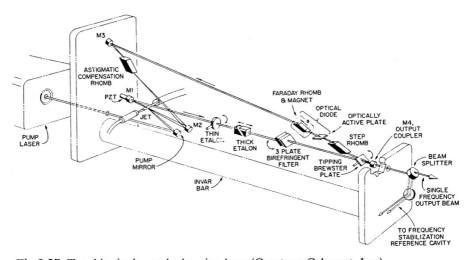

Fig.8.27. Tunable single-mode dye ring laser (Courtesy: Coherent, Inc.)

close to a maximum of the Lyot filter some 10 modes may lase. Adjacent etalon maxima occur where the filter has a substantially reduced transmission. If an additional, thicker etalon is introduced, having a substantially reduced free spectral range, say 10 GHz, single-mode operation can be achieved. In Fig.8.26 the transmission profiles of the different resonator components are shown. In order to achieve a high single-mode power, a ring laser cavity is frequently used instead of the usual standing wave linear cavity. This avoids the problem of the linear cavity of high non-utilized gain in the standing wave nodes, that will ultimately be utilized by an unwanted second cavity mode, that for high pumping power can no longer be suppressed. A ring laser arrangement is shown in Fig.8.27. Tuning curves for a single-mode dye laser are shown in Fig.8.28. Dye lasers have been discussed in detail in [8.54, 55], and laser dyes are covered in [8.56].

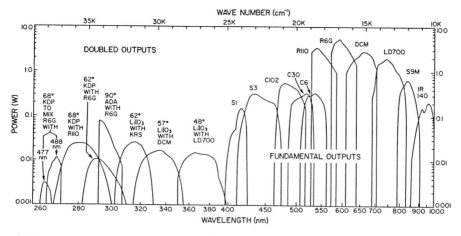

Fig.8.28. Tuning curves for a CW single-mode dye laser (Courtesy: Coherent, Inc.)

8.5.2 Colour-Centre Lasers

In the near-IR region colour-centre (F-centre) lasers can be used [8.57]. These lasers resemble dye lasers with regard to function and construction. Instead of the dye solution a cooled F-centre crystal is used as the active, laser-pumped medium. The colour centre is a defect in a crystal lattice, e.g. KCl. The defect consists of an ion vacancy that has trapped an electron. If impurity ions are close to the vacancy the optical properties are changed. F-centre crystals absorb visible radiation and fluorescence in the near-IR region. Using different crystals, the range 1-3.5 μm can be covered. In Fig.8.29 F-centre structures, absorption and lasing curves are shown as well as a practical arrangement in a commercial F-centre laser.

8.5.3 Tunable Solid-State Lasers

Certain solid-state materials have rather broad gain curves and can thus be tuned over a certain range. This is the case for the Nd:Glass laser, which can be tuned in the region 1.0-1.1 μm. The alexandrite (Cr^{3+}: $BeAl_2O_4$) and emerald (Cr^{3+}: $Be_3Al_2Si_6C_{18}$) lasers that bear a close resemblance to ruby lasers can, in the same way, be tuned in the regions 730-800 nm and 700-850 nm, respectively. These lasers can be pumped directly with a flash-lamp. Finally, the transition element lasers should be mentioned. $Co:MgF_2$ can be pumped by a simple, non-Q-switched Nd:YAG laser and yields tunable radiation in the region 1.7-2.3 μm. If the material is instead doped with vanadium, it can be tuned in the region 1.0-1.1 μm. A further interesting laser material is $Ti:Al_2O_3$ providing tunable radiation between 660 and 1100 nm. Direct generation of tunable laser radiation without using laser systems for pumping is very attractive. In order to extend the tuning range, frequency doubling and stimulated Raman scattering can be used (Sect.8.6). Tunable solid-state lasers have been discussed in [8.23, 58-60].

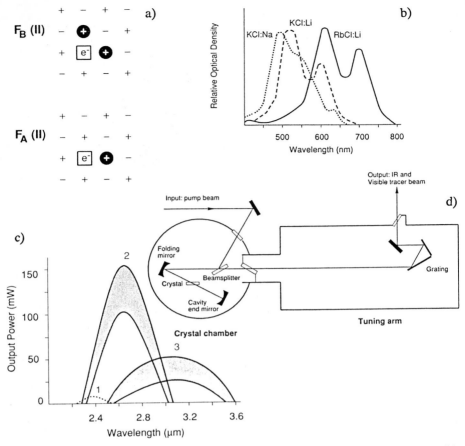

Fig.8.29a-d. Basics of F-centre lasers. The figure shows (**a**) F-centre structures, (**b**) absorption, (**c**) laser-emission curves, and (**d**) the cavity arrangement for a commercially available unit (Courtesy: Burleigh Instruments, Inc.)

8.5.4 Tunable CO_2 Lasers

The carbon dioxide laser is the most efficient gas laser with a wall plug efficiency of up to 20%. It works in the IR region around 10 µm. In numerous applications a non-tuned CO_2 laser is used. However, from a spectroscopic point of view, the fact that it can be line-tuned and continuously tuned at high gas pressures is very useful. In Fig.8.30 a level diagram and a practical arrangement for a tuned CO_2 laser are shown, together with a diagram of the available lines. The CO_2 molecule has three fundamental modes of vibration (Sect.3.4). In the mode that corresponds to symmetric stretching (ν_1), the two oxygen atoms move in opposite directions while the carbon atom is stationary. In the bending mode (ν_2), the molecule is bent back and forth. Finally, in the asymmetric stretching mode (ν_3), the oxygen atoms move together while the carbon atom moves in the opposite direction.

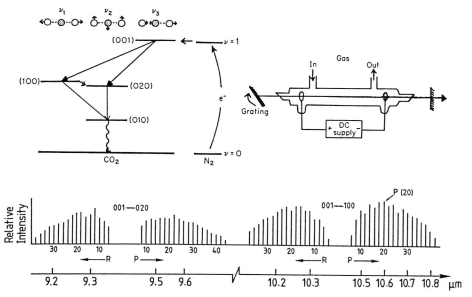

Fig.8.30. Energy-level diagram relevant to the CO_2 laser with individual vibrational-rotational laser lines and a cavity arrangement illustrated [8.22]

The vibrational state of the molecule is specified by a set of vibrational quantum numbers (v_1, v_2, v_3), one for each vibrational mode. In the CO_2 laser the level corresponding to asymmetric stretching (001) is populated efficiently by almost resonant collisions with excited molecules of nitrogen that have been added to the CO_2 gas. An electric discharge populates the first excited vibrational level in nitrogen. A strong population inversion with respect to the lower-lying CO_2 states (100) and (020) occurs. In a suitably arranged laser cavity, laser emission can occur in the bands 10.2–10.8 μm and 9.2–9.7 μm using a large number of rotational lines. If no special precautions are taken, lasing of the strongest line occurs, P(20) at 10.59 μm.

Using a Littrow-mounted grating other individual P and R branch lines can be chosen. Within every line, fine tuning can be accomplished within about 50 MHz, corresponding to the Doppler width at 10 μm for the CO_2 molecules. By utilizing isotopic molecules, like $^{13}CO_2$, further fixed wavelengths can be chosen. A continuous CO_2 laser of this type normally has an electric discharge along the tube. However, it is also possible to work at high gas pressures (atmospheric pressure and up to 10 atmospheres) if a transverse, pulsed discharge is used (TEA, Transverse Electric Atmospheric laser, Sect.8.4.3). At 10 atmospheres the pressure broadening is about 2 cm^{-1} (60 GHz) and the individual rotational-vibrational lines that are typically separated by 1–3 cm^{-1} merge. Continuous laser action within the band is then possible. Continuous laser action with a certain tunability can be obtained in *waveguide lasers*, working at 1 atmosphere pressure.

Non-spectroscopic high-power CO_2 lasers have been constructed with a continuous output up to tens of kW. Such lasers are used industrially for cutting, welding, hardening, etc. High-power pulsed CO_2 lasers are also used for fusion research. CO_2 lasers were discussed in [8.61-63].

Lasers with CO as an active medium can also be constructed. This laser type yields a large number of lines in the region 5.1-5.6 μm. In order to achieve efficient operation, the discharge tube must be cooled to low temperatures which complicates practical use. It is simpler to use HF or DF lasers, which give lines in the 2.8-4.0 μm region. Such lasers are examples of chemical lasers for which the active molecules are formed in the discharge tube from the supplied gases H_2/D_2 and SF_6.

As we have seen above, lasing has been achieved for a very large number of atomic and molecular lines. A list of these lines can be found in [8.64].

8.5.5 Semiconductor Lasers

In 1962 several American researchers discovered, essentially simultaneously, that laser action can be achieved in certain semiconductor diode arrangements. The physics behind this process is different from that discussed so far. However, the fundamental requirement of the creation of a population inversion still exists. A typical diode laser material is gallium arsenide (GaAs) which has been strongly doped. Laser action occurs in the transition zone between p- and n-doped material in a diode subject to a voltage applied in the forward direction. In Fig.8.31 the basic energy-level diagrams for the cases of voltage off and on are shown. The voltage forces electrons as well as holes to the transition region. Here the conduction-band states

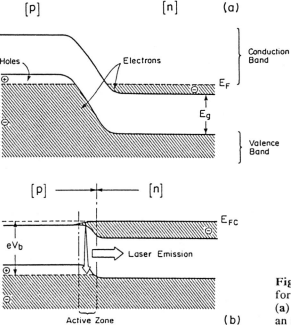

Fig.8.31. Energy-level diagram for a semiconductor diode laser (a) without voltage and (b) with an applied voltage [8.65]

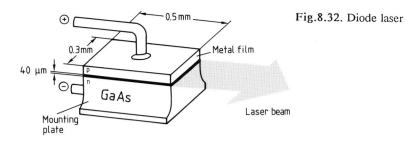

Fig.8.32. Diode laser

have excess population with respect to the empty states of the valence band. Stimulated light emission with a photon energy corresponding to the band gap will result upon the recombination of electrons and holes. Semiconductor lasers are very small; with typical dimensions of less than one mm. In Fig.8.32 the geometry of a typical laser diode is shown. Two opposite sides are polished and serve as a cavity. Due to the smallness of the cavity, single-mode oscillation is frequently obtained. Diode lasers with an external cavity can also be used.

Semiconductor lasers can be used in the red and particularly in the IR spectral region. Most materials must be cooled for operation, while GaAs diodes can be used at room temperature. The wavelength of the emitted light can be tuned between 800 and 900 nm by changing the temperature. The frequency change is due to a change in the energy gap as well as in the refractive index. Other possible ways of changing the wavelength of semiconductor lasers include the application of a magnetic field or hydrostatic

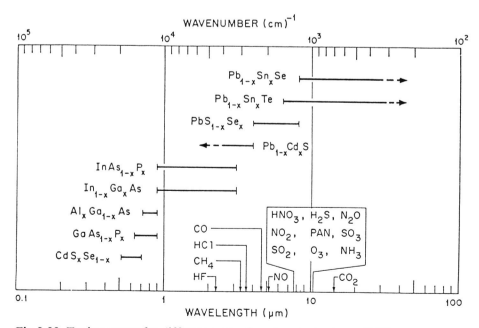

Fig.8.33. Tuning ranges for different types of semiconductor lasers [8.65]

pressure. Using several materials, generally made up of three elements such as Pb, Sn and Te ($Pb_{1-x}Sn_xTe$) or In, Ga and As ($In_xGa_{1-x}As$), the whole IR region up to about 40 μm can be covered. In Fig.8.33 the ranges of some diode lasers are shown. Intense research aimed at achieving diode laser light through the visible region is being pursued.[1] Laser diodes can be used in pulsed or continuous operation. A continuous power of 1 W or more can be obtained from certain diodes arrangements. Tunable diode lasers have been discussed in more detail in [8.66-70]. Efficient frequency doubling of diode-laser radiation to the blue spectral region has recently been accomplished.

Laser diodes have a very high efficiency (up to 80%). Since the light intensity can be modulated very quickly (up to 10^{10} Hz) by voltage variations, semiconductor lasers are of great interest for communication. Since the best transmission properties in fibres are obtained at 1.3 μm, great effort has been put into optimizing lasers at this wavelength.

8.6 Nonlinear Optical Phenomena

Several nonlinear optical phenomena can be utilized to extend the available wavelength range for a certain laser type. First we will consider *frequency doubling*. In an electric circuit, frequency doubling of an input signal can be achieved by using nonlinear components. The corresponding optical phenomenon, which is an example of *nonlinear optical effects*, can be observed in certain crystals that are traversed by very intense laser light. Light propagation through a transparent medium can be considered as a process in which electrical dipoles (electrons bound to a nucleus) are made to oscillate by the radiation field, and thus to re-emit the light. Through interference, light in the beam direction only is obtained. As long as the oscillation amplitude is small the dipoles can follow the applied oscillation, but at higher amplitudes nonlinearities occur when the dipoles can no longer reproduce the applied oscillation. Harmonics then occur. If $\mathcal{E}$ is the applied electric field and P is the polarization for the optical medium we can write

$$P = \chi^{(1)}\mathcal{E} + \chi^{(2)}\mathcal{E}^2 + \chi^{(3)}\mathcal{E}^3 + ... \tag{8.6}$$

where $\chi^{(i)}$ are polarizability constants (susceptibilities). From the second term in (8.5) we obtain for $\mathcal{E} = \mathcal{E}_0 \sin\omega t$

$$P_2 = \chi^{(2)}\mathcal{E}_0^2 \sin^2\omega t = \frac{1}{2}\chi^{(2)}\mathcal{E}_0^2(1 - \cos 2\omega t) .$$

This term is responsible for the generation of frequency-doubled light. A prerequisite is that the crystal has a non-central symmetric structure. Using symmetry arguments, it can be shown that for a material with a centre of symmetry all even coefficients $\chi^{(2)}$, $\chi^{(4)}$ etc. must vanish. Since $\chi^{(2)}$ is always small it is necessary for $\mathcal{E}_0$ be sufficiently high. With pulsed lasers electric field strengths of a sufficient magnitude (~10^5 V/cm) can easily be

---
[1] 670nm diode lasers are commercially available.

obtained so that the terms $\chi^{(1)}\mathscr{E}$ and $\chi^2\mathscr{E}^2$ become comparable in magnitude. The generation of frequency-doubled radiation is hampered by wavelength dispersion for the two waves at frequencies ω and 2ω. Since the waves do not normally propagate with the same velocity, destructive interference occurs with a resulting low yield. By utilizing doubly refractive materials, for which the velocity of propagation for the ordinary ray at ω coalesces with the velocity for the extraordinary ray at 2ω, "*phase matching*" can be obtained. The phase-matching condition also expresses the requirement for conservation of momentum. By tilting the crystal, and thus selecting the direction of propagation through it, phase matching can be achieved within a certain wavelength range. In Fig.8.34 the generation of a frequency-doubled wave and the achievement of phase matching are illustrated.

Temperature also influences the phase matching. It is especially advantageous if phase matching for a direction at 90° to the optical axis can be achieved by temperature adjustment. Then the extraordinary ray is not displaced with respect to the ordinary ray which is otherwise the case, reducing the beam overlap. KDP (potassium dihydrogen phosphate) and KPB (potassium pentaborate) are frequently used crystals. An energy conversion efficiency of tens of per cent can be achieved for the former material, whereas the latter has a substantially smaller efficiency. β-Barium-Borate (BBO, BaB_2O_4) is a new high-efficiency material, which is replacing KPB. Because of the nonlinear nature of the frequency-doubling process it is normally used in connection with pulsed lasers. Using tunable dye lasers, doubling down to about 200 nm can be achieved, with the limit set by material absorption. In Fig.8.35 frequency-doubling curves for a Nd:YAG-pumped dye laser are given.

By using an intracavity frequency-doubling crystal or an external enhancement cavity [8.71, 72] continuous frequency doubling and mixing can also be achieved for cw dye laser radiation. Powers of several mW can then be achieved (Fig.8.28).

Frequency doubling can be seen as a *mixing* of two waves of the same frequency in a nonlinear medium. The more general processes, sum and difference-frequency generation, are also possible and can be used to generate new frequencies using two lasers. The process can be explained expressing the light field as

$$\mathscr{E} = \mathscr{E}_1 \cos\omega_1 t + \mathscr{E}_2 \cos\omega_2 t . \tag{8.7}$$

The quadratic term in (8.6) can then be written

$$\begin{aligned}P_2 &= \chi^{(2)} [\mathscr{E}_1^2 \cos^2\omega_1 t + \mathscr{E}_2^2 \cos^2\omega_2 t + 2\mathscr{E}_1\mathscr{E}_2 \cos\omega_1 t \cos\omega_2 t \\ &= \chi^{(2)} [\tfrac{1}{2}(\mathscr{E}_1^2 + \mathscr{E}_2^2) + \tfrac{1}{2}\mathscr{E}_1^2 \cos2\omega_1 t + \tfrac{1}{2}\mathscr{E}_2^2 \cos2\omega_2 t + \\ &\quad + \mathscr{E}_1\mathscr{E}_2 \cos(\omega_1 + \omega_2)t + \mathscr{E}_1\mathscr{E}_2 \cos(\omega_1 - \omega_2)t] .\end{aligned} \tag{8.8}$$

It can be seen that apart from the doubled frequencies, sum and difference frequencies are obtained. The phase matching is chosen to strongly enhance

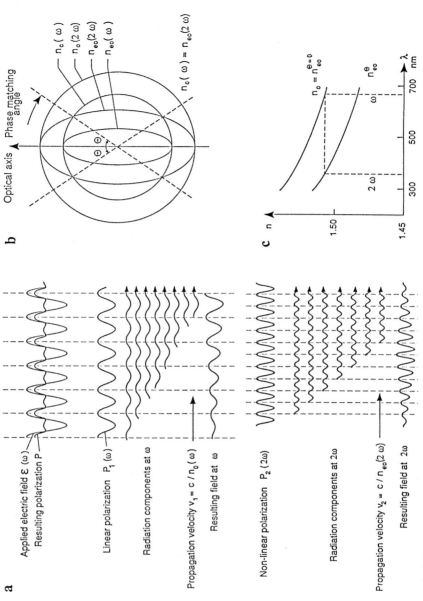

Fig. 8.34a–c. The generation of a frequency-doubled wave in a nonlinear crystal. (a) Fields and polarizations in the crystal. (b) The concept of phase matching is illustrated. (c) Dispersion curves for KDP

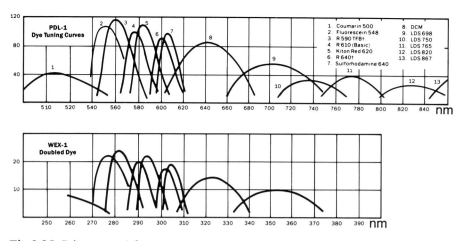

Fig.8.35. Primary, and frequency-doubled tuning curves for a Nd:YAG-pumped dye laser (Courtesy: Spectra Physics, Inc.)

one of the terms. In Fig.8.36 curves for sum generation employing a dye laser and the fundamental frequency of the Nd:YAG laser are given.

For difference-frequency generation materials that are transparent in the IR region are needed. In Fig.8.37 IR generation in $LiNbO_3$ (lithium niobate) is illustrated employing single-mode Ar^+ and dye lasers, and output curves for a Nd:YAG-based difference-frequency generator are also given. Crystals of $CdGeAs_2$, $AgGaS_2$ and $AgGaSe_2$ have recently been grown allowing frequency doubling and sum-frequency generation for

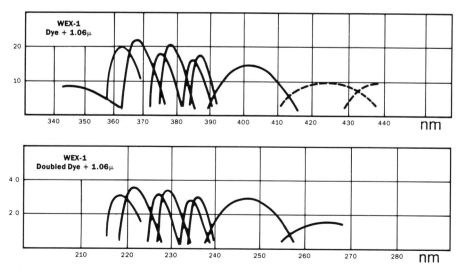

Fig.8.36. Sum-frequency generation curves for a Nd:YAG-based tunable dye laser (Courtesy: Spectra Physics, Inc.)

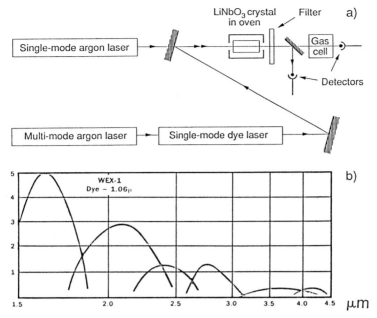

Fig.8.37a,b. IR radiation by difference frequency generation in $LiNbO_3$: (a) A set-up for CW generation [8.73] and (b) curves for pulsed IR radiation obtained using a Nd:YAG-based tunable laser (Courtesy: Spectra Physics, Inc.)

pulsed CO_2 lasers. Nonlinear materials for frequency doubling and mixing have been discussed in [8.74-8.76].

The *Optical Parametric Oscillator* (OPO) process should also be mentioned. Here a nonlinear crystal in a cavity is used to generate two new frequencies (ω_1 and ω_2) out of a single one (ω) that is used to pump the crystal. Energy conservation requires $\omega_1 + \omega_2 = \omega$. The frequency division between the two new waves (the *signal* and the *idler*) is chosen by the phase matching condition. The parametric process can also be used in *Optical Parametric Amplifiers* (OPA). Parametric laser light generation was reviewed in [8.77, 78].

Frequency mixing can also be achieved in mixtures of metal vapours and inert gases. Since asymmetry is not present in a gas, terms depending on $\chi^{(2)}$ are excluded. Third-order processes ($\chi^{(3)} \neq 0$) can be utilized. Direct frequency tripling and four-wave mixing processes are of this type. By changing the pressures of the metal vapour and the inert gas, appropriate phase matching for the desired process can be achieved in the presence of anomalous dispersion close to the metal atom lines [8.79]. Degenerate four-wave mixing, where all beams have the same frequency, is a very useful technique for achieving optical phase conjugation [8.80, 81]. Widely tunable short-wavelength radiation can be obtained by mixing in inert gases [8.82, 83]. Tuning curves for nonlinear mixing in Kr are shown in Fig.8.38. Laser spectroscopy in the vacuum ultraviolet (VUV) region is a rapidly growing field. Sources and techniques have been discussed in [8.84-87].

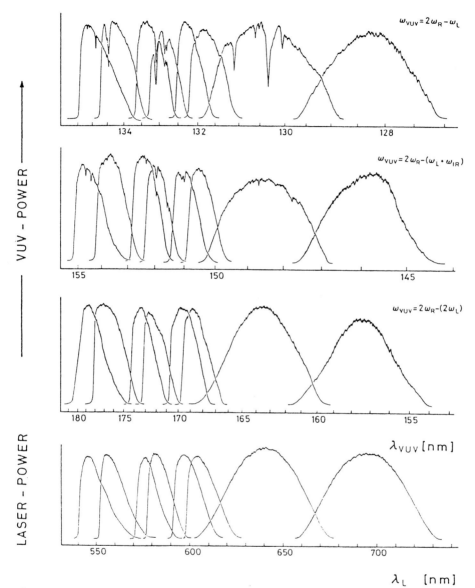

Fig.8.38. Tuning curves in the vacuum-ultraviolet spectral region obtained using different mixing processes in Kr [8.83]

Stimulated Raman scattering processes yield further possibilities. If a gas is irradiated with laser light of sufficiently high power, stimulated Raman scattering occurs. The Stokes-shifted light propagates in the same direction as the pump radiation. In hydrogen gas a vibrational shift of 4155 cm^{-1} (~0.5eV) is obtained. The process is illustrated in Fig.8.39 together with an illustration of how higher Stokes components are obtained by

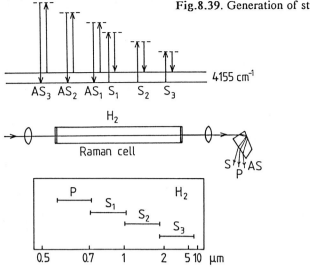

Fig.8.39. Generation of stimulated Raman scattering

further Stokes shifts. Radiation of shorter wavelength can also be obtained (anti-Stokes components) by four-wave mixing between the pump radiation and Stokes-shifted components. A gas pressure of 10 atmospheres is common. There is no phase-matching condition. By using a small number of very efficient dyes in a pulsed dye laser, wavelengths from the vacuum ultraviolet to 10 μm can be achieved by Raman shifting. The efficiency in the conversion to the first Stokes component can exceed 10% whereas higher-order components are much weaker. In Fig.8.40, the relation between the primary radiation and the generated wavelengths is illustrated. Besides H_2, other gases, in particular D_2 (2987 cm^{-1}) and CH_4 (2917 cm^{-1}), can be used in Raman shifters. Raman shifting was discussed in [8.88, 89].

We will end this section on frequency extension techniques by discussing the CARS (Coherent Anti-Stokes Raman Scattering) process, which is a special case of four-wave mixing [8.90]. The process is useful for the generation of new frequencies (1st anti-Stokes component in the stimulated Raman scattering discussed above) and also for powerful spectroscopic applications (Sect.10.1.4). In Fig.8.41 basic diagrams for the process are given illustrating energy and linear momentum conservation.

In the CARS process the sample is irradiated by two laser beams and the frequency difference between the beams is chosen to correspond to the vibrational (rotational) splitting of the irradiated molecules. The beams are denoted the *pump beam* (at frequency ω_P) and the *Stokes beam* (at frequency ω_S). Two photons of frequency ω_P are mixed with a photon of frequency ω_S, through the third-order susceptibility $\chi^{(3)}$, to generate a stimulated anti-Stokes photon of frequency ω_{AS} (in the anti-Stokes position with regard to the pump beam)

$$\omega_{AS} = 2\omega_P - \omega_S . \tag{8.9}$$

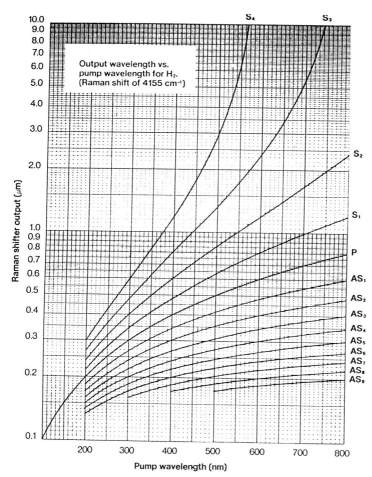

Fig.8.40. The relation between applied wavelength and generated coherent Raman components using H_2 gas (G. Bjorklund, private communication)

The phase-matching condition can be written

$$k_{AS} = 2k_P - k_S \tag{8.10}$$

where **k** is the wavevector with $|\mathbf{k}| = \omega n/c$.

In liquids and solids with a strong wavelength dispersion the two primary laser beams must be crossed at a small angle and the generated CARS beam will emerge at still another angle to fulfil the phase matching condition (8.10). In a gas with negligible dispersion ($n \simeq 1$), collinear phase matching is possible. In diagnostic applications, the so-called BOXCARS geometry is frequently used to achieve improved spatial resolution [8.92]. The CARS beam is generated only from the region in which the incoming laser beams cross. In this case two pump beams and one Stokes beam are used.

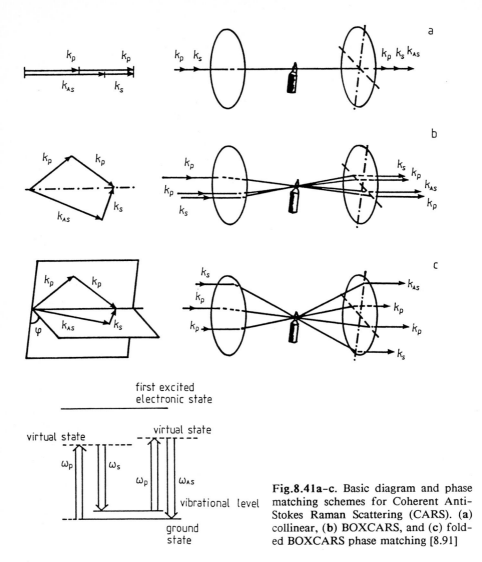

Fig.8.41a-c. Basic diagram and phase matching schemes for Coherent Anti-Stokes Raman Scattering (CARS). (a) collinear, (b) BOXCARS, and (c) folded BOXCARS phase matching [8.91]

When the phase-matching condition is fulfilled the power P_{AS} of the generated beam is related to the powers P_P and P_S of the pump and Stokes beams, respectively, through the expression

$$P_{AS} \simeq |\chi^{(3)}|^2 \cdot P_P^2 P_S . \tag{8.11}$$

Since $\chi^{(3)}$ is proportional to the molecular number density, N, we make the interesting (and unusual) observation that the generated beam strength is proportional to N^2 rather than to N.

The basis for most nonlinear optical processes has been discussed in [8.93]. Several recent monographs cover the field extensively [8.94-97].

9. Laser Spectroscopy

The wide applicability of lasers in spectroscopy is due to several factors. As we have seen, very high intensities can be obtained in a small frequency interval. The favourable spatial properties of laser beams with the possibility of very good focussing is also of great importance. With the advent of tunable lasers, completely new types of experiments have become possible and investigations that were only barely possible with conventional light sources can now be readily performed. It is fair to state that tunable lasers have revolutionized optical spectroscopy. Several monographs and review articles on laser spectroscopy have been published [9.1-18]. A wealth of material is presented in the proceedings of international laser spectroscopy conferences [9.19-27].

9.1 Basic Principles

9.1.1 Comparison Between Conventional Light Sources and Lasers

A comparison is made in Table 9.1 between a conventional line light source and a continuous single-mode dye laser; both sources with representative data. In many spectroscopic experiments a decisive factor is by what power per unit area and spectral interval a sample can be irradiated. The designation $I(\nu)$ is used for this power density/frequency unit. In the comparison in

Table 9.1. Comparison between a conventional light source (RF discharge lamp) and a single-mode dye laser

	Conventional line source, RF discharge lamp	Continuous single-mode dye laser
Linewidth	1000 MHz	1 MHz
Total output of line	10^{-1} W	10^{-1} W
Power within a useful solid angle	10^{-2} W	10^{-1} W
Irradiated area (depends on focussing)	10 cm^2	10^{-4} cm^2
$I(\nu)$ power density per unit frequency	10^{-6} W/(cm$^2 \cdot$MHz)	10^{3} W/(cm$^2 \cdot$MHz)

Table 9.1 a 10^9 times higher value of $I(\nu)$ is obtained for the laser. With a pulsed laser $I(\nu)$ may be increased by many more orders of magnitude.

9.1.2 Saturation

A very high population in short-lived excited states can be obtained using the high values of $I(\nu)$ that are available with lasers. For a two-level system (Sect.4.1) with populations N_1 and N_2 in the lower and upper states, respectively, we obtain using (4.21 and 22)

$$\frac{N_2}{N_1 + N_2} = \frac{1/2}{1 + A/2B\rho(\nu)} \ . \tag{9.1}$$

For $B\rho(\nu) \gg A$ we have

$$\frac{N_2}{N_1 + N_2} = \frac{1}{2} \ , \tag{9.2}$$

i.e. half the atoms are found in the excited state, and half in the lower state. This situation is called *saturation*. (If statistical weights (g = 2J+1) are taken into account, the expression $g_2/(g_1+g_2)$ is obtained instead of 1/2). We will now calculate which values of $I(\nu)$ are required to achieve this situation. $I(\nu)$ is related to the energy density $\rho(\nu)$ through

$$I(\nu) = c\rho(\nu) \ . \tag{9.3}$$

Using this equation and (4.27) we find the following condition for saturation at a chosen wavelength of 600 nm:

$$I(\nu) \gg c\frac{A}{B} = 16\pi^2 \frac{\hbar c}{\lambda^3} = \frac{16 \cdot 3.14^2 \cdot 1.05 \cdot 10^{-34} \cdot 3 \cdot 10^8}{6^3 \cdot 10^{-21}}$$

$$= 2 \cdot 10^{-5} \ \frac{W}{m^2 \cdot Hz} = 2 \cdot 10^{-3} \ \frac{W}{cm^2 \cdot MHz} \ . \tag{9.4}$$

Thus, a laser power of the order of mW/(cm²·MHz) is needed to achieve saturation. This condition is easily met even with a continuous laser.

The first excited $^2P_{1/2}$ state of the sodium atom (3p) can only decay back to the 3s $^2S_{1/2}$ ground state (the D_1 line). The natural radiative lifetime of the 3p $^2P_{1/2}$ state is 16 ns. At saturation $(1/2)(1/16 \cdot 10^{-9}) \simeq 3 \cdot 10^7$ scattered fluorescence photons per second are obtained. This means that it is in principle possible to "see" individual atoms (Sect.9.5.4).

Two types of laser spectroscopy can be distinguished. In the first kind, the laser is adjusted to the desired wavelength. The experiment is then performed at this fixed wavelength, either by pulsing the light, or influencing the atoms or molecules in other ways. In this kind of measurement a good frequency stability of the laser is desirable. In the second type of spectroscopic experiment, the laser wavelength is varied continuously. Most frequently, single-mode operation is required. A continuous sweep without mode hopping must then be achieved.

9.1.3 Excitation Methods

Several different excitation schemes can be used in laser spectroscopy. This is illustrated in Fig.9.1 for the case of alkali atoms.

a) **Single-step excitation** - Atoms are transferred directly from the ground state to the excited state using an allowed electric dipole transition. This means an S-P transition for an alkali atom.

b) **Multi-step excitation** - Since tunable lasers have high output powers enabling the saturation of optical transitions, step-wise excitation via short-lived intermediate states is possible. A 2-step process has been indicated in the figure. For an alkali atom this may mean S-P-D transitions. Step-wise excitations give access to states that cannot normally be reached.

c) **Multi-photon absorption** - At high laser powers higher-order optical absorption processes become non-negligible. Thus, it becomes possible for an atom to simultaneously absorb two photons, thus bridging energy differences between two states without utilizing real intermediate states. The theory of two-photon absorption processes was presented by M. Goeppert-Mayer as early as 1931, but only in 1961 could such transitions be observed with lasers. The transition probability R_{gf} from the ground state g to the final state f can be expressed in matrix elements involving non-resonant intermediate states i

$$R_{gf} \propto \sum_i \left| \frac{\langle g|er|i\rangle \langle i|er|f\rangle}{\delta E_i} \right|^2 P^2 , \qquad (9.5)$$

where the summation is performed over all allowed intermediate states i. δE_i is the energy difference between the virtual and the real intermediate states i, and P is the laser power. Frequently, a single real state close to the virtual state dominates the expression. A quadratic laser power dependence

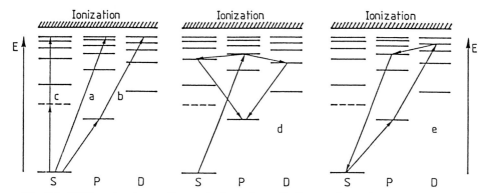

Fig.9.1. Different schemes for excitation of an alkali atom including (*a*) single-step, (*b*) two-step, (*c*) two-photon, and (*d,e*) cascade excitation

is obtained. The two-photon absorption method connects states of the same parity. An S-S transition is indicated in the figure. Similarly, more than 2 photons can be absorbed simultaneously. For example, molecules can be made to absorb 10's of IR photons in this way (Sect. 10.4.2).

In the cascade decay of a laser-excited state, lower-lying states are populated as illustrated in the figure for the case of alkali atoms. This provides further possibilities for laser spectroscopy investigations.

9.1.4 Detection Methods

There are several methods of detecting whether an excited state has been populated. This is particularly true for highly excited states, i.e. *Rydberg states*, that have been the object of many laser spectroscopy investigations. Rydberg atoms are essentially hydrogen-like and some of their properties are given in Table 9.2. The spectroscopy of Rydberg states has been discussed in [9.28-31].

Table 9.2. Properties of Rydberg atoms [9.1] (R: Rydberg constant, a_0: Bohr radius, n: principal quantum number)

Quantity	n dependence	Numerical examples for	
		Na(10d)	H(n=50)
Binding energy	$-Rn^{-2}$	0.14 eV	0.027 eV
Orbital radius	$a_0 n^2$	147 a_0	2500 a_0
Geometrical cross section	$\pi n_0^2 n^4$	$7 \cdot 10^4 a_0^2$	$6 \cdot 10^6 a_0^2$
Dipole moment	$\propto n^2$	143a_0	
Polarizability	$\propto n^7$	210 kHzV^{-2}cm^2	
Radiative lifetime	$\propto n^3$	10^{-6} s	10^{-3} s
Stark splitting in electric field E = 1kV/cm	$\Delta w \propto n(n-1)E$	~15 cm^{-1}	~10^2 cm^{-1}
Critical field strength E_c for field ionization	$E_c = \pi \epsilon_0 R^2 e^{-3} n^{-4}$	$3 \cdot 10^6$ V/m	$5 \cdot 10^3$ V/m

Considering these properties we will now discuss a number of detection methods.

a) **Fluorescence** – The most direct way of detecting whether an atom has been excited to an upper state is to detect the fluorescence light released when the atom decays. The atom can return to the state from which it was excited via an allowed electric dipole transition. With increasing excitation energy the number of possible decay paths increases corresponding to many spectral lines. Since the detection must frequently be limited to a narrow wavelength band, only a small fraction of the decays will be observed. Further, the lifetime increases as n^3. If pulsed excitation is used the detection temporal interval must be successively increased in order to be able to detect a reasonable fraction of the photons. This leads to an increase in the background signal. These factors make fluorescence detection of highly excited states less and less attractive as n increases.

b) Photoionization - An excited atom can be photoionized by absorbing a further photon of sufficient energy to bring the atom above the ionization limit. One way of detecting the excited atom is to detect the released electron with an electron multiplier, which is essentially a photomultiplier tube without a photocathode. For highly excited states an IR photon, e.g. from a CO_2 laser, can be used (~0.1eV). If a laser of sufficiently high power is used for the excitation process ionization can occur using another photon from the laser beam. The cross-section for photoionization is low but is strongly enhanced in the presence of resonant auto-ionizing states above the ionization limit.

c) Collisional ionization - Highly excited atoms have a high probability of colliding because of their considerable size. Since the thermal energy of the atom (~kT) may be larger than the energy required to reach the ionization limit, there is a significant probability that a collision results in the formation of an ion. A *thermo-ionic detector* can be used to detect the presence of the ions. This technique is extremely sensitive [9.32,33].

d) Field ionization - Highly excited atoms are very sensitive to electrical fields. Field ionization (Sect.2.5.2) can be brought about by populating a well-defined highly excited state using step-wise, pulsed excitation in the absence of an electric field and then applying an electric field pulse [9.34,35]. The electrons can be detected with a suitably located electron multiplier.

Ionization detection is further illustrated in Sect.9.2.6.

9.1.5 Laser Wavelength Setting

Before discussing different types of laser spectroscopy experiments, we will consider the question of how the frequency of the laser is made to coincide with the absorption frequency of the atoms or molecules of interest. The problem is not at all trivial considering the sharpness of both the laser and the absorption lines and the mode structure of both laser resonators and intracavity etalons. A high-resolution grating monochromator can be used but it is considerably more convenient to use digital laser wavelength meters. In Fig.9.2 two systems are shown. Figure 9.2a shows a wavelength meter for continuous single-mode lasers, and Fig.9.2b shows a system that can also operate with a pulsed laser beam.

The wavelength meter for continuous light is essentially a double Michelson interferometer (Sect.6.2.4) in which the light from the "unknown" laser and the light from a reference laser are directed in counterpropagating paths [9.36,37]. The movable mirror is a retro-reflector ("corner cube", Sect.6.2.1) which is self-adjusting with regard to the beam direction. It can be mounted on a carriage, which moves on a frictionless air track. Interference fringes from each laser are recorded at different de-

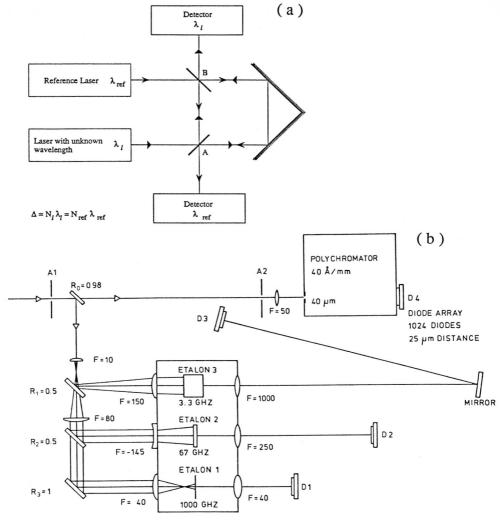

Fig.9.2. (a) Wavelength meter for continuous lasers [9.36]. (b) Wavelength meter for pulsed or continuous lasers [9.38]

tectors during a certain time. The optical path difference Δ can be expressed in two ways

$$\Delta = N_\ell \lambda_\ell = N_{ref} \lambda_{ref} , \quad \text{i.e.,} \quad \lambda_\ell = \frac{N_{ref}}{N_\ell} \lambda_{ref} , \tag{9.6}$$

where N_{ref} and N_ℓ are the numbers of counted interference fringes for the reference laser and the unknown laser, respectively. Since the laser beams overlap in the instrument, the method is quite insensitive to air turbulence, etc. Normally, a single-mode HeNe laser is used as the reference source. If this laser is stabilized at a saturation dip (Sect.9.5.2), very high accuracy and stability can be obtained. It is practical to stop the fringe counting in the electronic unit when N_ℓ equals the stored digital number for the wavelength of the HeNe laser. Then the corresponding number N_{ref} can be directly displayed as λ_ℓ, see (9.6). Since only whole fringes can be electronically counted the accuracy can be considerably increased if the fringe counting frequency is multiplied by a factor (i.e., 10) using a *Phase-Locked Loop* (PLL).

A wavelength meter that also works for pulsed laser light employs a small spectrometer and Fabry-Pérot interferometers with different free spectral ranges [9.38]. The image plane of the spectrometer and the ring systems of the interferometers are detected by diode arrays (Sect.6.3). Since the laser light is intense no amplification is needed, in contrast to the case of an OMA (Sect.6.3). The information is handled by a microprocessor which calculates the wavelength. A HeNe laser can be used as a reference. However, if the free spectral ranges of the interferometers are constant and well known a reference laser is not needed for moderate accuracies. Several other techniques for accurate laser wavelength determination have been developed [9.39-41]. The accuracy of laser wavemeters has been discussed in [9.42].

In spectroscopy using continuous laser tuning it is very important to be able to monitor the change in wavelength. For this purpose, a stable multipass interferometer (Sect.6.2.3) can be used from which the fringes are recorded together with the spectrum [9.43].

We will now consider different laser spectroscopy methods. We will first discuss techniques in which a reasonable resolution is sufficient and where relatively broad-band lasers (0.1-0.01Å) and Doppler-broadened samples are used. We will then describe how resonance techniques can be combined with laser excitation. In the following sections, time-resolved spectroscopy will be considered. Lifetime measurements and structure determinations using such techniques will be described. In this context a comprehensive survey of methods for determining the radiative properties of atoms and molecules is made.

While the laser is used essentially as a bright lamp in the above-mentioned contexts, a survey of the most important methods of Doppler-free spectroscopy, employing the extremely small linewidth of a single-mode laser, is made in Sect.9.5.

9.2 Doppler-Limited Techniques

In this section we will discuss measurement techniques in which Doppler broadening, caused by the thermal movements of the atoms or molecules, limits the achievable resolution. Lasers with a comparatively large linewidth (0.1-0.01Å) are therefore normally employed. Experiments within this category are aimed at the determination of primary energy-level structure without regard to finer details, such as hyperfine structure. Several important analytical applications are also discussed.

9.2.1 Absorption Measurements

A tunable laser is very useful for measuring the absorption of a sample as a function of the wavelength. In this case the laser replaces the normal light source as well as the monochromator in a spectro-photometer (Sect.6.5.6). Because of the good geometrical properties of the laser beam it is possible to work with small samples or with long absorption paths and low concentrations. Dual-beam techniques are generally employed. With continuous lasers, lock-in techniques are used for recording. When pulsed lasers are used gated integrators (boxcar integrators, see Sect.9.4.3b) are employed. Sample and reference signals are divided for normalizing purposes. In Fig.9.3 a commercial diode spectrometer for the IR region is shown. In this particular case, a grating monochromator has been included to suppress unwanted laser modes.

If gaseous samples of low concentration are to be investigated a *White cell* can be used [9.44,45]. This is a multi-pass device in which a well-defined beam path is obtained by repeated focusing using spherical mirrors of suitable curvature. The number of passes can easily be chosen by adjusting

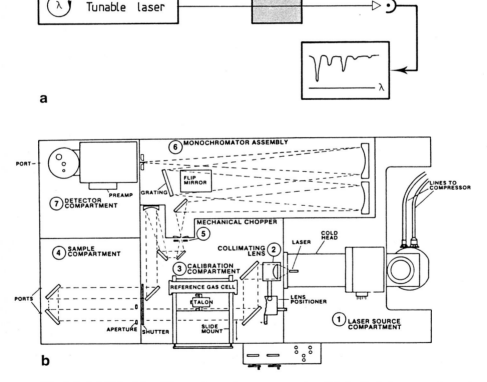

Fig.9.3. (a) Principle of laser absorption spectroscopy. (b) A diode laser spectrometer (Courtesy: Spectra Physics)

the mirrors. 50-100 double passes can be achieved. The transmitted intensity is largely determined by the mirror reflectivities. For absorption experiments on metal atoms, for which a high temperature is required to obtain a sufficient vapour pressure (Fig.7.2), a *heat-pipe oven* can be used [9.46,47]. This consists of an evacuated metal tube, in which the central inner part is covered by a fine metal mesh. The metal is placed in the centre of the tube and a few torrs of inert gas are frequently added. The central part of the tube is resistively heated while the end parts, with optical windows, are water-cooled. A micro-climate develops in the tube, and metal atoms evaporate out into the cold zones where condensation occurs. The liquid metal is then sucked back towards the centre by surface tension forces between the metal atoms and the mesh. The gas is fractioned so that the inner zone contains metal atoms while the windows are protected against deposits, due to metal evaporation, by the inert gas in front of them. Different aspects of laser absorption spectroscopy have been discussed in [9.48].

9.2.2 Intra-Cavity Absorption Measurements

In the absorption measurements described above an external laser beam was used. A great increase in sensitivity can be achieved if the sample is placed inside the laser cavity [9.49, 50]. This is due to the mode competition in a multi-mode laser. The principle of intra-cavity absorption measurements is given in Fig.9.4.

A multi-mode dye laser is run in the wavelength region where the species under investigation absorbs. As an example we will consider the case of molecular iodine (I_2). If the intra-cavity cell is first taken out and the laser is adjusted to the desired wavelength region, laser-induced fluorescence will be observed in an external iodine cell. Normally, several molecular transitions are induced. If the intra-cavity iodine cell is now introduced the

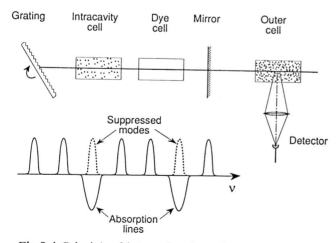

Fig.9.4. Principle of intracavity absorption measurements

fluorescence light will disappear or be strongly reduced. The reason is that in the cavity, the modes that excited the molecules in the external cell will be greatly reduced in intensity compared with other modes due to the intra-cavity absorption. Since we have multiple passages through the internal sample and the laser is very sensitive to a small imbalance, very few molecules are needed to strongly influence the fluorescence light in the external cell. The sensitivity when using an intra-cavity sample is about 10^5 times greater than when an external cell is used.

9.2.3 Absorption Measurements on Excited States

Classical absorption measurements with a continuum light source and a photographically recording spectrograph only display absorption lines originating in the ground state or in low-lying, thermally well-populated metastable states. Through the possibility of saturating an optical transition with

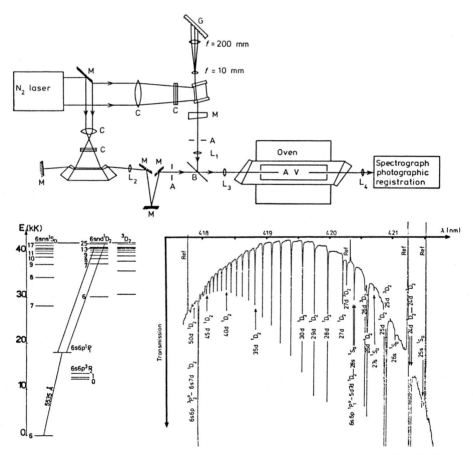

Fig.9.5. Illustration of absorption spectroscopy from an excited state. A level scheme for Ba, an experimental set-up and a resulting absorption spectrum are shown [9.52]

a laser, thus transferring essentially half the number of atoms into a short-lived excited state, it is now possible to extend absorption experiments to spectral series originating in excited states [9.51,52]. In Fig.9.5 an example for Ba is given. A nitrogen laser is used for synchronous pumping of two dye lasers. One has a narrow bandwidth and is used to saturate the $6s^2$ $^1S_0$-$6s6p$ $^1P_1$ transition at 5535 Å. The barium vapour is contained in a heat-pipe. The second laser has a normal incidence mirror instead of a Littrow-mounted grating and it yields an intense continuum at the same time as the excited state is populated. As illustrated in the recording, light is absorbed from the continuum at wavelengths corresponding to the $6s6p$ $^1P_1$-$6sns$ $^1S_0$ and $6snd$ $^1D_2$ series. The trace in the figure has been obtained from the photographic plate using a microdensitometer.

9.2.4 Level Labelling

The analysis of molecular spectra is complicated because of the very large number of lines that is obtained simultaneously in normal excitation or absorption experiments. With narrow-band laser excitation an individual excited rotational-vibrational level can be populated selectively and only the decays originating in the excited state are observed. A similar simplification in absorption measurements is very desirable. Through the possibility of saturating optical transitions, a certain lower level can be "labelled" by depleting the population with a laser (pump laser). If this laser is switched on and off repetitively, all absorption lines originating in the labelled level will be modulated when induced with a second (probe) laser [9.53,54]. A number of schemes for modulation detection are indicated in Fig.9.6. Several schemes can be used to ascertain that absorption has occurred, as discussed in Sect.9.1.4. A still more powerful technique using equipment similar to that described in the preceding section is shown in Fig.9.7. The probe laser

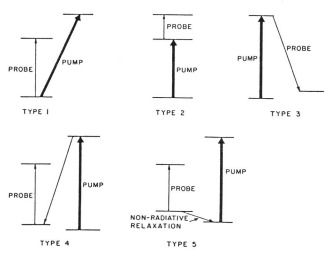

Fig.9.6. Different schemes for level labelling [9.54]

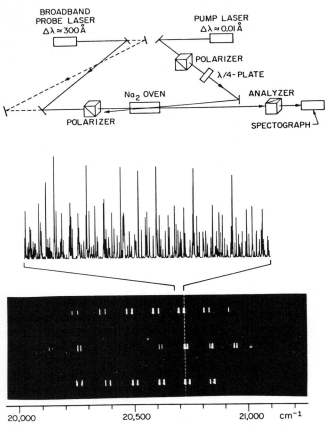

Fig.9.7. Principles of level polarization labelling. The experimental set-up and a recording for Na$_2$ molecules, displaying $\Delta J=+1$ and $\Delta J=-1$ transitions from the labelled level are shown [9.55]

is a broad-band laser and can normally not illuminate the entrance slit of the recording spectrometer since it is blocked by crossed polarizers. The pump laser is circularly polarized and induces an optical anisotropy by molecular orientation of the sample, which is placed between the crossed polarizers [9.55]. For lines coupled to the oriented molecules the plane of polarization is rotated for the probe laser light and the entrance slit is illuminated. On the photographic plate all absorption lines originating in the labelled level give rise to bright lines.

9.2.5 Two-Photon Absorption Measurements

Series of levels of the kind discussed in connection with absorption measurements from excited states (Sect.9.2.3) can also be investigated in two-photon absorption experiments. Then only a single pulsed dye laser is needed. Amplifier stages may be used to achieve sufficient power (Sect.

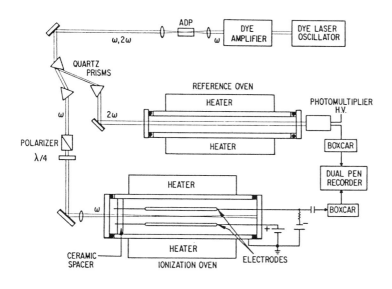

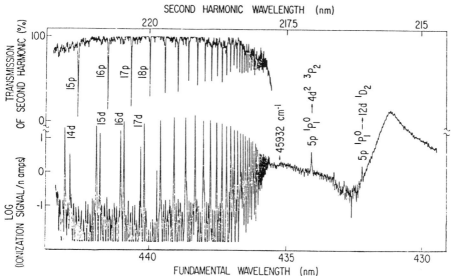

Fig.9.8. Experimental set-up, ionization recording of two-photon transitions, and absorption recording of single-photon transitions in strontium vapour [9.56, 21]

8.5.1). When the laser has been tuned to bridge the energy difference between the ground state and an excited state of the same parity by means of two simultaneously absorbed laser photons, this can be detected by using still another photon to photoionize the excited atoms. The ion current can be detected by electrodes at a low dc voltage inserted into the heat-pipe oven. The ion transients are averaged in a boxcar integrator. In Fig.9.8 two-photon transitions to 5sns $^1S_0$ and 5snd $^1D_2$ states in Sr are shown. Simultaneous single-photon absorption detection of $5s^2\ ^1S_0 \to 5snp\ ^1P_1$ transitions is illustrated using frequency-doubled laser radiation. In the figure the increase in ionization current above the ionization limit of Sr is shown, as well as a dispersion-shaped broad resonance due to an auto-ionizing state (Fano resonance).

9.2.6 Opto-Galvanic Spectroscopy

In the previous section we described how *optical* resonance could be detected by direct observation of *electrical* phenomena. From this point of view we described a special type of *opto-galvanic* spectroscopy. When electrical detection of optical resonance in connection with measurements on electrical discharges or flames is performed the term *Laser-Enhanced Ionization* (LEI) spectroscopy is frequently used. In Fig.9.9 opto-galvanic spectroscopy (LEI) on a gas discharge is illustrated. The laser beam is directed into the discharge and when it is tuned to an optical transition the discharge current is changed, since the probability of collisional ionization is different for

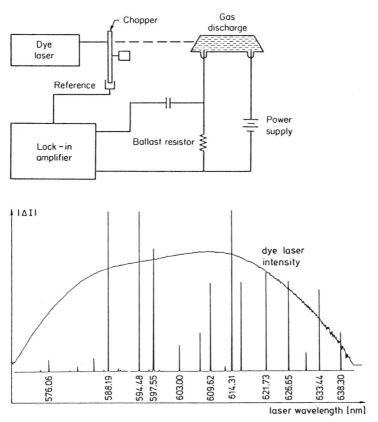

Fig.9.9. Experimental arrangement for opto-galvanic spectroscopy on a gas discharge [9.57] and a recorded spectrum for a Ne discharge [9.1]

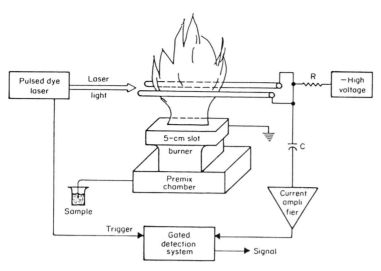

Fig.9.10. Set-up for opto-galvanic flame spectroscopy [9.59]

high-lying levels compared with low-lying ones. Current changes ΔI can be detected as a voltage change, $\Delta V = R\Delta I$, over a ballast resistor R in the discharge circuit. An opto-galvanic spectrum of a neon discharge is also given in the figure. The technique can conveniently be used for wavelength calibration of a tunable laser. A small part of the laser light is directed into a hollow cathode containing, e.g. thorium or uranium, which both have a large number of lines distributed over the visible and ultraviolet spectral regions [9.58]. A very simple pure electrical method of detection of the reference lines can thus be obtained. Optical resonance in radio-frequency discharges can also be detected electrically by observing an imbalance in the RF circuitry used for the oscillator (Sect.9.5.2).

Opto-galvanic spectroscopy of flames has important analytical applications. If an atomic absorption flame (Sect.6.5.3) is irradiated by a tunable laser a change in the current between two electrodes, placed in the outer parts of the flame, can be detected. A typical arrangement is shown in Fig.9.10. A sensitivity exceeding that obtainable in atomic absorption spectroscopy (Sect.6.5.2) can be achieved for elements seeded into the flame. Two-step excitation improves both sensitivity and background rejection. The opto-galvanic technique can also be used for studying normal flame constituents such as O, H and OH [9.60].

As we have seen, collisions are important for the signal generation in LEI. In low-pressure experiments photoionization instead is the principal origin of the signal. The term *Resonance Ionization Spectroscopy* (RIS) is then frequently used. Several examples of opto-galvanic detection schemes for different atoms are shown in Fig.9.11. If multi-photon excitation of the atoms to be studied is used the technique is referred to as REMPI (*REsonance Multi-Photon Ionization*) spectroscopy. The selectivity of RIS and REMPI can be further enhanced by using a mass spectrometer to ana-

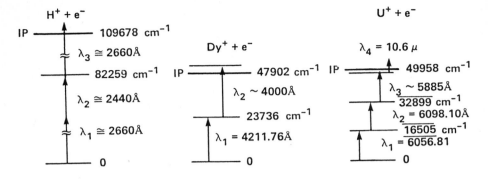
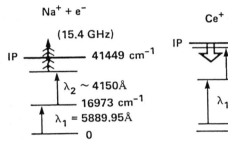
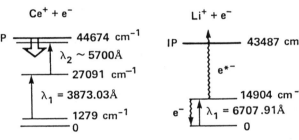
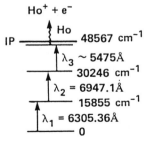
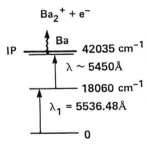

Fig.9.11. Examples of resonance ionization pathways [9.61]

lyse the produced ions. This technique is referred to as RIMS (*Resonance Ionization Mass Spectroscopy*). The numerous aspects of optogalvanic spectroscopy were covered in [9.61-70].

9.2.7 Single-Atom Detection

We have already mentioned that single atoms or ions can, in principle, be detected by the emission of a large number of photons per second at saturation. We will come back to this in Sect.9.5.4. The resonance ionization techniques discussed in the previous section also have the potential of single-atom detection. The energy density of the laser beams is chosen in such a way that the probability of photoionization of an atom in the beam is unity. The released electron can be detected with 100% probability if the atom is located inside a proportional counter. It has been demonstrated that a single Cs atom in a background of 10^{19} Ar atoms can be detected in this way. Single-atom detection is of great interest, e.g., for studying rare reactions such as those induced by neutrinos from the sun: $^{37}Cl(\nu,e)^{37}Ar$. Other applications include dating of the polar ice caps, studies of ocean mixing by monitoring of atmospherically produced ^{39}Ar etc. Single-atom detection using resonant photoionization is reviewed in [9.71-72]. Different techniques for ultra-sensitive laser spectroscopy were discussed in [9.73-76]. The general field of analytical laser spectroscopy was treated in [9.77-79].

9.2.8 Opto-Acoustic Spectroscopy

In this section we will describe another non-optical technique for the detection of optical absorption: *opto-acoustic spectroscopy*. The principle of a laser-based opto-acoustic spectrometer (spectrophone) is given in Fig.9.12. A molecular sample at relatively high pressure is contained in a closed volume and is irradiated with a chopped cw laser beam tuned to resonance. The excited molucules are mainly de-excited radiationlessly at the prevailing pressure. The excitation energy is transferred to translational energy

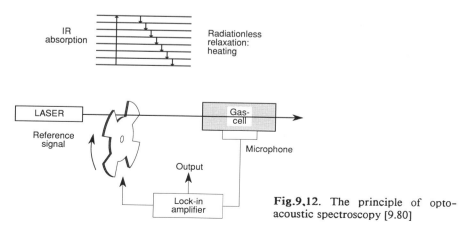

Fig.9.12. The principle of opto-acoustic spectroscopy [9.80]

upon collision. Thus the pressure rises and falls periodically in the cell. The pressure variations can be detected as a sound wave by a microphone mounted in the cell wall. Weak signals can be detected by lock-in techniques. If the cell is shaped as a resonant cavity for the chopping frequency a very high sensitivity can be achieved, better than 1 ppb (1 part in 10^9) in the measurement of, e.g., air pollutants. CO_2 or diode lasers can be used to induce vibrational transitions. Since the detector is only sensitive to the absorbed part of the radiation, very small absorptions can be measured. This is not possible in normal optical absorption measurements where the difference between two almost equally large quantities is to be measured. Another feature of opto-acoustic spectroscopy is that it is insensitive to stray light. The technique has further been discussed in [9.81-85].

The absorption of solid or liquid materials can also be measured using these techniques [9.86]. The laser illuminates the sample and the heat variations are coupled to a non-absorbing gas, which is in contact with the microphone. Piezo-electrical detection can also be used with solid or liquid samples. Elastic waves in the material are then detected. In Fig.9.13 an example of an acousto-optic spectrum of a powder is given.

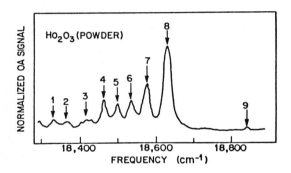

Fig.9.13. Example of an acousto-optic spectrum from a solid [9.86]

9.3 Optical Double-Resonance and Level-Crossing Experiments with Laser Excitation

In this section we will deal with experiments in which, similarly to the conditions in the previous sections, Doppler-broadened media and rather broad-band lasers are used. In spite of this fact, a resolution is obtained that is only limited by the Heisenberg uncertainty relation. This is due to the fact that the signals are resonant in nature with half-widths limited essentially only by the lifetime of the excited state. Optical double-resonance and level-crossing experiments (Sects.7.1.4,5) with laser excitation utilize the high intensity of polarized laser light in order to populate levels that are normally not accessible for investigation for intensity reasons. Especially experiments employing step-wise excitation become possible [9.87]. For the alkali atoms, a large number of investigations on fine structure, hyperfine structure and the Stark effect have been performed using such techniques

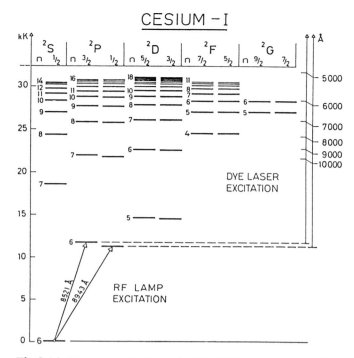

Fig.9.14. Energy level scheme for Cs. Rf-lamp excitation of the first excited ^{2}P states is indicated and absorption wavelengths for subsequent ^{2}S and ^{2}D state excitation are given [9.90]

[9.88-92]. As an example, the energy-level diagram of Cs is given in Fig.9.14. Step-wise excitation of S and D levels via the first P-doublet is illustrated in the diagram. At the early stage of development of cw dye lasers only a narrow wavelength region in the yellow and red spectral range was available with Rhodamine dyes. Thus the spectroscopic applicability for such lasers was quite limited. However, if the laser excitation is initiated from the first P state, many highly excited states become accessible for the given spectral range. This is true not only for Cs but also for the other alkali atoms. The P states can be weakly populated using the very strong resonance lines (D_1 and D_2) in the alkali atoms. A sealed RF lamp (Sect.6.1.1) is very suitable for the efficient generation of these lines. At a vapour pressure of 10^{-3} torr strong multiple scattering is obtained in the resonance cell for these lines, and thus the average population of the P levels is increased. Reemitted photons are reused to bring the atoms into the P state, which has a lifetime of some tens of ns. Still only about 1 atom in 10^4 is in the P state. The RF lamp cannot, by a long way, saturate the optical transition. A multi-mode dye laser is used to induce the second excitation step. By piezoelectrically vibrating one of the laser mirrors the mode structure can, on the average, be washed out and a "white" excitation is obtained. In Fig.9.15 an experimental arrangement for optical double-resonance and level-crossing spectroscopy is shown.

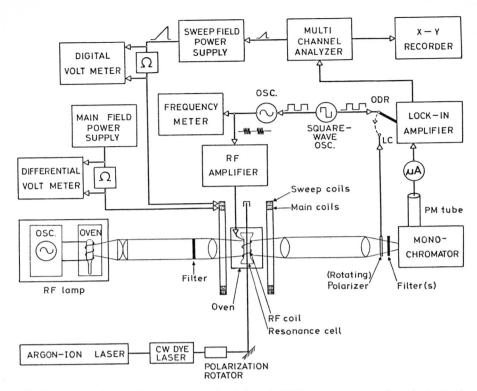

Fig.9.15. Experimental arrangement for LC and ODR spectroscopy of excited alkali atom levels, populated using stepwise excitation [9.89]

Lock-in detection is used for signal detection and the lock-in signals are further averaged using a multichannel analyser, which has a channel advancement which is synchronized with a repetitive magnetic field sweep. In ODR experiments the RF transmission is pulse-modulated, while LC signals are detected by recording the fluorescent light intensity through a rotating linear polarizer. Normally, circularly polarized light components are obtained in the direction of the magnetic field (Fig.4.9). The only occasion for which linearly polarized light can be obtained in the direction of the magnetic field is at a coherently excited and coherently decaying level crossing, where the σ^+ and σ^- light is phase-related and results in linearly polarized light. This light is then modulated by the rotating linear polarizer and this modulation is detected by the lock-in amplifier.

In Fig.9.16 a fluorescence light spectrum for Cs is given as recorded with the apparatus shown in Fig.9.15. The 12 $^2D_{5/2}$ state is populated by step-wise excitation. In cascade decays, as illustrated in Fig.9.1, lower-lying P and F levels are populated and fluorescence lines are obtained in the blue and UV, and in the red spectral regions, respectively.

If the primary excitation is performed, e.g., with π light (Fig.4.9) the magnetic sublevels of the $^2D_{5/2}$ state will be non-uniformly populated. The

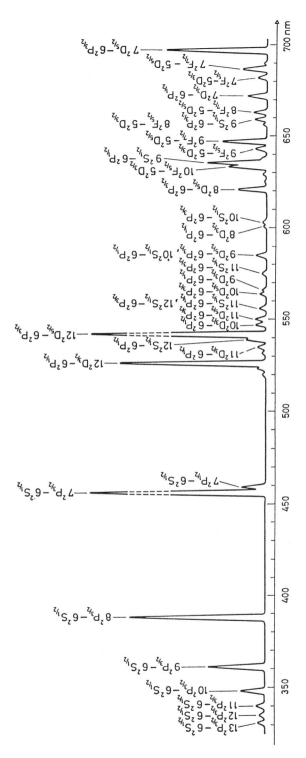

Fig.9.16. Fluorescence spectrum from Cs atoms following stepwise excitation to the 12 $^2D_{5/2}$ state [9.90]

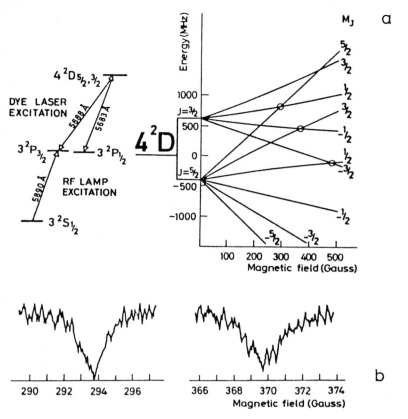

Fig.9.17. Illustration of fine-structure measurement for an alkali atom D doublet using level-crossing spectroscopy [9.93]

fluorescence in the different decay lines will then also be linearly polarized. Radio-frequency transitions, that are induced in the primary state or in cascade states, will give rise to a depolarization and redistribution of the light that can be detected. Thus the properties of many P, D and F states can be investigated. An example of ODR signals in an excited S state, populated by step-wise excitation, was given in Fig.7.11.

In Chap.7 we have discussed how hyperfine structure can be determined by level-crossing spectroscopy. Clearly, alkali atom ^{2}D states can readily be studied using this technique after step-wise excitation. We will here instead choose an example illustrating fine-structure measurements. In Fig.9.17 the example of the inverted sodium 4d $^2D_{5/2,3/2}$ state is given. From the measured level-crossing positions the fine-structure splitting can be calculated using the Breit-Rabi formula for the fine structure, (2.31).

We will conclude this section by illustrating how the tensor Stark polarizability constant can be measured using level-crossing spectroscopy. As illustrated in Fig.9.18, for the case of the potassium 5d $^2D_{3/2}$ state the unknown Stark effect is measured in terms of the well-known Zeeman effect

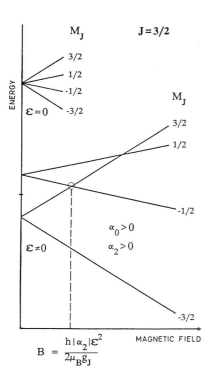

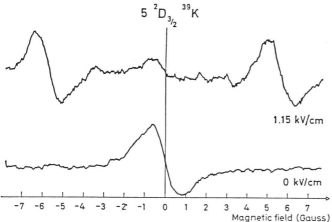

Fig.9.18. Illustration of the determination of the tensor polarizability of an excited alkali atom D state using level-crossing spectroscopy in parallel electric and magnetic fields [9.88]

(g_J is well-defined in LS coupling). In the absence of an external electric field the normal Hanle signal (differentiated) is recorded. If, in addition, a fixed electric field $\mathcal{E}$ is applied over the atoms a high-field level-crossing will be produced when scanning the electric field. The simple relation given in Fig.9.18 is easily verified from (2.29) and (2.35).

9.4 Time-Resolved Spectroscopy

In this section we will study time-resolved laser spectroscopy and generally discuss radiative properties of atoms and molecules, and methods of studying these properties. Since very short laser pulses with a power density sufficient to well saturate optical transitions can be obtained, a large fraction of the irradiated ground-state atoms can be transferred to the excited state. Using step-wise excitations with synchronized lasers a large number of atoms can be excited into very highly excited states. When the laser pulse ceases the exponential decay of the excited state can be monitored. Note, that primarily the population number N(t) decays exponentially, i.e.,

$$N(t) = N_0 e^{-t/\tau} . \tag{9.7}$$

This decay can be monitored by observing the decay of the fluorescence light in an arbitrary spectral line originating in the state. For the light intensity I(t) we have

$$I(t) \sim -\frac{dN(t)}{dt} = \frac{N_0}{\tau} e^{-t/\tau} = I_0 e^{-t/\tau} . \tag{9.8}$$

Because of the special properties of the exponential function the light decays with the same time constant τ as the population decay. The light decay can be followed by a fast detector connected to fast, time-resolving electronics. If the excited state has a substructure, e.g. because of the Zeeman effect or hyperfine structure, and an abrupt, coherent excitation is made, oscillations (quantum beats) in the light intensity will be recorded. The oscillation frequencies correspond to the energy level separations and can be used for structure determinations. We will first discuss the generation of short optical pulses and measurement techniques for fast optical transients.

9.4.1 Generation of Short Optical Pulses

Broad-band flash-lamps with specially designed driving circuits can be used to produce light pulses of a length of the order of 1 ns. Synchrotron radiation can be produced as pulses of ~0.1 ns duration [9.94]. A suitable spectral region can be filtered out from the broad-band distribution using a monochromator and be used for exciting species with broad absorption bands, i.e. liquids and solids. For free atoms or molecules absorbing in a very narrow spectral region, the high monochromaticity of lasers is normally needed for efficient excitation. In Chap.8 we described a variety of pulsed lasers with a pulse duration in the 3-20 ns range. The repetition rate of such lasers seldom exceeds a few hundred Hz and is frequently substantially lower (e.g., 10 Hz). Pulses at a higher repetition rate can be obtained by modulating a cw laser beam by means of a Pockels cell or a Bragg diffraction cell (acousto-optic modulation). In the latter device, an RF oscillator coupled to a piezo-electric transducer induces density variations in a crystal causing a deflection of the beam into an aperture because of grating diffraction. While a rise-time below 10 ns can be achieved in this way the peak power of the pulses is, by necessity, minute. A much more efficient way of producing laser pulses of short duration and high peak power is by *mode-locking*.

9.4.2 Generation of Ultra-Short Optical Pulses

A short description of techniques for generating very short laser pulses by mode-locking will be given. More detailed accounts of such techniques can be found in [9.95-97]. If no frequency discriminating element is present in the cavity, laser operation on an atomic or ionic line will yield only a few cavity modes, whereas a corresponding dye laser with a very broad gain medium will feature a large number of modes. The modes are separated by $\Delta\nu = c/2\ell$, or, in angular frequency $\Delta\omega = 2\pi(c/2\ell) = \pi c/\ell$, where ℓ is the cavity length. The phases of these modes are completely unrelated. If we assume an idealized situation, in which 2N+1 modes of equal amplitude $\mathcal{E}_0$ oscillate with a constant phase relation

$$\phi_k - \phi_{k-1} = \alpha \tag{9.9}$$

between adjacent modes (mode-locking), the resulting light field $\mathcal{E}(t)$ is given by

$$\mathcal{E}(t) = \sum_{k=-N}^{N} \mathcal{E}_0 \exp[i(\omega_0 + k\Delta\omega)t + k\alpha] . \tag{9.10}$$

Here ω_0 is the centre angular frequency. The evaluation of this expression involves the same type of mathematics as that used for calculating the spectral intensity distribution from a grating, and we have

$$\mathcal{E}(t) = A(t)\exp(i\omega_0 t) \tag{9.11}$$

with

$$A(t) = \mathcal{E}_0 \frac{\sin[(2N+1)(\Delta\omega t+\alpha)/2]}{\sin[(\Delta\omega t+\alpha)/2]} . \tag{9.12}$$

Eq.(9.11) expresses an optical sinusoidal carrier wave of frequency ω_0, which is amplitude-modulated in time according to (9.12). The light intensity is proportional to $A^2(t)$, and in Fig.9.19 the intensity for the case of 7 locked modes is given.

The phase relation (9.9) causes the continuous partial waves to interfere constructively to give short pulses separated in time by

$$\Delta t = 2\ell/c . \tag{9.13}$$

(The maxima occur when the denominator in (9.12) approaches 0). The pulse separation is the time taken for light to go back and forth in the cavity ("cavity round-trip time") and is typically 10 ns (1 ns ↔ 30 cm). From

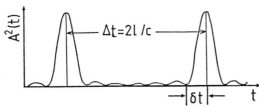

Fig.9.19. Temporal structure of 7 locked modes of equal amplitude

(9.12) we also obtain the time δt from the pulse maximum to the first zero point

$$\delta t = \frac{2\pi}{(2N+1)\Delta\omega} = \frac{1}{\nu_{osc}}. \qquad (9.14)$$

Thus, δt is the inverse of the oscillation bandwidth ν_{osc} of the laser and we note that a large number of locked modes is needed in order to obtain a short pulse length. Clearly, this is entirely consistent with the uncertainty relation requiring a large frequency uncertainty if the time uncertainty is small. It also reflects the general properties of the Fourier transform between the time and the frequency domains. In a dye laser with a broad frequency selective element (e.g., a single-plate Lyot filter), resulting in a bandwidth of ~5 Å, mode-locking results in pulses of a few picoseconds (1 ps = 10^{-12} s) duration. From (9.12) it follows that the pulse peak intensity $A^2(t)$ is proportional to $(2N+1)^2 \mathscr{E}_0^2$, whereas the light intensity for non-locked modes would be proportional to $(2N+1)\mathscr{E}_0^2$. This is consistent with the observation that $\Delta t/\delta t = (2N+1)$. The total time-averaged intensity remains constant but the distribution in time is entirely different. We note that very high peak powers can be obtained with a dye laser that has a high N value.

We will now describe how mode-locking according to (9.9) can be accomplished. Two different approaches can be used: active or passive mode-locking. *Active* mode-locking is obtained by modulating the gain or loss of the laser at a frequency that matches the mode separation $\Delta\omega$. This can be accomplished using a Bragg cell as illustrated in Fig.9.20a. Optical sidebands are then generated for every mode. These side-bands coalesce with the neighbouring mode positions and impose their phase on them. In this way all modes are successively locked in phase with each other, i.e. mode--locking. In an alternative way of describing the phenomenon the loss modulation can be seen as a shutter in the cavity. The shutter is only open during certain intervals separated by the cavity round-trip time. Then a multi-mode oscillation with unrelated phases cannot occur since each mode would be strongly attenuated. On the other hand, if the phases are locked as in (9.9) the energy distribution in the resonator is a short pulse that passes the shutter when it is open, which happens at exactly the right times. Because of the "regenerative" nature of the laser action the phases tend to lock in a way leading to low effective losses (at pulse formation).

A common way to mode-lock a dye laser that is pumped by an ion laser is to use *synchronous pumping*. An actively locked ion laser with the same cavity length as the dye laser is then used (the length of the dye-laser

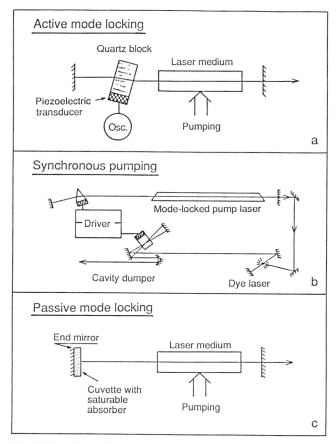

Fig.9.20. (a) Active mode locking by acousto-optic modulation. (b) Synchronous pumping of a dye laser equipped with a cavity dumper. (c) Passive mode-locking using a saturable absorber

cavity is accurately adjusted). Then the laser medium is inverted only during short time intervals, and only a light pulse that bounces back and forth in the cavity passing the gain medium at exactly the right time can be amplified (Fig.9.20b). By using a second Bragg cell in the dye laser a bouncing pulse can be deflected out of the resonator when it has gained in intensity in the low-loss cavity. Using this *cavity-dumping* technique, stronger picosecond pulses at longer time separations than those from normal mode-locking are obtained. Single picosecond pulses can then be further amplified in successive stages using amplifier cells with dye that is pumped by a high-power pulsed laser, e.g., a frequency-doubled Nd:YAG laser.

In *passive* mode-locking a thin (0.1mm) *saturable absorber* (special dyes) is used in the cavity. For sufficiently high light intensities the absorption of the substance is reduced (bleaching). Such intensities are obtained at a mode-locking with associated pulse formation, whereas all unrelated cavity modes are strongly absorbed and quenched. The laser modes then tend to arrange themselves so that oscillation and intensity can build up and passive mode-locking is achieved. The best efficiency is obtained if the absorber is placed in direct connection to the high reflector of the cavity so that a still higher intensity is obtained in the turning pulse (Fig.9.20c). In a technique using *"colliding" pulses* a ring cavity is used with a very thin saturable absorber (normally a bleachable dye jet). In the ring cavity light circulates in both directions. Only if extremely short pulses from both directions simultaneously pass the saturation medium is sufficient bleaching obtained and thus high transmission. In this way pulse lengths of 100 femtoseconds (1fs = 10^{-15}s), corresponding to a spectral width of about 5 nm, can be achieved. The formal mathematical treatment for active and passive mode-locking is rather complex and we give only the above rather crude phenomenological descriptions.

Further reductions in pulse length can be accomplished by *pulse compression*. A femtosecond pulse is then transmitted through an optical fibre, which has an index of refraction that depends on the intensity. The spectral distribution of the pulse is then broadened by self-phase modulation and the pulse length is shortened by a subsequent passing of the pulse through a wavelength-dependent optical retarder consisting of two gratings, as illustrated in Fig.9.21. Six-femtosecond pulses corresponding to only three optical periods can be obtained in this way [9.98].

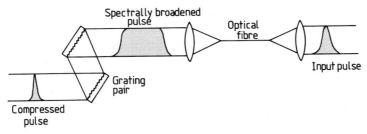

Fig.9.21. Pulse compression using spectral broadening in an optical fibre and subsequent wavelength-dependent retardation in a grating pair

9.4.3 Measurement Techniques for Optical Transients

In this subsection we will discuss various techniques for recording a decaying light intensity. A short laser pulse is used to excite the atoms at time t=0. Fluorescence photons are detected by a fast photomultiplier tube. The corresponding electrical signal can be handled in different ways which can conveniently be discussed with reference to Fig.9.22.

a) **Transient-digitizer technique** - The most direct way of capturing the transient signal from the photomultiplier tube is to use a normal fast oscilloscope and take a photograph of the screen with open camera shutter while the electron beam, which is trigged by the laser, sweeps the screen drawing the decay curve. Clearly, it is inconvenient to further process data stored as photographs. A *transient digitzer* instead captures the transient

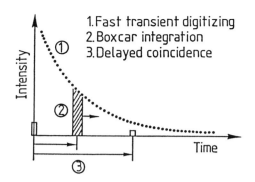

Fig.9.22. Illustration of methods to measure optical transients

digitally by successively recording the input voltage at given time intervals, e.g. 10 ns, and generating numbers that are proportional to the input voltage. The whole sequence of numbers is stored digitally and the memory can then be read out repetitively as an analogue signal at a low rate. Thus a permanent oscilloscope trace of high brightness is obtained. Alternatively, the sequence of numbers can be transferred to a mini-computer and an arbitrary number of individual transients can then be averaged. This method of measurement is very efficient since all the signal information from each transient is captured. A time resolution below 1 ns can be obtained. As a result of the digitization, data can easily be further processed.

b) **Boxcar technique** - In a *boxcar integrator* a time window of a certain length is set for the input signal at a certain delay after the trigger signal, and the corresponding electrical charge is stored in a capacitor with a suitable time constant. Transients are repeatedly recorded and the capacitor is charged to a stationary value. The voltage over the capacitor is read out to a recorder. If the time delay is slowly increased to longer and longer values (with a speed that is compatible with the chosen discharge constant) the exponential decay will be reproduced on the recorder. An ADC (Analogue-to-Digital Converter) at the output of the boxcar integrator can be used to adapt the signal for computer processing. The boxcar technique is relatively simple, but clearly not very efficient, since only photons at a certain time delay are used while the others are discarded. In this respect, the cases of transient digitizer vs. boxcar integrator, and optical multichannel analyser vs. photoelectrically recording slit monochromator are analogous. The transient digitizer, as well as the boxcar technique, requires strong optical transients consisting of a large number of individual photon pulses. Possible nonlinearities in the photomultiplier, which has to record very strong and very weak signals at short time intervals, constitute a limitation to both techniques as regards the recording accuracy.

c) **Delayed-coincidence techniques** - This method operates in the extremly low intensity regime, in which single photon counts are recorded. The principle is illustrated in Fig.9.23.

A single photon gives rise to an electric pulse, which in good photomultipliers has a length of only a few nanoseconds. A Time-to-Amplitude pulse height Converter (TAC) is a critical component when using this technique. A clock is started when the excitation pulse' is fired. The clock runs

Delayed Coincidence Technique

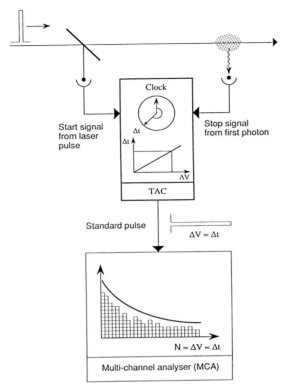

Fig.9.23. Principle of lifetime measurements using the delayed-coincidence technique

until the photomultiplier records the first fluorescence photon. A standard electronic pulse, the height (=voltage) of which is proportional to the time difference is generated and is fed to a Multi-Channel Analyser (MCA). An ADC at the input of the MCA converts the voltage into a channel address number, which is proportional to the pulse height. One unit is then added to the contents of this channel. Since the decay is exponential it is likely that the first photon will arrive early. It is important that the probability of two or more photons reaching the detector is kept low since later photons cannot be recorded. In such cases small time delays would be strongly over-represented, giving rise to the so-called *pile-up* effect. If the probability of detecting a single photon is kept below 1:30 the pile-up correction will be negligible [9.99]. A histogram is then built up in the multichannel analyser. The histogram displays the exponential decay when a sufficient number of counts have been accumulated. The microscopic probability function is then transferred into a macroscopic intensity function. Clearly, a high repetition rate is necessary for the excitation source in order to provide an acceptable measurement time. Thus, a pulse-modulated cw laser or a mode-locked laser is normally required. Very accurate results are obtained with this

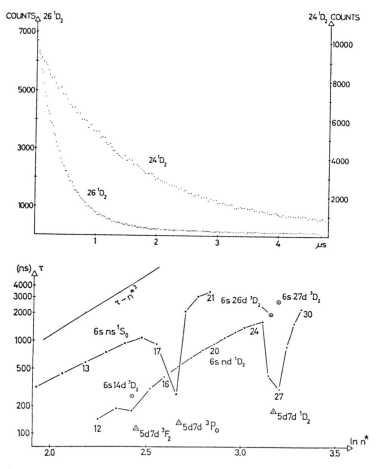

Fig.9.24. Delayed-coincidence recordings of decay curves for excited barium 6snd $^1D_2$ states and plot of lifetime results for perturbed barium Rydberg state sequences. An excitation scheme can be found in Fig.9.5 ([9.101, first Ref.] and [9.102])

method since there are no linearity problems. It is simply a matter of *when* the photon comes, not how large the corresponding pulse is. When using high repetition rate sources it is useful to reverse the role of the start and stop pulses so that the TAC is activated by a fluorescence photon and is stopped by the next excitation pulse. In this way dead time in the electronics is reduced. Delayed-coincidence measurements with pulse-modulated or mode-locked cw lasers have been discussed in [9.100].

In Fig.9.24 examples of decay curves recorded using delayed-concidence techniques with an acousto-optic modulator are shown for excited Ba states. Two-step excitation is used and only the second laser is pulse-modulated. Lifetime results are also given for the 6sns $^1S_0$ and 6snd $^1D_2$ sequences of Ba in the figure. An overall trend of $\tau \sim n_{\text{eff}}^3$ is found, but strong localized perturbations are also evident. These are due to admixtures into

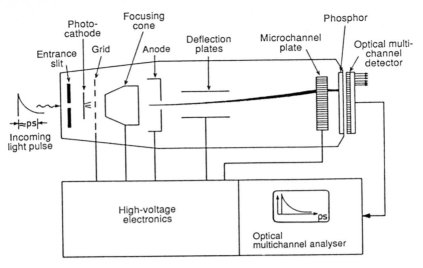

Fig.9.25. The construction of a streak camera

the wavefunction of a doubly excited state of the same parity, belonging to the 5d7d configuration. Doubly excited states have very short lifetimes in comparison with the highly excited states of the Rydberg sequences. Thus the experimental results can be used to investigate the composition of the atomic wavefunctions [9.101].

d) *Streak-camera techniques* - For the recording of very fast phenomena a *streak camera* can be used [9.103]. The principle is illustrated in Fig.9.25. The incoming transient light impinges on a photo-cathode. The released electrons are accelerated, focused and deflected by plates subject to a rapidly rising voltage. The deflected electron beam hits a microchannel plate (Sect.6.3) in which every electron gives rise to an electron burst. The temporal electron distribution gives rise to a streak of light on a phosphor screen. The intensity on the screen is proportional to the time-resolved light intensity. The streak is read off by a sensitive optical multichannel system (Sect.6.3) and the intensity is displayed on an oscilloscope or relayed to a computer. The time resolution in commercially available systems is of the order of ~1ps.

e) *Pump-probe techniques* - The decay of an excited state can also be monitored using a second laser tuned to a transition starting in the excited state. The time-integrated fluorescence induced by this second laser is proportional to the number of excited-state atoms. Thus, by successively measuring the fluorescence intensity as a function of the delay of the second laser pulse with regard to the pump pulse the excited-state lifetime can be measured. Note that in this case, Eq.(9.7) is directly applicable, rather than (9.8). In picosecond experiments with mode-locked lasers pump-probe techniques using a single laser frequency are used. This principle is shown in Fig.9.26. Each pulse is divided into two parts by a beam-splitter, and a variable delay line is used to produce a probe pulse which

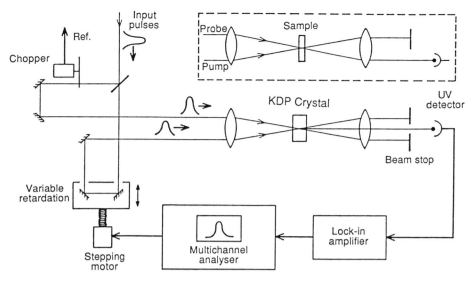

Fig.9.26. The principle of pump-probe experiments using picosecond pulses (*insert*) and pulse-length determinations using autocorrelation with frequency doubling

follows the pump pulse. When the pump pulse depletes the lower state the probe pulse can monitor how quickly the population is restored through relaxation processes. The figure also demonstrates how the pulse length can be determined with an autocorrelation technique employing frequency doubling. Picosecond and femtosecond spectroscopic methods are very useful for investigating ultrafast processes in solids, semiconductors and liquids. Much literature can be found on this topic [9.104-109].

9.4.4 Background to Lifetime Measurements

On several previous occasions we have discussed measurement techniques yielding information on natural radiative lifetimes. We will now discuss the motivation behind such measurements. In Fig.9.27 some important relations between radiative properties are presented.

In many contexts, transition probabilities A_{ik} (Sect.4.1) are the quantities of primary interest. As we will see in the next section, relative transi-

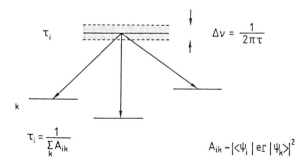

$$\Delta\nu = \frac{1}{2\pi\tau}$$

$$\tau_i = \frac{1}{\sum_k A_{ik}}$$

$$A_{ik} \sim |\langle \psi_i | e \mathbf{r} | \psi_k \rangle|^2$$

Fig.9.27. Relations between radiative properties

tion probabilities, a_{ik}, can be determined by several methods. A direct, absolute determination is considerably more difficult. However, lifetimes can be used to determine the relevant constant c, used to normalize relative transition probabilities to absolute ones

$$\frac{1}{\tau_i} = \sum_k c a_{ik} = \sum_k A_{ik} \; . \tag{9.15}$$

The reasons for studying radiative properties are many:

1) The natural lifetime determines the *fundamental limit* of resolution $\Delta\nu_N = 1/2\pi\tau$ in spectroscopic investigations.

2) Transition probabilities can be used for sensitive *testing of atomic wavefunctions*. A_{ik} is related to the matrix element of the electric dipole operator er between the two wavefunctions (Sect.4.1). Note that A_{ik} is especially sensitive to the outer part of the wavefunction (at larger r values) because of the r weighting.

3) As we have mentioned in Sect.6.7, transition probabilities and the related oscillator strengths f_{ik} are of utmost importance for *astrophysics*, e.g. for calculations of the relative abundances of the elements in the sun and stars.

4) The radiative properties of atoms and ions are also of great importance in *plasma physics*, e.g. for temperature determination and for the calculation of the concentrations of different constituents. In fusion research, determination of the concentrations of contaminating species that tend to cool the plasma are of great interest. The study of the properties of refractory elements that are used for plasma confinement (Mo, Va etc.) is especially important [9.110].

5) In *laser physics* lifetimes and transition probabilities are decisive for predictions of potential laser action in specific media.

9.4.5 Survey of Methods of Measurement for Radiative Properties

We will now discuss a number of measurement techniques for radiative properties. We will start with methods of lifetime determination, and will continue with techniques for the measurement of transition probabilities. Detailed descriptions can be found in [9.111-114].

a) **Linewidth measurements** - The natural linewidth of a short-lived state is $\Delta\nu_N = 1/2\pi\tau$. If this constitutes the only broadening mechanism, the measured linewidth yields the lifetime directly. For very short lifetimes the technique is quite useful. The lifetimes of inner-shell states connected by X-ray transitions can be determined in this way (Sect.5.2.1). Short-lived states for external electrons can also be determined from the linewidth if the Doppler effect is totally eliminated and a single-mode laser of small linewidth is employed [9.115].

b) **ODR and LC** - Here the natural radiative width is also determined directly. The contribution from Doppler broadening is negligible since the measurement is performed in the radio-frequency regime ($\sim 10^7$ Hz). Thus the Doppler broadening is about 10^8 times smaller than for optical transitions. In radio-frequency measurements the lifetime is given by $\tau = 1/\pi\Delta\nu$ in the limit of vanishing RF field strengths. At higher field strengths a broadening of the signal occurs, since the perturbation is no longer negligible. In level-crossing experiments the lifetime is directly obtained from high-field crossing signals. If $I = 0$ or $I = J$ the Hanle signal is also easy to analyse, since only signal contributions of the same half-width occur ($d\nu/dB$ is the same for all contributing level pairs; $g_F = g_J/2$ for all F if $I = J$). In the general case a superposition of signal contributions of different intensities and half-widths has to be analysed [9.116].

c) **Beam-foil techniques** - The beam-foil method has been discussed in Sect.6.1. It is a very general method for measuring lifetimes of atoms and ions. However, the nonselective excitation, leading to cascading decays, places heavy demands on the data analysis and sometimes a detailed study of the different cascade channels is necessary for reliable lifetime evaluations. While the nonselective excitation frequently constitutes a problem, it is also an advantage of the method since a multitude of excited states are populated. For measurements of (multiply) charged ions in particular, the technique provides unique measurement possibilities where other techniques are not applicable.

d) **Beam-laser techniques** - By exchanging the exciting foil in the beam-foil technique for a focused laser beam, selective excitation can be obtained and the problem of cascades is eliminated [9.117]. By directing the laser beam at a certain angle θ, with regard to the ion beam, considerable Doppler shifts in the interaction with the fast beam are obtained. The relativistically correct formula for the interaction wavelength λ_i experienced by the ions that are illuminated by a laser of wavelength λ_ℓ is given by

$$\lambda_i = \lambda_\ell \frac{\sqrt{1 - v^2/c^2}}{1 - (v/c)\cos\theta} . \tag{9.16}$$

For $v/c = 1/100$ we note that Doppler shifts of tens of Å (see also Fig.6.82) are easily accomplished by varying θ. Measurements of this kind are illustrated in Fig.9.28. A fixed-frequency laser (e.g., an Ar$^+$ laser) can be used

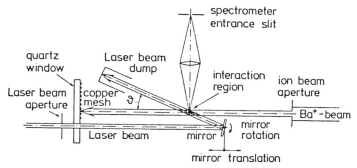

Fig.9.28. Lifetime measurement with the beam-laser technique using Doppler tuning [9.117]

due to the possibility of Doppler-tuning. The first experiments of this kind were performed for the resonance line of Ba$^+$ 6s $^2S_{1/2} \rightarrow$ 6p $^2P_{3/2}$ at 4554 Å. The 4545 Å Ar$^+$ line could be utilized by chosing $\theta = 23°$ for an ion velocity of $0.63 \cdot 10^6$ m/s. Since the transition wavelengths for ions of different charge states rapidly shift into the ultraviolet and VUV regions, due to the electronic shell contraction, and as cw lasers are scarce at such wavelengths, the applicability of this high-precision technique is unfortunately limitied.

e) **Time-resolved spectroscopy with pulsed lasers** - We have recently discussed these techniques in the sections above.

f) **Time-resolved spectroscopy with pulsed electron beam excitation** - Abrupt excitation of atoms and molecules can be accomplished with a pulsed electron beam. In the same way as for the beam-foil interaction, non-selective excitation is obtained. One of the most efficient varieties of this technique is the *high-frequency deflection* method [9.118,119], which is illustrated in Fig.9.29. An electron beam of high power (~20keV, 0.5A) is

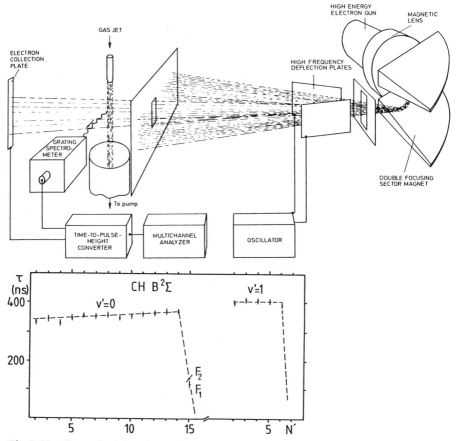

Fig.9.29. The principle of the high-frequency deflection method for lifetime measurements, and experimental lifetime data for the CH radical [9.118, 120]

deflected using RF deflection plates, as in an oscilloscope. The beam is swept over a slit and the repetition rate and pulse length can easily be selected. Delayed-coincidence techniques (Sect.9.4.3c) are used to record the decay. Because of the high light intensity obtained with this technique a high-resolution monochromator (with a narrow slit) can be used for the detection. This is of great importance, particularly in studies of molecules, for which individual rotational-vibrational lines can be isolated without blending. For molecules, cascade problems are frequently less severe than for atoms. Lifetime results for the CH molecule are shown in the figure illustrating the shortening of lifetimes for higher rotational levels due to predissociation at potential curve crossings.

g) **Phase-shift method** - In the phase-shift method atoms or molecules are excited by a sinusoidally modulated light or electron beam, as illustrated in Fig.9.30. Fluorescence light is recorded with a photomultiplier

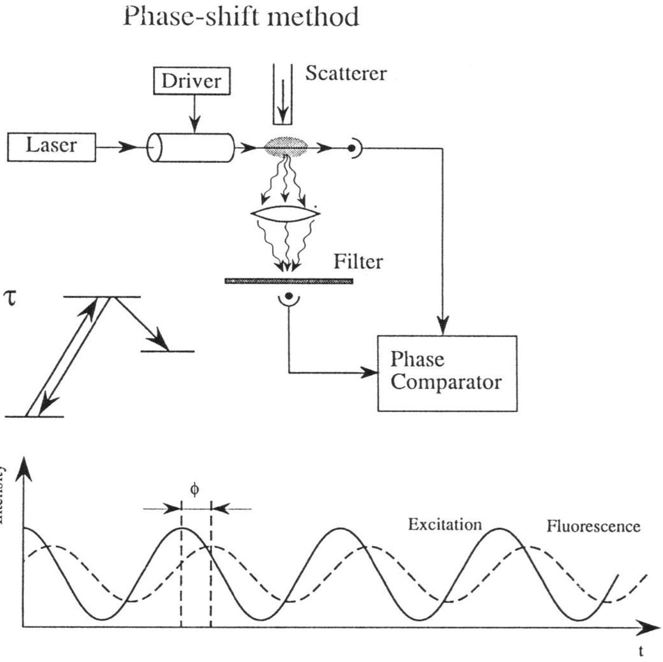

Phase-shift method

$$I_{exc} = I_0 (1 + a \sin \Omega t) \cos \omega t$$

$$I_{fl} = bI_0 \left[1 + \frac{a}{\sqrt{1 + \Omega^2 \tau^2}} \sin (\Omega t + \phi) \right] \cos \omega t$$

$$\tan \phi = \Omega \tau$$

Fig.9.30. The principle of lifetime measurements with the phase-shift method

tube. This light will then also be modulated, but because of the delay in the excited state a phase-shift has been introduced [9.121]. At the same time, the contrast in the modulation is reduced. If light is used for the excitation, the phase shift can be determined by comparison with the signal that is obtained when stray light is sent directly into the detector by inserting a small scattering object into the light beam at the point of interaction. If the exciting light is described by the expression

$$I_{exc} = I_0(1 + a\sin\Omega t)\cos\omega t , \tag{9.17}$$

it can be shown, by integrating the exponential decays from the different excitation function time elements, that the fluorescence light intensity is given by

$$I_{fl} = bI_0\left[1 + \frac{a}{\sqrt{1 + \Omega^2\tau^2}}\sin(\Omega t + \phi)\right]\cos\omega t \tag{9.18}$$

with

$$\tan\phi = \Omega\tau . \tag{9.19}$$

If multiple scattering processes occur an erroneously long lifetime will be obtained, as for several of the above-mentioned methods. The opposite effect is obtained if, in addition to the fluorescence light, non-shifted stray light from the modulated light source is recorded. If the modulation is not perfectly sinusoidal, the first Fourier component can be isolated and the phase shift for this component will still yield the lifetime.

We will now turn to the determination of transition probabilities.

h) **The emission method** - If the effective temperature in local thermodynamic equilibrium is known, the relative strength of spectral lines from a light source can be used to calculate *relative transition probabilities*. Clearly, corrections for the wavelength dependence of the spectrometer transmission and the photomultiplier sensitivity must be applied. This can be done by replacing the actual light source by a standard tungsten lamp (Planck radiator) with a known emissivity. A large number of elements have been investigated by *Corliss* and *Bozman* using a standardized arc discharge [9.122]. Actually, absolute values are given, but a correction factor of up to 5, as determined from normalizing lifetime measurements, must frequently be applied.

i) **The hook method** - The hook method is based on the Kramer-Kronig dispersion relation, which relates the refractive index of a gas to transition probabilities [9.123, 124]. At the same time as an atomic vapour absorbs at the transition frequency, an anomalous dispersion is obtained. (The increase in refractive index of transparent materials towards the UV region is related to the UV absorption of the material, see Figs.6.18 and 48.) In Fig.9.31 the relation between the absorption of and dispersion at the sodium D lines is illustrated. In the hook method, light from a continuum source is sent through a Mach-Zehnder interferometer, where a heat-pipe oven with the metal vapour is inserted in one arm. The interference pattern is imaged as a function of wavelength through a spectrometer on a photo-

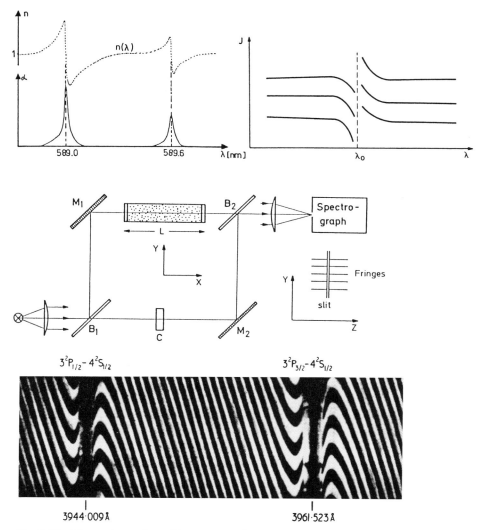

Fig.9.31. Illustration of oscillator strength measurements [9.1]. An experimental recording for the AlI resonance doublet is included

graphic plate. The fast change in refractive index at a line gives rise to the "hooks" in the recording. It can be shown, that the product of the atomic density and the transition probability is proportional to Δ^2, where Δ is the hook separation. By measuring absorption as well as dispersion, absolute transition probabilities can, in principle, be determined. The hook method yields very accurate, reliable relative transition probabilities. Recently a laser-based related technique, "the inverse hook method", has been introduced [9.125].

9.4.6 Quantum-Beat Spectroscopy

We have already discussed quantum-beat spectroscopy (QBS) in connection with beam-foil excitation (Fig.6.6). There the case of abrupt excitation upon passage through a foil was discussed. Here we will consider the much more well-defined case of a pulsed optical excitation. If two close-lying levels are populated simultaneously by a short laser pulse, the time-resolved fluorescence intensity will decay exponentially with a superimposed modulation, as illustrated in Fig.6.6. The modulation, or the quantum beat phenomenon, is due to interference between the transition amplitudes from these coherently excited states. Consider the simultaneous excitation, by a laser pulse, of two eigenstates, 1 and 2, from a common initial state i. In order to achieve coherent excitation of both states by a pulse of duration Δt, the Fourier-limited spectral bandwidth $\Delta\nu \simeq 1/\Delta t$ must be larger than the frequency separation $(E_1 - E_2)/h = \omega_{12}/2\pi$. If the pulsed excitation occurs at time t = 0, the wavefunction of the excited state can be written as a linear superposition of the eigenstates

$$|\psi(0)\rangle = \sum_k a_k |\phi_k(0)\rangle , \qquad (9.20)$$

where the coefficients a_k are probability amplitudes for finding the atom in level k. Due to the exponential decay to the final level f, the time-dependent wavefunction is given by

$$|\psi(t)\rangle = \sum_k a_k |\phi_k(0)\rangle \exp(-iE_k t/\hbar) e^{-t/2\tau} . \qquad (9.21)$$

The time-dependent fluorescence light intensity from the excited states is determined by the transition matrix element, see (4.28) and (7.26).

$$I(t) = C |\langle\phi_f | e_g \cdot r | \psi(t)\rangle|^2 , \qquad (9.22)$$

where e_g is the polarization vector for the detected light. Inserting (9.21) in (9.22) yields

$$I(t) = C e^{-t/\tau}(A + B\cos\omega_{12} t) \qquad (9.23)$$

with

$$A = a_1^2 |\langle\phi_f | e_g \cdot r | \phi_1\rangle|^2 + a_2^2 |\langle\phi_f | e_g \cdot r | \phi_2\rangle|^2 , \qquad (2.24)$$

$$B = 2a_1 a_2 |\langle\phi_f | e_g \cdot r | \phi_1\rangle| |\langle\phi_f | e_g \cdot r | \phi_2\rangle| . \qquad (9.25)$$

We note, that a modulation is obtained (B≠0) only if the matrix elements for the transitions 1 → f and 2 → f are non-zero at the same time. A quantum-mechanical interpretation of the beats is based on the observation that it is impossible to determine whether the atom decayed via the transition 1 → f or 2 → f. Then the total probability amplitude is the sum of the two corresponding amplitudes and the observed intensity is the square of this sum and the cross-term gives rise to interference. If it had been possible to detect the transitions separately (spectrally, or by means of the polarizations), then the interference would have been lost. The situation here is analogous to that in Young's double-slit experiment. If attempts are made to determine through which of the slits the photon passes, the interference is lost. Note the very close relationship between quantum-beat spectroscopy and level-crossing spectroscopy. If time integration of the QBS phenomenon is performed, the LC phenomenon is obtained, for which continuous excitation and detection are used. Quantum beats are induced by a tunable laser with a short pulse duration (~5ns). If very fast beats are to be studied, the pulse must be correspondingly shorter (in order to obtain Fourier components spectrally overlapping both levels, or, expressed differently; to ensure a sufficiently small initial spread in the wavefunction phase factors). With a mode-locked laser, pulse lengths of the order of 1 ps can be obtained (Sect.9.4.2).

In Fig.9.32 an example of Zeeman quantum beats is shown. The geometries for excitation and detection are the same as for a recording of the Hanle effect (Sect.7.1.5). The signal is also a $\Delta M = 2$ phenomenon. Zeeman quantum beats can also be explained semi-classically using the same model as was used for the Hanle effect (Fig.7.15). The linearly polarized light (which has σ^+ as well as σ^- nature, Fig.4.9) induces an oscillating *electric dipole* in the electron cloud, which will radiate with a dipole distribution, shrinking with a time constant of τ. Since the atom has a *magnetic dipole moment* it will Larmor precess (Sect.2.3.1) in the external magnetic field and a radiation lobe is turned in towards the detector twice every full revolution. The excited atom is a microscopic "light-house". The Hanle effect is the part of the Zeeman quantum beat that "survives" the time integration because of the common initial condition and slow lobe rotation in the field region of the Hanle effect. The situation is illustrated in Fig.7.15.

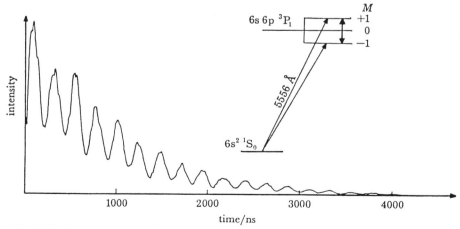

Fig.9.32. Zeeman quantum beats for the resonance line in ytterbium [9.126]

For high fields the lobes pass in and out of the detector direction many times during the decay of the dipolar distribution. Thus, the fraction of time the detector is illuminated will become essentially independent of the field strength and the signal levels off.

From the semi-classicial picture it is also evident that the phase of the beat signal is changed by 180° if the linear polarizer for the case of signal detection in the direction of the magnetic field is turned by 90° (this corresponds to a sign change of the Hanle effect). Because of the phase inversion it is possible to eliminate the primary exponential decay by subtracting two signals of equal unmodulated light intensity but recorded for perpendicular detection polarizer settings. When pure exponentials are desirable for lifetime evaluation a sufficiently high magnetic field can be applied, corresponding to frequencies well above the cut-off frequency of the detection system, to wash out the quantum beats.

Although the general features of the quantum-beat phenomenon can be understood in a simple semi-classical model, a full quantum-mechanical description is required for calculating the correct relation between B and A in (9.23) to determine the beat contrast in the decay curve. Quantum-beat spectroscopy has been discussed in detail in [9.127, 128].

Fine- and hyperfine splittings also give rise to quantum-beat signals. For example, the hfs beats can be semi-classically understood considering the radiating electronic shell precessing at hyperfine frequency in the internal field due to the nucleus. $\Delta F = 1$ and $\Delta F = 2$ beats can be obtained in zero magnetic field. In Fig.9.33 hfs quantum-beat signals corresponding to two polarizer settings are given. It can be shown that the beat amplitude is proportional to $3\cos^2\theta - 1$, where θ is the angle between the polarizers in the exciting and detection beams. (Thus, by making $\theta = \arccos(1/\sqrt{3}) \simeq 54.7°$ ("magic angle") it is possible to suppress beats all together). When many beat frequencies occur simultaneously they can be isolated by a Fourier transformation from the time to the frequency domain. In Fig.9.34 an example of hfs beats from the two stable gallium isotopes is shown and a corresponding Fourier transform is also given.

Quantum-beat spectroscopy as well as level-crossing and optical double resonance measurements using pulsed lasers for high-resolution investigation of small energy intervals in a very wide wavelength region [9.131]. This is because pulsed lasers can be shifted to UV/VUV or IR wavelengths using nonlinear optical techniques (Sect.8.6).

As a final point in this section, we will discuss the question of increasing the spectral resolution beyond the natural radiation width $\Delta\nu_N = 1/2\pi\tau$. Different experiments have been performed, in which the observation after pulsed excitation is limited to "old" atoms, i.e. the detection system is activated after a certain delay such that only the atoms that have survived longest in the statistical process will contribute to the signal [9.132, 133]. Normally, the whole exponential decay is utilized and in the integration from 0 to ∞ a Lorentzian of half-width $1/2\pi\tau$ is obtained in frequency space (Fig.4.4). The Breit formula (7.25) describing the level-crossing phenomenon has been obtained through such a complete integration. Lorentzian crossing signals are then obtained. If the integration is performed from

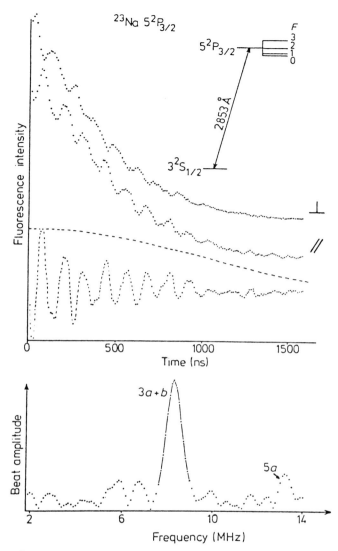

Fig.9.33. Hyperfine quantum beats for a $^2P_{3/2}$ state in ^{23}Na recorded for two different polarizer settings [9.129]

a delayed time T to ∞, oscillations in the signal tails are obtained but the central peak becomes narrower. The same is true for quantum-beat signals, for which the Fourier transform also narrows when integration is performed after a time delay. Clearly, a very strong loss in intensity is obtained. For example, after 4τ the intensity has been reduced to 1% of the original intensity. Thus it is not at all clear that the information content has been improved by restricting the observation to "old" atoms. In special cases, e.g. if the signals are non-symmetric or overlapping, reduced system-

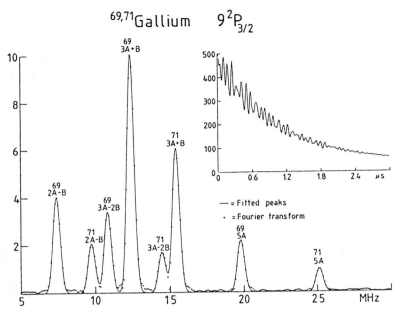

Fig.9.34. Superimposed hyperfine quantum beats for the two stable gallium isotopes and the corresponding Fourier transform displaying the beat frequencies [9.130]

atic errors can be obtained if the linewidth is reduced in spite of the significant loss in intensity.

9.5 High-Resolution Laser Spectroscopy

In this section we will describe a number of high-resolution methods, in which the extremely narrow linewidth of single-mode lasers is utilized. Various ways of eliminating Doppler broadening have been investigated during recent years, leading to the development of Doppler-free laser spectroscopic techniques. The effective linewidth that is experimentally obtained is determinated by a number of effects:
- The natural radiation width.
- Residual Doppler broadening, in particular the second-order Doppler effect (due to the time dilatation effect $(1-v^2/c^2)^{1/2}$ in (9.16), which is present irrespective of the direction of motion.
- The laser linewidth.
- The transit time broadening, due to the finite time that an atom dwells in the laser beam because of its motion. (See also Sect. 7.1.2.).

We will first describe spectroscopy on *collimated atomic beams* and on *kinematically compressed ion beams*. Two groups of nonlinear spectroscopic techniques will be discussed: *saturation techniques* and *two-photon absorp-*

tion techniques. We will also deal with the optical analogy to the Ramsey fringe technique (Sect.7.1.2). Finally, the *atom-* and *ion-trap* techniques will be discussed. Here the atoms are brought to rest by *laser cooling*, and the Doppler effect, as well as the transit time broadening, is almost totally eliminated.

9.5.1 Spectroscopy on Collimated Atomic Beams

As we have already noted (Sect.6.1.1), a well-collimated atomic beam displays a very small absorption width perpendicular to the atomic beam. As shown in Fig.9.35, the collimation ratio C for an atomic beam is defined as

$$C = s/d \ . \tag{9.26}$$

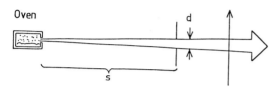

Collimation ratio $C = \frac{s}{d}$ Fig.9.35. Definition of the collimation ratio

The residual (first-order) Doppler broadening $(\Delta\nu_D)_{res}$ for a collimated atomic beam is given by

$$(\Delta\nu_D)_{res} \simeq \frac{\Delta\nu_D}{C\sqrt{2}} \ , \tag{9.27}$$

where $\Delta\nu_D$ is the normal Doppler broadening. High-resolution spectroscopy can be performed by irradiating such a beam at right angles with a narrow-band, single-mode laser. With a collimation ratio of 100 a typical residual Doppler broadening of 5 MHz is obtained.

The atomic beam technique is a very versatile one. Atomic beams can be produced for essentially any element, whereas conventional cell techniques are limited to a temperature interval up to about 1000°C. Besides reducing the Doppler width, collisional effects are significantly reduced with the atomic-beam technique compared with cells. The possibility of utilizing spatially separated interaction regions along the beam is also valuable in certain cases. We note that the most probable velocity v of a thermal atomic beam emerging from an oven is

$$v = \sqrt{\frac{3kT}{M}} \ , \tag{9.28}$$

where T is the absolute temperature, M the mass number and k Boltzmann's constant. This formula results in typical thermal velocities of about 300

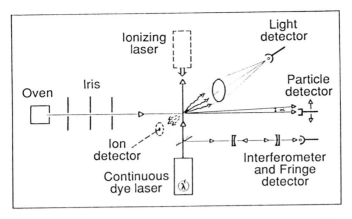

Fig.9.36. Laser spectroscopy on a collimated atomic beam. Three different detection methods are illustrated

m/s. Typical mean free paths, between collisions with residual gas in the vacuum system, are about 10 m at 10^{-5} torr. There are several methods for detecting the narrow resonance induced by the laser. In Fig.9.36 a few methods are illustrated:

a) **Detection through fluorescence** - The most direct way to study optical resonance is to observe the *fluorescence light* that is released after the excitation. In Fig.9.37 a schematic spectrum for the D_2-line in ^{23}Na is shown (3s $^2S_{1/2} \rightarrow$ 3p $^2P_{3/2}$). Two well-separated component groups, corresponding to the ground-state hyperfine splitting occur. The components corresponding to the small excited-state splittings are well resolved and the a and b constants for the hyperfine structure can be evaluated. The natural radiation width corresponding to τ = 16 ns for the 3p $^2P_{3/2}$ state is 10 MHz.

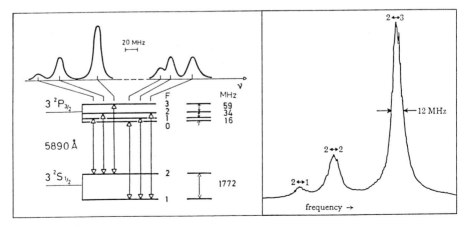

Fig.9.37. Schematic representation of the hyperfine structure in the sodium D_2 line. An experimental curve from a well collimated sodium beam is included [9.126]

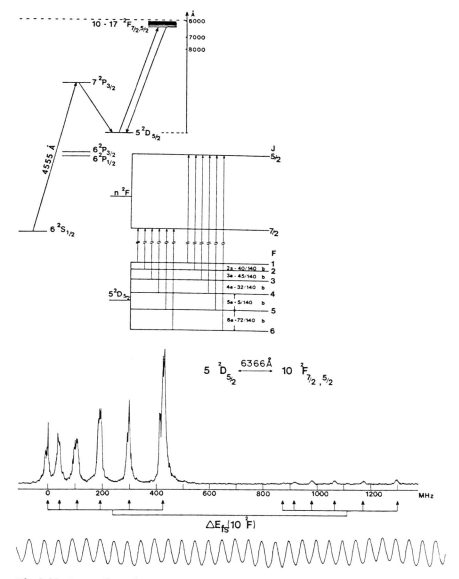

Fig.9.38. Recording of a D - F caesium transition obtained in a collimated atomic beam experiment using stepwise excitations [9.134]

As can be seen from the inserted experimental curve, such a linewidth can be approached. In Fig.9.38 a further example of this type of spectroscopy is given. Here a broadband cw laser has been used to populate the rather long-lived 5d $^2D_{5/2}$ (τ~1 μs) state in ^{133}Cs in the cascade decay of the primary excited 7p $^2P_{3/2}$ state. A narrow-band laser is swept through the components of the 5d $^2D_{5/2}$ → 10f $^2F_{7/2,5/2}$ transition. The inverted hyperfine structure of the lower state and the inverted fine structure (evident

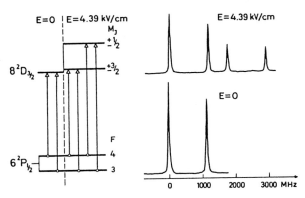

Fig.9.39. Stark effect determination in the transition $6\,^2P_{1/2} - 8\,^2D_{3/2}$ in ^{133}Cs. The lower state has too low a polarizability to be affected by the electric field [9.135,91]

from the intensity ratio between the low- and high-frequency groups) of the highly excited state are shown. The hyperfine structure of the F-state is too small to be resolved.

An example of the Stark effect is given in Fig.9.39. Note, that while only the tensor Stark interaction constant α_2 (Sect.2.5.2) can be determined in an LC experiment (Sect.7.1.5; Fig.9.18), the scalar interaction constant α_0 can be obtained in this type of experiment as well as α_2. In the same way, isotope shifts can be measured by direct optical high-resolution methods while resonance methods and quantum-beat spectroscopy can only be used for measurements of splittings within the *same atom*.

b) **Detection by photoionization** - In this method the atomic beam is simultaneously irradiated by a narrow-band laser and an intense laser. The latter has a sufficient photon energy to photoionize the atoms that are transferred to the excited state, while ground-state atoms cannot be ionized. The photo-electrons are detected in an electron multiplier, which is constructed like a photomultiplier tube but without a photo-cathode. For the case of very long-lived Rydberg atoms, field ionization can also be applied as an efficient detection method. During a period of 10 µs atoms move several mm, and it is possible to physically separate a field-free laser interaction region from the field-ionization region [9.136].

c) **Detection by the recoil effect** - When an atom absorbs a photon impinging perpendicularly to the direction of flight of the atom the photon momentum

$$p = h\nu/c \qquad (9.29)$$

is transferred to the atom. For a sodium atom absorbing a D-line quantum the transverse velocity change is about 3 cm/s. The atom is then deflected by an angle given by

$$\alpha \simeq p/mv \simeq 10^{-5} \text{ rad}, \qquad (9.30)$$

where m and v are the mass and the velocity of the atom, respectively. If the atom is de-excited by stimulated emisson the absorbed perpendicular momentum is cancelled by the recoil of the atom when the stimulated photon is emitted in the same direction as the incoming light. On the other hand, if spontaneous decay occurs, the direction of the recoil will be randomized. Thus, on average, a transverse momentum is transferred to the atomic beam, which is deflected and also broadened. At optical resonance the intensity at a particle detector placed in the path of the original atomic beam is reduced. The deflection of the beam depends on the intensity of the laser and the length of the interaction region. Deflections are normally small (~1°). In principle, isotopes can be separated in this way by isotope-selective excitation. However, more efficient methods have been developed (Sect. 10.4.2).

d) **Detection by magnetic deflection** - In one version of this kind of experiment the atomic beam in an ABMR apparatus (Sect.7.1.2) is irradiated by a sharply tuned laser in the C region. At optical resonance, a flop-in signal is obtained due to the pumping of atoms with $\mu_{eff} < 0$ into states with $\mu_{eff} > 0$. A single inhomogeneous magnetic field, e.g. from a sextupole magnet, can also be used for non-optical signal detection. The atomic beam emerging from the source is perpendicularly irradiated by the laser beam before it enters the sextupole magnet, which focuses $\mu_{eff} < 0$ atoms and deflects $\mu_{eff} > 0$ atoms. The focused atoms hit the detector, which can be the entrance slit of a mass spectrometer. Such a detection system is particularly valuable when dealing with small amounts of radioactive atoms. When the laser beam pumps atoms from one F-state into another in a field-free region, a flop-in (increase) or a flop-out (decrease) signal is recorded depending on which Paschen-Back group a particular F-level communicates with. In Fig.9.40 an example of this type of experiment for radioactive alkali atoms is shown. Extensive measurements of short-lived radioactive alkali isotopes of the alkali atoms, including francium, have been performed [9.138, 139].

As we have seen, magnetic deflection can be used for the detection of optical resonance. However, laser beams can also be used to replace the inhomogeneous A and B magnets in ABMR (Fig.7.5). Then the laser is tuned to substantially reduce the population of a particular F-level while the atoms are in the A region. A beam from the same laser is used in the B region, where the fluorescence induced by the beam is monitored. Here the light is reduced as a result of the low F-state population. However, if $\Delta F = \pm 1$ transitions from neighbouring, well-populated levels are induced in the C region, the population of the state initially chosen is increased, which is observed in the B region as an increase in the fluorescence light [9.140]. With the A and B magnets retained, laser-induced fluorescence can be used to detect the transmitted atoms [9.141].

We will conclude this section by describing the special conditions pertaining to fluorescence spectroscopy on fast ion beams. As we have noted in connection with beam-foil spectroscopy, the transverse Doppler broadening of an ion beam is substantial due to the high ion velocities. Thus, high-resolution spectroscopy cannot be accomplished by perpendicular laser beam radiation. However, it is possible to obtain a high resolu-

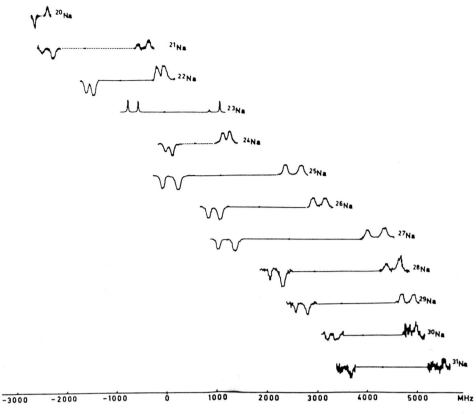

Fig.9.40. Recordings illustrating hfs and isotopic shifts between different isotopes of sodium. The recordings were made on collimated beams and detection by magnetic deflection was used [9.137]

tion by directing the laser beam along the ion beam, which at first sight might seem very surprising. The effect is due to a *kinematic compression*, which occurs in the ion acceleration process [9.142, 143]. For a particle in movement the non-relativistic kinetic energy is given by

$$E = \tfrac{1}{2}mv^2 \ . \tag{9.31}$$

By differentiating we obtain

$$\Delta E = mv\Delta v \ . \tag{9.32}$$

For a given energy spread ΔE of the ions, mainly due to the thermal velocity spread in the ion source we obtain, since ΔE is the same after the acceleration,

$$\Delta v \propto 1/v \ . \tag{9.33}$$

The velocity spread Δv that gives rise to the Doppler broadening, is thus inversely proportional to the ion velocity. At v = 0.001c, which should be compared with the thermal value of v ≃ 300 m/s, we have a reduction of the Doppler width by a factor of 1000. This technique can also be applied to neutral atoms. An ion beam can be neutralized by sending it through a charge-exchange gas. Many studies of isotope shifts and hfs have been performed for longitudinally excited fast beams. In particular, short-lived radioactive nuclei have been studied. Reviews of determinations of nuclear properties using various laser spectroscopic techniques are given, e.g., in [9.144-149].

9.5.2 Saturation Spectroscopy and Related Techniques

We have earlier discussed how a multi-mode laser reacts in the presence of an intracavity absorption cell (Sect.9.2.2). We will now consider the corresponding situation when the laser is forced to run in a single-cavity mode through the action of an intra-cavity etalon (Fig.9.41). The atoms or molecules in the intra-cavity absorption cell have a Doppler broadening of the order of 1000 MHz. Different velocity groups of the atoms are responsible for the absorption in the different parts of the absorption curve. The narrow-band laser light will be absorbed at two different positions in the Doppler profile, symmetrically placed around the peak. Light moving to the right is absorbed at one position while left-moving light is absorbed at the other position. Through the action of the intense light, essentially half the number of the atoms with the corresponding velocity will leave the ground state; "holes" are burnt in the velocity distribution (saturation, Sect.9.1.2). If the laser is tuned towards the centre of the Doppler-broadened absorption line the output power of the laser will drop successively. When the laser is tuned exactly to the peak of the absorption curve both laser beams will be absorbed at the same position (atoms moving perpendicularly to the laser beam are affected). Since light of *one* direction of propagation can saturate the transition for the particular velocity group, the total absorption will *decrease* when light of *both* directions of propagation interacts with the same atoms. Compared with the situation just before the laser tuning had reached

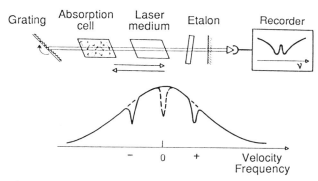

Fig.9.41. Inverted Lamb-dip spectroscopy

the line centre only half the number of atoms are available as ground-state absorbers. The output power of the laser will be significantly *increased* at the centre of the absorption line. The width of the corresponding signal is no longer limited by the Doppler broadening. We note that the reason why the signal at the line centre is obtained, is that the absorption is not linear, due to the saturation phenomenon. If the laser intensity is weak and we are far from saturation, very weak hole burning occurs and we have the linear absorption situation. Then no narrow signal is observed at the line centre, since the absorptions add linearly at the line center as well as off the line centre. In contrast, the *inverted Lamb-dip* experiment [9.150] we have just described constitutes an example of *nonlinear laser* spectroscopy. The *normal Lamb-dip* effect [9.151], which is closely related to the above described process, produces a *decrease* in the output power of a single-mode gas laser (without absorption cell) when the frequency is tuned to the centre of the laser gain profile. The decrease occurs since only one population-inverted velocity group is available at the line centre, whereas two velocity groups contribute to the amplification off the line centre. The sharp Lamb dip or inverted Lamb dip can be used for frequency stabilization of a laser. The output power is sensed and a servo system adjusts the length of the cavity. A small modulation is introduced into the length regulation and the derivative of the light intensity is monitored using lock-in techniques. Thus the servo system adjustment can be performed with the correct phase. In this way a HeNe laser at 6328 Å can be stabilized on an I_2 absorption line or a methane line at 3.39 μm. The frequency of the HeNe transitions, and some I_2 lines, has been measured directly with the Cs clock as the reference, using a special technique employing ultrafast mixing photodetectors and several intermediate stabilized lasers [9.152-155]. Further, the wavelength of the stabilized light has been determined with high precision using interferometric techniques. The orange (Doppler broadened) line at 6058 Å from ^{86}Kr, which until recently defined the metre, has been used as the reference. By multiplying frequency and wavelength, a very accurate value for the velocity of light could be calculated. A best value of c = 299, 792, 458 m/s has been established from measurements at different laboratories. An error is associated with this result, mainly reflecting the uncertainty in the definition of the metre, which has posed a significant problem. At the meeting of "Conférence Internationale des Poids et des Mesures" in 1983, it was decided that the metre should be directly connected to the much more precise time (second) definition by *defining* the velocity of light as the above-mentioned value (it no longer has an error) resulting in a definition of the metre as the distance light travels in vacuum during 1/299, 792, 458 s [9.156, 157]. Thus, one basic unit definition has effectively been eliminated and length is now expressed in terms of time. For practical purposes, secondary standards must, of course, be used, and the most obvious ones are the Lamb-dip stabilized laser lines used in the process of establishing the above-mentioned value of c. Authorized data for a few lines are given in Table 9.3. Lamb-dip studies of transitions in gas lasers have been performed for a long time. With the advent of the dye laser a renewed interest in spectroscopy utilizing saturation effects has evolved. The first saturation

Table 9.3. Secondary length standards based on saturation spectroscopy using HeNe laser oscillations locked to molecular absorption lines [9.157]

Transition	Frequency	Wavelength
CH_4, ν_3, P(7) comp.$F_2^{(2)}$	88,376,181,608 kHz	3,392,231,397.0 fm
$^{127}I_2$, 17-1, P(62), comp. o	520,206,808.51 MHz	576,294,760.27 fm[a]
$^{127}I_2$, 11-5, R(127), comp. i	473,612,214.8 MHz	632,991,398.1 fm
$^{127}I_2$, 9-2, R(47), comp. o	489,880,355.1 MHz	611,970,769.8 fm

[a] Frequency doubled 1.15 μm HeNe transition

spectroscopy experiments using a tunable narrow-band laser were performed by *Hänsch, Schawlow* and co-workers [9.158, 159] and by *Bordé* [9.160]. The atomic sample is normally placed outside the cavity for practical reasons. An experimental set-up for studying the sodium D_1 line (3s $^2S_{1/2} \rightarrow$ 3p $^2P_{1/2}$) is shown in Fig. 9.42.

The primary laser beam is divided into two beams using a partially reflecting mirror (beam splitter). The two beams then pass the sodium vapour cell in opposite directions and with overlapping paths. One of the beams, the saturating beam or *pump beam*, has a high intensity while the other, the detection or *probe beam*, is weak. The intensity of the probe beam is measured after passing through the cell. When the laser is tuned towards the centre of the Doppler-broadened absorption line, the strong laser beam saturates the transition for atoms in a certain velocity group. At the same time, the probe beam experiences an increasing absorption. When the laser is tuned to the line centre, the probe beam "detects" the hole that has been burnt by the saturating beam and thus its transmission is increased. In order to isolate the part of the signal that depends on the saturation (the Doppler-free signal) a lock-in amplifier, which is synchronized with a light-chopper in the saturating beam, is used. The lock-in amplifier records the amplitude of the ac component that can be obtained when the saturating beam is switched on and off. Out in the line wings the two beams interact with completely different velocity groups and therefore the transmission of the probe beam does not depend on the presence or not of a saturating beam. However, at the line centre the transmisson strongly depends on whether a hole has been burnt or not and therefore an ac signal is obtained at the chopper frequency. Information is then transferred from one beam to the other, counter-propagating one with the atoms as a transfer medium. In the lower part of Fig. 9.42 the Doppler-free structure of the D_1 line is schematically shown when the laser is tuned through the line. The Doppler-free signal is given by the expression

$$S(\nu) = f(I/I_{sat}) \frac{1}{1 + [(\omega-\omega_0)/\gamma_s]^2}, \tag{9.34}$$

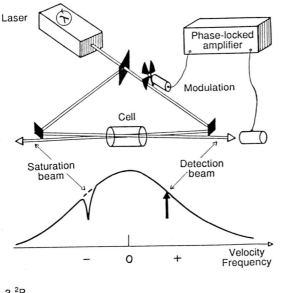

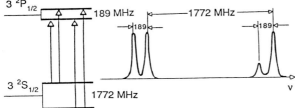

Fig.9.42. Saturation spectroscopy of the sodium D_1 line [9.159]

where the signal amplitude f depends on the pump-beam intensity in relation to I_{sat}, which is the intensity required to saturate the transition.

The half-width of the signal increases with pump beam intensity according to

$$\gamma_s = \tfrac{1}{2}\gamma_0(1 + \sqrt{1 + I/I_{sat}}) \tag{9.35}$$

where

$$\gamma_0 = \pi\Delta\nu_N . \tag{9.36}$$

A particularly simple form of saturation spectroscopy is obtained if a sample that is sufficiently dense to completely block a probe beam is used. Only at the line centre does the bleached path induced by a strong counter-propagating pump beam allow the probe beam to emerge from the cell and hit the detector. A simple set-up and a schematic curve are shown in Fig.9.43 for this type of high-contrast transmission spectroscopy [9.161].

In normal saturation spectroscopy the detector essentially "looks" straight into the laser and a substantial absorption is needed in order to observe a signal. This means that only strong spectral lines, originating in

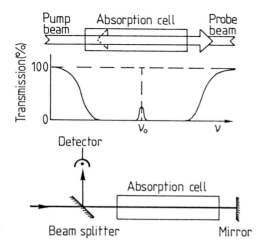

Fig.9.43. High-contrast transmission spectroscopy [9.161]

well-populated states, can be investigated by this method. In cases where less favourable conditions prevail, the so-called *polarization spectroscopy* method can instead be used [9.162]. The experimental set-up is shown in Fig. 9.44.

The set-up is similar to the one used in normal saturation spectroscopy but the probe beam is now blocked by crossed polarizers that are placed at opposite sides of the absorption cell. The polarizer in front of the cell is used to increase the linear polarization of the laser beam, which is frequently already well polarized. A matched pair of polarizers (frequently Glan-Thompson prisms, Fig.6.46) can have a rejection ratio of 10^6 to 10^7 in the crossed position. The pump beam is circularly polarized and induces an anisotropy in the gas for the affected velocity group. At the line centre these atoms can, considered as an optically active medium, turn the plane of polarization so that probe beam light can pass the analysing polarizer (Sect.9.2.4). If the relative orientation of the polarizer is off set by a small angle θ, a constant background results, but at the same time signal amplification is also obtained. It can be shown that the intensity of the polarization spectroscopy signal is given by

$$I = I_0 \left[\theta^2 + \frac{1}{2}\theta s \cdot \frac{x}{1+x^2} + \frac{1}{16}s^2 \frac{1}{1+x^2} \right] \tag{9.37}$$

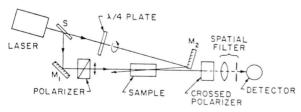

Fig.9.44. Arrangement for polarization spectroscopy [9.162]

where s is a factor that depends on the degree of saturation, which in turn is determinated by the transition probability and the laser spectral power density. The variable x is proportional to the laser frequency detuning from the line centre. For completely crossed polarizers ($\theta = 0$) the signal is a Lorentzian while the admixture of a dispersion-shaped curve increases with θ. The amplitude of the dispersion curve increases but so does the background θ^2. Thus, there exists a small angle θ, for which the signal-to-noise ratio has an optimum value. The dispersion-shaped signal can be very useful for locking a laser onto an atomic transition. A change in signal sign is obtained at the line centre without a need for frequency modulation of the laser.

One interesting application of Doppler-free laser spectroscopy is the precision determination of the Rydberg constant, which has been performed by *Hänsch, Schawlow* and co-workers [9.163, 164], and others [9.165-167]. The Rydberg constant can be evaluated if an accurate wavelength determination is performed for a suitable line in hydrogen [9.168, 169]. A large number of evaluations have been performed by conventional spectroscopy on the red Balmer line H_α at 6563 Å. The very large Doppler broadening (~6000 MHz) of the very light element hydrogen presents a significant problem because of the wide overlap of the fine structure components. In Fig. 9.45 the fine structure of the H_α line and a theoretically calculated Doppler-broadened line are shown. In the lower part of the figure the Doppler-free structure obtained by saturation spectroscopy is shown. The Lamb-shift can be directly observed. The individual positions of the components can now be well established and the uncertain deconvolution procedure that previously had to be used can now be eliminated. In this way, the Rydberg constant can be determined much more accurately than before. The best value for the Rydberg constant obtained so far is

$$R_y = 109737.31573(3) \text{ cm}^{-1}.$$

In Figs. 9.45 a so-called cross-over resonance is shown. This is an inherent phenonomenon in saturation spectroscopy and occurs when two lower or upper state sublevels have transitions to a common level in the other state. The two oppositely propagating laser beams can then interact at frequencies half-way between the normal resonances. Then atoms moving with a certain velocity along the laser beams are utilized [9.170, 171].

An experimental set-up similar to the one used in polarization spectroscopy is employed in certain parity-violation experiments. A small optical rotation is induced by interference between neutral weak and electromagnetic interactions in atoms [9.172-174].

A further method of monitoring Doppler-free signals using transmitted beams is also possible. In this technique (saturated interference spectroscopy) [9.175, 176], the change in refractive index for the atoms at the "hole" position is used to influence the light interference condition in a two-beam interferometer. If the set-up is initially adjusted for destructive interference an increase in light intensity will be observed at the line centre.

It is also possible to observe Doppler-free saturation signals in cell experiments without detecting the intensities of the transmitted laser beams.

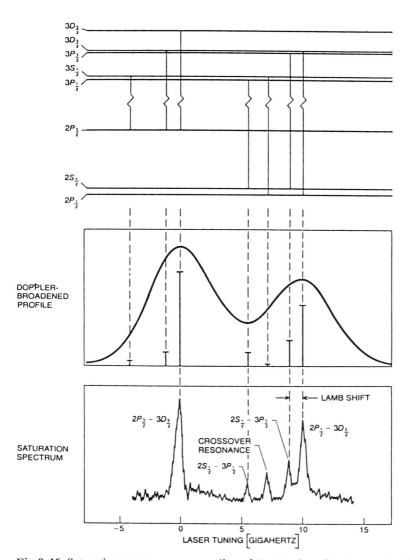

Fig.9.45. Saturation spectroscopy recording of the H_α line of hydrogen [9.163]

Fluorescence, opto-galvanic and opto-acoustic detection can all be used. However, since the signal information is no longer carried by a well-defined probe beam, which is coupled to the pump beam only at the line centre, a double modulation technique must generally be employed in order to isolate the Doppler-free signal from the Doppler-broadened one. This is because a modulation of the pump beam at frequency f_1 clearly causes a huge Doppler-broadened signal at frequency f_1 in these alternative detection schemes. If the oppositely directed beam (which in this case has an intensity similar to that of the pump beam) is modulated at a different frequency f_2 a modulation of the signal will also occur at this frequency.

Clearly, these modulations are normally quite independent of each other since they are coupled to different velocity groups. However, at the line centre the two modulations interact due to the nonlinear response of the medium resulting from the saturation. Thus, for the same reason as frequency sum and difference generation occurs in nonlinear crystals (Sect. 8.6), signals at the frequencies (f_1+f_2) and (f_1-f_2) will arise at the line centre in these *intermodulated* experiments. In Fig.9.46 arrangements for intermodulated fluorescence [9.177] and intermodulated opto-galvanic spectroscopy [9.178] are shown. Doppler-free signals can also be obtained using RF opto-galvanic spectroscopy [9.179] where amplitude variations in the RF oscillator driving a discharge in the sample are utilized. Doppler-free acousto-optic intermodulated spectroscopy is described in [9.180].

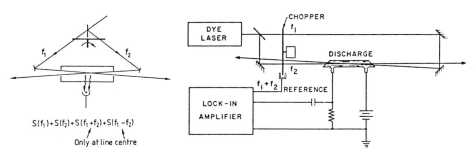

Fig.9.46. Experimental arrangements for intermodulated fluorescence spectroscopy (left) and intermodulated opto-galvanic spectroscopy (right) [9.178]

The need for intermodulation in the experiments discussed above arises from the fact that the beams are amplitude modulated. By instead utilizing beam overlap modulation, e.g. by reflecting the laser beam back through the cell from a vibrating mirror, both beams are present at all times and no modulation is normally obtained. However, when one beam is swept to overlap the other, the nonlinear interaction at the line centre changes the total signal at the vibration frequency f or 2f, depending on the sweep geometry [9.181]. By combining the saturation spectroscopy techniques discussed above with rotating polarizers, instead of choppers in the laser beams, problems due to (f_1+f_2) and (f_1-f_2) frequency generation caused by nonlinearities in the detectors, rather than true signals, can be eliminated. The technique is then called POLINEX (POLarization INtermodulated EXcitation) spectroscopy [9.182]. The polarization methods also have an additional advantage of being largely insensitive to velocity-changing collisions, which can cause broad signal pedestals under the narrow signals even if lock-in detection at the frequencies (f_1+f_2) or (f_1-f_2) is performed. (The collisions tend to spread the "hole" over a larger frequency range). Since the orientation of the atoms induced by the polarized light is normally changed by collisions, the POLINEX signal, which is sensitive to polarization rather than to light intensity, will essentially only exhibit the narrow signal contribution. The effect is illustrated in Fig.9.47 in which the reduction in collisional effects is shown for a Ne transition.

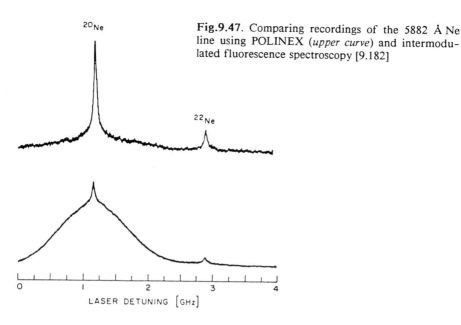

Fig.9.47. Comparing recordings of the 5882 Å Ne line using POLINEX (*upper curve*) and intermodulated fluorescence spectroscopy [9.182]

As a final method for sensitive detection of saturation spectroscopy signals we will mention *frequency modulation spectroscopy* [9.183]. Here a frequency modulation ν_1 is electro-optically imposed on a narrow-band laser beam of frequency ν_0 generating side-bands $\nu_0 \pm \nu_1$. The phase factors for the beat notes at ν_1 between the two sidebands and the carrier frequency ν_0 differ by 180° and thus normally no beat amplitude can be detected in the laser beam transmitted through the sample. However, if the sidebands are differently absorbed by the atoms the amplitude of the beat at ν_1 is non-zero.

Frequency-modulation spectroscopy can also be used for sensitive absorption measurements using pulsed lasers [9.184]. The sidebands are then generated by microwaves (~10GHz) to separate them from the rather broad-band pulsed radiation. Small differential absorptions from, e.g., pressure-broadened molecular transitions, can be detected. Other extremely sensitive laser absorption spectroscopy techniques have been described in [9.185, 186].

9.5.3 Doppler-Free Two-Photon Absorption

We have already discussed the two-photon absorption process [9.187] (Sect. 9.1.3c). Normally, the signals are Doppler broadened, and high resolution is not obtained even if narrow-band lasers are used. However, a very important observation was made as early as 1970 by *Chebotayev* and co-workers [9.188]. The Doppler broadening can be eliminated if the two photons are extracted in such a way that one is supplied from one of two counter-propagating laser beams and one from the other beam. For a certain atom with a velocity component v along one direction of propagation, the energy in-

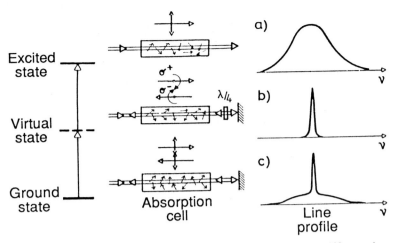

Fig.9.48. Two-photon absorption spectroscopy signals for different laser beam arrangements

terval ΔE can be bridged by two Doppler-shifted contributions

$$\Delta E = h\nu(1 + v/c) + h\nu(1 - v/c) = 2h\nu . \tag{9.38}$$

Thus, the first-order Doppler shift in one absorption is exactly cancelled by the corresponding shift in the second absorption. Since the velocity is eliminated, *all* irradiated atoms can contribute to the signal and not just a certain velocity group, as in saturation techniques. In S-S transitions we have the selection rules $\Delta F = 0$, $\Delta M = 0$ for the two-photon transition between the two states of equal parity. By making both laser beams either right-hand circularly polarized or left-hand circularly polarized it is possible to ensure that the two photons come from different laser beams ($\Delta M = +1$ and $\Delta M = -1$ make effectively $\Delta M = 0$). In Fig.9.48 this case is shown (b) as well as the case of a Doppler-broadened signal, which is obtained if only one laser beam (unpolarized or linearly polarized) is used (a). If two such oppositely propagating beams are used, a Doppler-free as well as a Doppler-broadened contribution will be obtained (c). Early experimental observations of Doppler-free signals were made by several groups [9.189-191]. Doppler-free two-photon absorption has been used in a large number of atomic level investigations. For highly excited states, thermionic diode detection is frequently employed. States up to n = 500 have been reached in step-wise excitations [9.192]. As an example, a recording is shown in Fig.9.49. Systematic studies of alkali atoms [9.193-195] and alkaline earth atoms have been performed. For the alkaline earth atoms the data obtained in two-photon and step-wise excitation experiments on sequences of Rydberg states have been analysed using multichannel quantum defect theory [9.196-200].

We will also discuss another series of investigations on hydrogen performed by *Hänsch* and co-workers, which is of fundamental importance

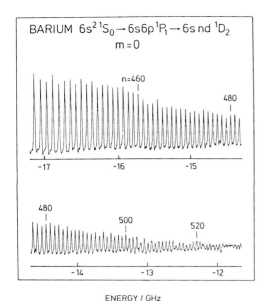

Fig.9.49. Doppler-free spectrum for high-lying Rydberg states in Ba [9.192]

[9.201-203]. The experiment deals with the determination of the Lamb shift in the ground state (n=1) (Fig.9.50). In contrast to the situation for the n=2 state, in which the position of the 2s $^2S_{1/2}$ state can easily be related to the 2p $^2P_{1/2}$ state position, there is no neighbouring state for the ground state. However, its energy position is still affected by QED effects, and in order to measure these, reference to the n=2 ↔ n=4 transition (H_β) can be made. The simple Balmer-Rydberg formula states that the n=2 ↔ n=4 interval should be 1/4 of the n=1 ↔ n=2 interval. Because of fine structure, relativistic and QED effects (the Lamb shift is the ground state QED effect which is to be determined) the simple energy interval relation does not hold exactly. Starting from a primary narrow-band laser at 4860 Å it is possible to simultaneously measure the H_β-line using polarization spectroscopy and the n=1 ↔ n=2 transition with two-photon spectroscopy, and compare the signal positions for the Lamb shift determination. Sometimes well-known lines from Te$_2$ molecules are used as reference lines. These molecules have many transitions in the blue spectral region which can be used as wavelength standards [9.204] in the same way as I$_2$ lines are used at longer wavelengths [9.205]. Light from the 4860 Å laser is amplified with pulsed amplifier stages pumped by a pulsed excimer laser and the pulses are frequency doubled to 2430 Å in a nonlinear crystal. The two-photon transitions between n=1 and n=2 are monitored using the Lyman line at 1216 Å, which is obtained after collisional transfer in the discharge from the 2s $^2S_{1/2}$ to the 2p $^2P_{1/2}$ state. The two-photon signals have a larger linewidth than the polarization spectroscopy signals because of the Fourier broadening of the pulsed amplification (pulse length 10ns → $\Delta\nu \simeq 1/10^{-8}$ Hz = 100MHz). The measurement yields a Lamb shift of the 1s state of about 8160 MHz, which is in close agreement with the theoretical predictions. Since the natural life-

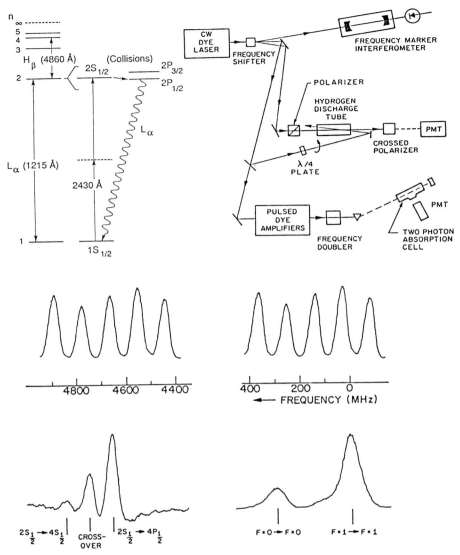

Fig.9.50. Experimental arrangement and recorded signals for simultaneous saturation spectroscopy (2s-4p) and two-photon absorption spectroscopy (1s - 2s) in atomic hydrogen [9.201]

time of the 2s state is about 10^{-1} s extremely narrow lines should, in principle, be attainable. Narrow-band 243 nm radiation has been generated using cw frequency mixing techniques yielding a resolution of $1:10^{10}$ [9.203, 206, 207].

Future possibilities for accurate H spectroscopy have been discussed in [9.208, 209]. The optical analogy to the Ramsey-fringe technique (Sect. 7.1.2) has been demonstrated for the case of two-photon absorption. In Fig.9.51 the inevitable Fourier broadening of a short pulse is illustrated.

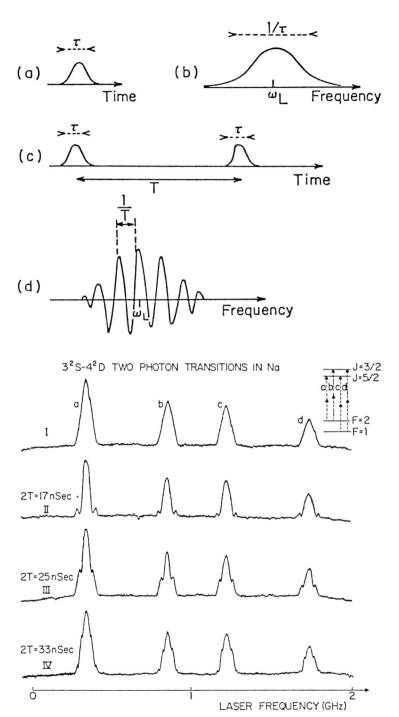

Fig.9.51. Illustration of the optical Ramsey fringe technique in the time domain [9.210]

However, by applying a further pulse at a delay T, a sharp interference pattern in the two-photon signal can be obtained. It is important that the signal is Doppler-free so that the phase memory from the first pulse is not lost through the velocity spread. Ordinary signals and Ramsey fringes for the two-photon transitions 3s $^2S_{1/2}$ → 4d $^2D_{3/2,5/2}$ in ^{23}Na are shown in the figure [9.210]. The signals display the hyperfine structure of the ground state (1772 MHz, Fig.9.42) and the inverted fine structure of the excited state (1028.8 MHz, see also Fig.9.17). The sharpness in the interference pattern can be increased using multiple pulses [9.211]. An analogy with the interference from a double slit and from a grating illuminates this effect.

The optical Ramsey effect has also been demonstrated for the case of saturation spectroscopy with cw laser beams [9.212]. An atomic beam is crossed by counter-propagating laser beams at three different locations along the atomic beam. In this way the transit-time broadening can be strongly reduced, which was the original reason for introducing the technique in the RF regime. The technique has primary importance in connection with the establishment of ultra-stable laser secondary wavelength standards. A special variety of the Ramsey fringe technique using a "fountain" of very slow atoms has been proposed and also been demonstrated [9.213].

9.5.4 Spectroscopy of Trapped Ions and Atoms

In principle, lasers can be stabilized to a linewidth in the Hz region. If transitions between the ground state and long-lived metastable states with a lifetime of the order of 1 ms are utilized, the natural radiation width is also very small (~100 Hz). On the other hand, transit-time broadening can be substantial if not reduced by the techniques discussed above. The transit time of a thermal atom through a 3 mm laser beam is about 10^{-5} s, which corresponds to a broadening of about 10 kHz. As we have seen, this effect can be handled by Ramsey-fringe techniques. The remaining fundamental broadening effect is due to the second-order Doppler effect which is present as soon as the atoms move. The effect amounts to about 100 Hz in the visible region. To achieve the ultimate resolution, the atoms must be brought to rest. Atoms can be slowed down using laser light. Basically, the transfer of momentum of a photon to an atom is utilized (Sect.9.5.1). The atomic motion can be decreased by directing the laser beam against the atomic beam. For each head-on absorption process, the velocity of a sodium atom is reduced by 3 cm/s. By absorbing (and emitting) photons some 20,000 times they can be brought to rest. Clearly, the fact that the atoms shift out of resonance with the counter-propagating single-mode laser beam as they slow down is a problem. By using the combined effect of optical pumping into the highest m quantum state (Sect.7.1.3) and Zeeman tuning in a magnetic field of changing strength along the atomic beam path, the atoms can be kept in resonance [9.214]. Alternatively, a frequency chirp can be applied to the tunable laser to bring a package of atoms to rest [9.215]. Actually, the atoms can even be reversed by this technique sending them back towards the oven. The techniques described above only affect the longitudinal velocity of the beam. In order to bring atoms or ions to com-

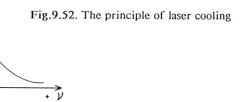

Fig.9.52. The principle of laser cooling

plete rest, all velocity components must be removed. This can be achieved if the atoms or ions are first confined in a trap. An electrically charged particle can be trapped in a potential well which is generated by oscillating RF fields. By irradiating the ion in the trap with laser radiation, tuned to the low-frequency side of the Doppler curve of the transition, the velocity of the ion will be reduced [9.216] as illustrated in Fig.9.52.

Because of the detuning of the laser, a photon can only be absorbed when the ion moves against the laser beam. The absorption of the momentum $p = h\nu/c$ of the photon will slow down the ion. The decay photon can be emitted in any direction. After some time, the ion has been brought to a very low velocity. A photograph of the fluorescence of a single Ba^+ ion is shown in Fig.9.53a. In Fig.9.53b the "crystallization" of a few Mg ions in a trap is illustrated. The atoms are cooled to such a low velocity that the electrostatic forces between the ions dominate and an ordered arrangement is obtained [9.218].

Atoms can also be confined in traps using magnetic forces, provided that the atoms are paramagnetic (Sect.7.1.6b). Clearly, the forces are much weaker than in the case of charged particles, resulting in very shallow traps [9.219]. Thus, the atoms have to be moving very slowly in order to be trapped. When trapped, they can be cooled down to very low effective temperatures using the techniques described above. It is also possible to cool a cloud of atoms without using a trap. The three spatial velocity components are "cooled" individually by using three mutually perpendicular standing-wave laser beams. In this way, sodium atoms have been cooled to an effective temperature of tens of μK [9.220]. The field of laser cooling and trapping is expanding very rapidly. Recent progress has been reviewed in [9.221-227].

An extremely high resolution ($1:10^{15}$) is obtainable if transitions to very long-lived (metastable) states are studied in cooled atoms by ultra-narrow-band laser light. A new problem then has to be dealt with, connected to the low fluorescence yield of such highly forbidden transitions. A technique in which a strong transition is used for the detection of the forbidden transition has been proposed and demonstrated [9.228-231]. The method applicable to a single atom or ion is illustrated in Fig.9.54 for the case of Hg ions. "Strong" fluorescence light induced by a laser tuned to the strong transition is normally detected. If another laser is tuned through the extremely narrow resonance of the "forbidden" transition a resonant transfer of the atom to the metastable state ("shelving") is detected through the

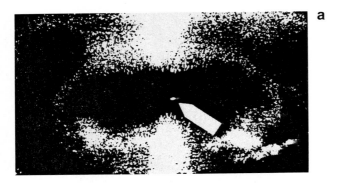
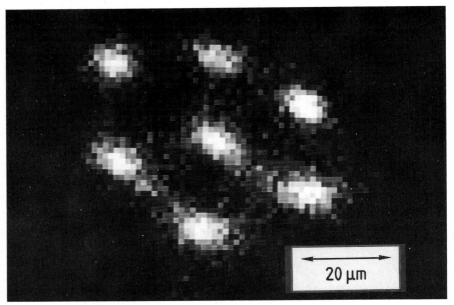

Fig.9.53. (a) Photograph of the fluorescence light from a single barium ion kept in an RF trap [9.217]. (b) Recording of the fluorescence of seven stored magnesium ions exhibiting "crystallization" behaviour [9.218, first ref.]

abrupt disappearance of the fluorescence light in the strong transition ("quantum jump"). An experimental recording of this phenomenon is included in Fig.9.54, as is a ultra-high resolution scan of the "forbidden" transition.

We have above discussed certain mechanical actions on atoms induced by laser light. Using resonance-radiation pressure atomic beams can also be focused and manipulated in interesting ways [9.233, 234]. Several other interesting mechanical effects exist and have been explored. Studied phenomena include light diffusive pulling [9.235], light-induced drift [9.236] and the optical piston [9.237].

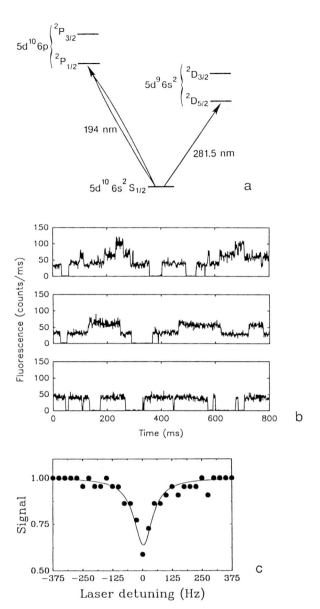

Fig.9.54. (a) Energy levels for Hg+, with the "strong" transition at 194 nm and the "weak" transition at 281.5 nm indicated (b). Fluorescence light recording at 194 nm illustrating quantum jumps in 3 ions (top), two ions (middle) and one ion (bottom) [9.230]. (c) An experimental scan of the "weak" transition using an electro-optically stabilized laser is shown, featuring a linewidths of 86 Hz. This corresponds to a spectroscopic resolution of $1:10^{15}$ [9.232]

10. Laser-Spectroscopic Applications

In the previous chapter we have seen how tunable lasers can be used in a multitude of ways to gain basic information on atomic and molecular systems. Thus, the laser has had a considerable impact on basic research, and its utility within the applied spectroscopic field is not smaller. We shall here discuss some applications of considerable interest. Previously, we have mainly chosen atomic spectroscopic examples rather than molecular ones, but in this chapter we shall mainly discuss applied molecular spectroscopy. First we will describe *diagnostics of combustion processes* and then discuss *atmospheric monitoring* by laser techniques. Different aspects of *laser-induced fluorescence in* liquids and solids will be considered with examples from the environmental, industrial and medical fields. We will also describe *laser-induced chemical processes and isotope separation with lasers*. Finally, *spectroscopic aspects of lasers in medicine* will be discussed. Applied aspects of laser spectroscopy have been covered in [10.1, 2].

10.1 Diagnostics of Combustion Processes

10.1.1 Background

Research in the field of combustion has been intensified recently because of the appreciation of the need for efficient combustion combined with low pollution. In order to obtain a deeper understanding of combustion processes it is necessary to perform the study on a molecular level. Laser spectroscopic techniques provide unique possibilities for non-intrusive measurements on the extremely aggressive media that burning or exploding gases constitute. Because of the unique properties of laser beams, both high spatial and temporal resolution can be achieved. Before we describe some measurement techniques we will give an elementary background to combustion processes [10.3].

A detailed understanding of combustion must start with simple processes such as hydrogen, methane or acetylene combustion in oxygen or air. Normal liquid hydrocarbons are considerably more complex and wood or coal combustion can hardly be attacked on a molecular level. Below we give some "effective" chemical reactions leading to a transformation of fuel and oxidant into carbon dioxide and water. The processes are strongly exothermic, which is, of course, a common feature for combustion processes (Table 10.1).

Table 10.1. Some effective chemical reactions in fuel combustion

Flame	Effective reactions	Temperature [K]	Energy release [J/g]
H_2/O_2	$2H_2 + O_2 \rightarrow 2H_2O$	3,100	24,000
CH_4/O_2	$CH_4 + 2O_2 \rightarrow CO_2 + 2H_2O$	3,000	10,000
C_2H_2/O_2	$2C_2H_2 + 5O_2 \rightarrow 4CO_2 + 2H_2O$	3,300	12,000

Combustion occurs with a large number of intermediate steps and even simple processes, such as the ones listed in Table 10.1, occur through dozens of coupled elementary reactions. With computer simulations it is possible to describe the interaction between the reactions, and concentration profiles can be calculated. In order to perform the computer calculations it is necessary to know the rate constants for the individual elementary reactions. Comparisons between theory and experiments are best made for a flat, premixed flame, which in its central part can be considered to have only one-dimensional (vertical) variation, allowing computer calculations to be performed comparatively easily. The most important reactions are included in the computer description. In Fig. 10.1 experimental and theoretically calculated concentration curves are given for the case of low-pressure ethane/oxygen combustion. As examples of important elementary processes we give the reactions

$$CO + OH \rightarrow CO_2 + H$$
$$H + O_2 \rightarrow O + OH$$

Reactive molecular fragments or radicals, such as OH, H and O are very important in combustion. The combustion zone of a stoichiometric CH_4/O_2 flame contains about 10% OH, and 5% each of H and O. In the second of the two reactions given above the number of radicals is doubled. A fast increase in radical formation frequently leads to explosive combustion. Because of the high reactivity of radicals they cannot be measured by probe (extraction tube) techniques, since wall reactions immediately eliminate them. Thus, laser techniques are particularly valuable for radical monitoring. Pollution formation in flames should also be considered. Nitrogen and sulphur oxides, incompletely burnt hydrocarbons and soot particles form important pollutants. It is of the utmost importance to understand which elementary reactions form and eliminate pollutants. The formation of nitric oxide is reasonably well understood. At temperatures above 2000 K the nitrogen in the air is attacked:

$$O + N_2 \rightarrow NO + N$$
$$N + O_2 \rightarrow NO + O$$

The two reactions constitute the so-called Zeldovich mechanism. NO is then oxidized to the toxic NO_2 by the oxygen in the air. NO can be elim-

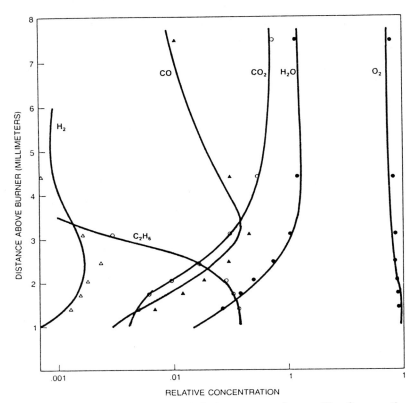

Fig.10.1. Experimental and theoretical concentration profiles for an ethane/oxygen flame [10.3, first Ref.] (Copyright 1982 by Scientific American, Inc.. All rights reserved)

ated from the post flame gases by the addition of NH_3. The following reactions occur [10.4]

$$NH_3 + OH \rightarrow NH_2 + H_2O$$
$$NO + NH_2 \rightarrow N_2 + H_2O$$

In the case of sulphur oxide SO it is immediately further oxidized to SO_2 in the flame. It seems that a reduction of SO_2 can only be obtained by using low-sulphur fuels. SO_2 and NO_2 are further oxidized to H_2SO_4 and HNO_3 in the atmosphere with subsequent acid rain formation.

Soot formation is the subject of many studies. Soot formation is enhanced at high fuel/air mixing ratios (rich flames) when using hydrocarbons with comparatively little hydrogen, and for bad mixing conditions. Since available fuels will become successively poorer in hydrogen, soot formation will become an increasing problem. The chemistry of soot formation is not well understood. Many processes, including polymerization of simple hydrocarbons to heaver ones and reactions with polyaromatic hydrocarbons, may be important. Basic flame combustion has been described in [10.3,5,6] and emission and absorption spectroscopy of flames discussed in [10.7].

Laser techniques have a great potential for studies of microscopic as well as macroscopic combustion in flames and engines. Combustion diagnostics with lasers was discussed in several reviews [10.8-13]. We will here give examples of measurements of concentrations and temperature (flame kinetics) using fluorescence, Raman and coherent Raman techniques. In practical combustion systems turbulence is extremely important and we will also briefly discuss laser techniques for flow and turbulence measurements.

10.1.2 Laser-Induced Fluorescence and Related Techniques

In laser-induced fluorescence (LIF) experiments a laser is normally tuned to an allowed dipole transition from a lower to an upper state of the species under consideration, and the fluorescence light that is released upon the subsequent decay is observed. We will start this section by considering the corresponding spontaneous emission process. At the high temperatures in a flame, upper levels become thermally populated and a natural emission giving the flame its colour occurs. In Fig.10.2 part of the emission spectrum from a C_3H_8/air Bunsen burner flame is shown, featuring strong bands due to the radical C_2. This emission was described by Swan as early as 1857 in one of the earliest molecular spectroscopy experiments. C_2 is responsible for the blue-green light from the lower parts of hydrocarbon flames. Schematic energy-level diagrams for C_2 and OH with the wavelengths of the individual bandheads are depicted in Fig.10.3. Hydrocarbon flames also exhibit strong bands due to the CH (~390 and 430 nm; especially in the flame front) and the OH radicals (~300 nm). In sooty flames the strong yellow light is due to incandescence of soot particles.

LIF yields a much more well-defined emission situation than the one pertaining to thermal emission. Different aspects of the use of LIF for combustion studies are treated in [10.16,17]. For quantitative LIF measurements it is necessary to consider and control the quenching of the fluorescence due to collisional, radiationless transitions. The quenching can be represented by a term Q to be accounted for on equal footing with A (describing spontaneous emission). The observed light intensity will be strongly reduced due to the strong but rather unpredictable degree of quenching. Eq.(9.1), describing saturation now must be written

$$\frac{N_2}{N_1 + N_2} = \frac{1/2}{1 + \frac{A+Q}{2B\rho(\nu)}} . \tag{10.1}$$

Since Q is frequently 10^3 times larger than A, a very high spectral energy density $\rho(\nu)$ is needed to obtain saturation with a laser beam. However, this can often be achieved and then the maximum fluorescence intensity is obtained despite the quenching, which no longer influences the measurement. Alternatively, the degree of quenching can be measured directly using time-resolved fluorescence spectroscopy. This is possible since the lifetime τ

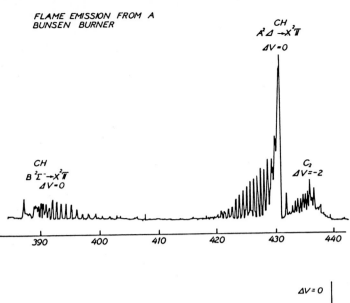
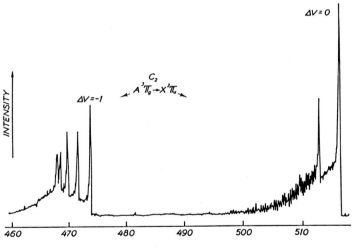
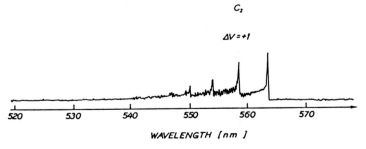

Fig.10.2. Emission from a propane/air Bunsen burner flame [10.14]

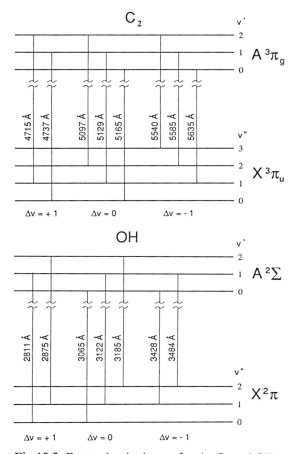

Fig.10.3. Energy level schemes for the C_2 and OH molecules [10.15]

is shortened from its natural value $1/A$ to $1/(A+Q)$. With a prior knowledge of the low-pressure "true" lifetime the observed value immediately yields the quenching at the particular pressure. Clearly, the quenched lifetimes are normally very short and have to be measured with picosecond laser techniques (Sect.9.4) [10.18].

An experimental set-up for studying LIF in flames is shown in Fig. 10.4. The output of a Nd:YAG pumped dye laser can be frequency doubled and, if needed, the doubled output can be mixed with residual 1.06 μm radiation to achieve still shorter wavelenghts. The beam is directed through the flame and the fluorescence can be spectrally analysed with the spectrometer shown in the upper part of the figure. The lower part of the figure shows how a diode-array detector can be used to obtain a one-dimensional image of the distribution of a radical across a flame [10.20]. The streak of LIF is imaged onto the detector, which is gated by the laser firing. Figure 10.5 shows distributions of OH fluorescence at various heights in a CH_4/O_2 flame. A single 10 ns pulse is used for each recording.

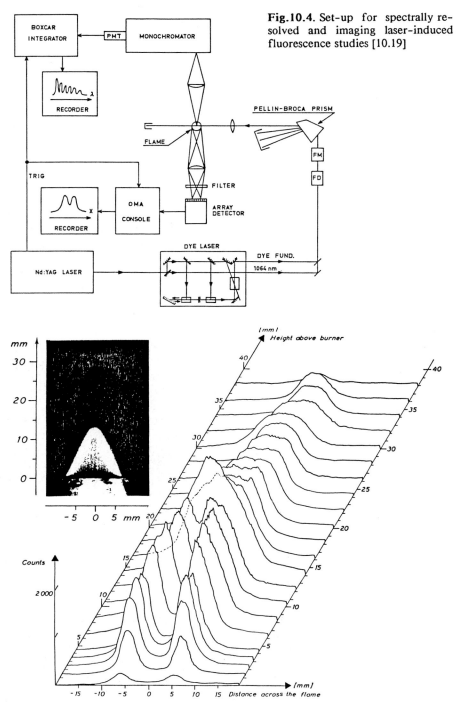

Fig.10.4. Set-up for spectrally resolved and imaging laser-induced fluorescence studies [10.19]

Fig.10.5. Spatial distributions of the OH radicals in a CH_4/O_2 flame [10.20]. The excitation wavelength was 281 nm and the detection wavelength 308 nm

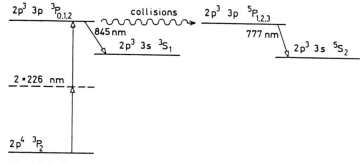

Fig.10.6. Two-photon excitation scheme for oxygen atoms [10.22]

Using matrix detectors or Vidicon tubes, LIF imaging techniques can be extended to two-dimensional imaging [10.21]. Many flame species, including OH, C_2, CN, CH and NO, have been studied using single-photon LIF. Several important flame species have their single-photon excitation wavelenghts in the VUV region where the flame gases absorb and tunable lasers are not readily available. It is then possible to use two-photon or even three-photon excitation (Sect.9.1.3). In Figs 10.6 and 7 an excitation scheme and a 2-photon LIF spectrum for oxygen atoms in a C_2H_2/O_2 welding torch are shown. The collisional transfer from the triplet to the quintet system in O should be especially noted.

Hydrogen atoms can be detected observing H_α or H_β emission from the n = 3 or 4 levels following two- [10.23] or three-photon excitation [10.24] or step-wise excitations [10.25]. CO molecules are also best detected using two-photon excitation [10.26]. The excitation of flame species to an upper level can also be detected by other means. Optogalvanic spectroscopy (Sect.9.2.6) [10.27] and photoacoustic spectroscopy (Sect.9.2.8) [10.29] employing pulsed lasers have been used. The former method is not non-intrusive in nature because of the need for electrodes. The latter technique util-

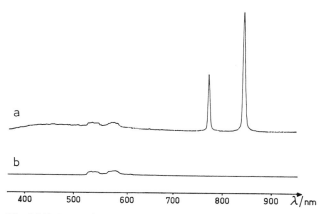

Fig.10.7. Laser-induced fluorescence spectrum from oxygen atoms in a C_2H_2 welding torch [10.22]

izes the local pressure increase following excitation, and a microphone close to the flame is used for detection. Spatial resolution is limited since the signal is collected along the laser beam. Ordinary absorption measurements clearly constitute the best examples of such line-of-sight measurements. Diode lasers, as well as dye lasers, have been used successfully in flame absorption measurements [10.30]. With tomographic techniques, similar to those used in medical X-ray imaging [10.31], spatially resolved information can be obtained from integrated absorption measurements in different directions through the flame [10.32]. By using tomography three-dimensional information can also be obtained utilizing thermal flame emission [10.33] and interferometry [10.34].

Laser beam deflection can also be used to detect optical resonance. In the region of optical excitation the index of refraction of the gas changes and a probing laser beam (frequently a He-Ne laser beam) crossing the excited region will be deflected [10.35].

10.1.3 Raman Spectroscopy

Because of its insensitivity to quenching (the lifetime of the virtual state is $\sim 10^{-14}$ s), Raman spectroscopy is of considerable interest for quantitative measurements on combustion processes. Further, important flame species such as O_2, N_2 and H_2 that do not exhibit IR transitions (Sect. 4.2.2) can be readily studied with the Raman technique. However, because of the inherent weakness of the Raman scattering process (Sect. 4.3) only non-luminous (non-sooting) flames can be studied.

Extractive Raman measurements on stable flame species can readily be performed. Here gases are transferred from the flame through a thin tube to the scattering cell of a laser Raman gas analysing system. In Fig. 10.8 Stokes Raman spectra, obtained using an Ar^+ laser operating on the 488 nm line are shown for the lower and upper part of a C_3H_8/air Bunsen-burner flame are shown. The conversion of fuel and O_2 into CO_2 and H_2O, constituting the over-all combustion process, is clearly demonstrated. The H_2O signal is prevented from increasing by a water vapour condenser in the gas feed-line. Soot particles, which give rise to a broadband LIF background, are also filtered away. The lower trace also displays signals due to CO and H_2, gases that will mostly burn up higher up in the flame. The weak signals in the shoulder of the strong Rayleigh line in the lower trace are due to pure rotational Raman transitions in the H_2 molecule, which, because of its small mass, has an exceptionally large rotation constant B (Sect. 3.2). The main hydrocarbon signal at about 570 nm has many components correponding to slightly different C-H stretch vibrational frequencies. Overtone and combination bands are also observed at smaller Raman shifts.

Using pulsed lasers and gated detection electronics, Raman measurements can also be performed for major species in flames that do not contain too many particles. Temperature measurements can then also be made using the Stokes/anti-Stokes signal asymmetry or the occurrence of slightly displaced Stokes hot-bands as discussed in Sect. 4.4.1. Flame Raman spectroscopy is discussed in further detail in [10.36, 37].

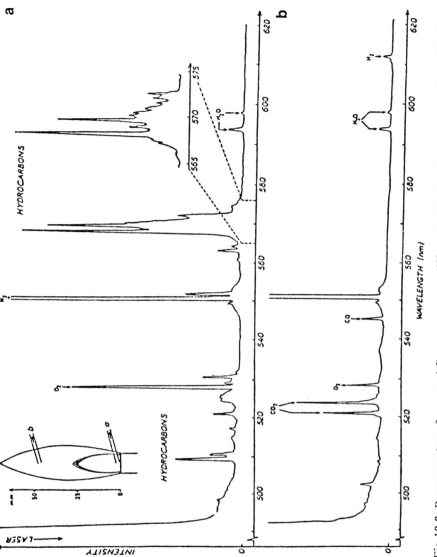

Fig.10.8. Raman spectra of extracted flame gases at two different heights in a Bunsen burner flame [10.14]

10.1.4 Coherent Anti-Stokes Raman Scattering

The CARS process has been described previously (Sect. 8.6). CARS spectroscopy is of particular interest for combustion diagnostics because of the strong signal available as a new laser beam emerging from the irradiated gas sample. Thus CARS is largely insensitive to the strong background light that characterizes practical combustion systems such as industrial flames and internal combustion engines. We recall that the spectroscopic information is contained in the third-order susceptibility term $\chi^{(3)}$. This term is given by the sum of a complex resonant term, (proportional to the concentration of the studied molecule) and a non-resonant background term. Thus, when the expression $|\chi^{(3)}|^2$ governing the signal strength is formed (Eq. 8.11) interference between the resonant signal and the background occurs, resulting in asymmetric signals, very much like the case of polarization spectroscopy, see (9.37). For a molecule with vibrational and rotational levels, whose populations are temperature dependent, there are many close-lying resonances and the signal shape has to be calculated with a computer program. In Fig. 10.9 theoretical curves for the N_2 molecule are shown for different temperatures. The occurrence of the first hot-band and the gradual widening of the vibrational peaks due to increasing rotational level population can clearly be seen for increasing temperatures. Figure 10.10 shows experimental spectra for a CH_4/air flame and for room temperature air recorded with a set-up of the type shown in Fig. 10.11.

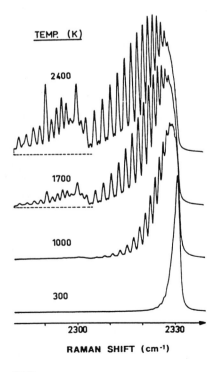

Fig. 10.9. Theoretically calculated CARS curves for N_2 molecules at different temperatures [10.38]

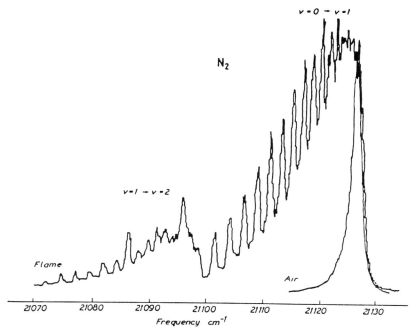

Fig.10.10. Experimental CARS recordings for room-temperature N_2 molecules and for N_2 molecules in a flame [10.38]

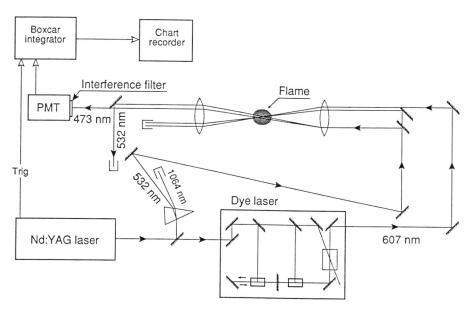

Fig.10.11. Experimental set-up for CARS spectroscopy using the BOXCARS phase matching scheme [10.15]

The BOXCARS phase-matching scheme (Fig.8.41) is implemented for the pump beams at ω_p and the Stokes beam at ω_s in the arrangement shown in Fig.10.11. After frequency doubling, the main part of the Nd:YAG laser output is used to pump the dye laser that generates the tunable ω_s beam, while the rest is used for the pump beam. The anti-Stokes beam at $\omega_{AS} = 2\omega_p - \omega_s$, which emerges when $\omega = \omega_p - \omega_s$ matches a rotational-vibrational transition for the molecule, is detected after proper spectral isolation from the strong pump beam. When the dye laser is slowly tuned through the signal region with the laser continuously firing, curves such as the ones shown in Fig.10.10 are recorded using a gated boxcar integrator. The resolution in the spectrum is given by the (small) laser linewidths and not by the resolution of a spectrometer, as in normal spontaneous Raman spectroscopy.

Because of the highly nonlinear nature of the CARS process [the signal is proportional to the square of the molecular number density and to $P_p^2 P_s$, see (8.11)], signal averaging does not yield a true value for rapidly varying, turbulent media. It is thus desirable to be able to perform the measurement using a single laser pulse. This is possible using an amplified broad-band dye laser (Sect.9.2.3) for the Stokes beam, which, in conjunction with a narrow-band pump-laser beam, will cover all the difference frequencies $\omega_p - \omega_s$ of interest for a specific molecular spectrum. The anti-Stokes signal frequency components are then all generated simultaneously through the action of the third-order susceptibility $\chi^{(3)}$. A gated and intensified linear diode array is used to capture the single-pulse spectrum. The technique can be used for measurements in specially adapted internal combustion engines. The laser firing can be strobed on a particular crank angle and the temperature at various time intervals can then be determined [10.39,40]. An experimental spectrum from N_2 molecules in an internal combustion engine is shown in Fig.10.12. CARS techniques have also been applied to full-scale coal furnaces [10.41] and other practical combustors [10.42,43]. CARS is especially useful for remote thermometry but species concentrations can also be determined, especially for major species. CARS techniques have been

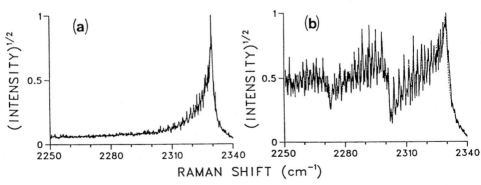

Fig.10.12. N_2 CARS spectrum from a firing engine [10.40] (a: 10 degrees after ignition, 7.1 atm. b: 40 degrees after ignition, 15.6 atm)

discussed in more detail in [10.44, 37, 38]. For minor species, LIF is frequently the most suitable technique. Since the LIF signals also depend on the temperature-dependent distribution of population on levels, the temperature can also be determined in such experiments if excitations from two metastable levels with different energies are used (two-line fluorescence method [10.45]).

Recently, degenerate four-wave mixing (Sect. 8.6) has also been applied to combustion studies [10.46]. In this technique the sensitivity of LIF is combined with the advantages of a coherent signal beam characteristic to CARS.

The techniques discussed here in connection with combustion diagnostics can clearly also be used for the monitoring of other reactive media. The techniques have been found to be valuable in the characterization of chemical vapour deposition (CVD) processes for semiconductor fabrication [10.47]. The examples mentioned here illustrate the power of laser spectroscopic techniques in studying chemical processes. Numerous other examples of chemical applications of laser spectroscopy can be found. The field was covered in [10.48, 2].

10.1.5 Velocity Measurements

Laser Doppler Velocimetry (LDV) is an important non-spectroscopic laser technique for intrusion-free measurements of velocities in liquid or gaseous flows, including combustion flows. In this technique two laser beams (frequently from an Ar^+ laser) are crossed at a small angle in the medium to be studied, as shown in Fig. 10.13. A standing interference pattern with bright and dark fringes is then formed. If a small particle being carried by the flow passes the interference pattern it will produce periodic glimpses of light that can be detected by a photomultiplier tube. The frequency f_D of the periodic signal (the Doppler burst) can be analysed by Fourier transformation, and the velocity v can then be determined since the fringe separation d is given by the laser wavelength λ and the beam crossing angle θ. We have

$$v = \frac{\lambda f_D}{2\sin(\theta/2)}. \qquad (10.2)$$

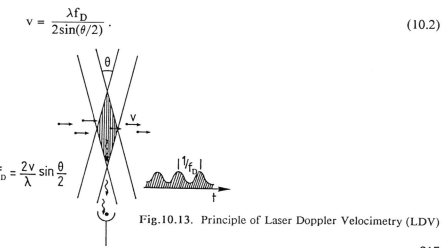

Fig. 10.13. Principle of Laser Doppler Velocimetry (LDV)

Clearly, if the particle is larger than the fringe separation, the contrast is reduced and thus LDV measurements also provide information on the particle sizes.

With one pair of crossing laser beams only one velocity component in the flow field can be determined. Furthermore, it is not possible to determine the sign of that velocity component since either flow direction produces the same Doppler burst. However, by frequency shifting one of the crossing beams with a Bragg cell (acousto-optic modulator), moving interference fringes are produced. It is then possible to decide from which direction a particle passed the interference field by noting if the burst frequency was shifted upwards or downwards from the frequency from a fixed object. A second velocity component can be measured by using two additional laser beams propagating in a plane perpendicular to the first laser beam plane and crossing at the same point. In order to be able to distinguish the Doppler burst from this interference pattern, different wavelengths are used for the laser beams. Thus, it is customary to use the 5145 Å line of an Ar^+ laser for one velocity component and the 4880 Å line for the other. The elastically scattered light is then detected through sharp interference filters in front of individual photomultiplier tubes. The third velocity component (in the direction of the bisectrix of the crossing laser beams) is harder to measure. However, by using three crossing beams in the same plane and extracting the information pair-wise from the central beam and one or other of the external beams, the third velocity component can be projected out, although at lower accuracy. When the scattered intensity is low, e.g. because of small particles, it is still possible to extract velocity information from the time correlation of the recorded individual photons. Special autocorrelation techniques have been developed for optimum information extraction [10.49]. It is frequently necessary to seed the flow with small particles. These particles do not necessarily truly follow the gas flow, which constitutes a complication with this technique.

We should conclude this description of LDV techniques with a comment on its common name. In our description of the technique we have not used "Doppler language" at all but rather expressed the observed phenomena in terms of a spatial interference pattern. However, we can alternatively consider the Doppler shift in the scattered light frequency caused by the motion of the particle. This shift is detected as a beat frequency against the light scattered from the other crossing beam. The beat frequency corresponds exactly to the frequency produced by a particle crossing the fringe pattern, so the two pictures are actually equivalent. LDV techniques have been discussed in detail in the monographs [10.50, 51].

A further class of velocity measurement techniques uses molecular "tagging". A "package" of molecules is first tagged with a pump pulse, and the movement of the molecules is then monitored with a second laser beam [10.52].

We will also consider a technique for velocity measurements which is based on Doppler shifting of absorption lines of atoms or molecules in the flow [10.53, 54]. The principle of the technique is illustrated in Fig. 10.14. A narrow-band cw laser is tuned to a frequency ν in the shoulder of a Dop-

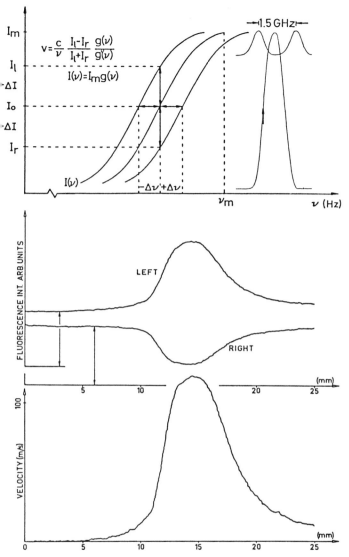

Fig.10.14. Spectroscopic flow imaging for an I_2 seeded gas jet [10.54]

pler-broadened absorption line, where fluorescence light of half the maximum intensity can be induced. If the gas molecules are displaced by a flow the whole Doppler-broadended profile will shift in frequency. Depending on the direction of this flow the fluorescence light intensity will increase or decrease. In order to measure the presence of a net flow, two alternate laser beams are sent through the medium in opposite directions. The fluorescence intensities are balanced out for a static gas, by adjusting the relative powers in the two beams. If the gas is moving, a modulation will be observed at the frequency of the beam direction shifting. By normalizing the signal it becomes independent of the number density of the molecules. We have

$$v = \frac{c}{\nu} \frac{I_{left} - I_{right}}{I_{left} + I_{right}} \frac{g(\nu)}{dg/d\nu} \,, \qquad (10.3)$$

where I_{left} and I_{right} are the two detected fluorescence intensities and $g(\nu)$ is the value of the line shape function at the chosen frequency ν.

A single-mode cw dye laser can be used for measurements of this kind. The experimental set-up is similar to the one used in Doppler-free intermodulated fluorescence measurements (Sect.9.5.2), but now the beams are presented to the atoms/molecules one at a time. For measurements of this kind, the flow can be seeded with sodium atoms or I_2 molecules. Imaging measurements using array detectors, as discussed above, can be performed. In [10.19-21,54] imaging measurements of species concentrations, temperatures and flows utilizing LIF have been described.

10.2 Laser Remote Sensing of the Atmosphere

Different aspects of remote sensing have been discussed in Sections 6.6 and 7.2. In this section we will describe how laser techniques can be used for monitoring of the atmosphere and its pollutants. General information on the atmosphere and its optical properties is given in [10.55-57]. Laser beams are particularly useful for monitoring over large distances because of the low divergence of the beam. Typically, a divergence of 0.5 mrad is obtained, corresponding to a spot diameter of 0.5 m at a distance of 1 km. Dry air contains 78.1% N_2 and 20.9% O_2. The rare gases Ar, Ne and He are present at levels of 9300, 18 and 5 ppm (parts per million), respectively. The CO_2 content is presently about 350 ppm, a value which increases annually by 0.5%. This may give rise to an increase in the average global temperature because of the change in the atmospheric radiation budget, the so-called *greenhouse* effect [10.58,59]. Ordinary air also contains widely varying amounts of water vapour. The gases N_2O, H_2, CH_4, NO_2, O_3, SO_2, CO, NH_3 etc. are also naturally present in the atmosphere in concentrations that vary from several ppm to fractions of ppb (parts per billion). If such gases are found in higher concentrations in the air because of human activities they are considered as pollutants [10.60-63]. Measurements in the troposphere, as well as in the stratosphere, are of great interest. In the troposphere, monitoring of industrial emissions, as well as ambient air quality, is needed for environmental protection purposes. Stratospheric measurements are important, e.g., for assessing possible long-term changes in the absorption characteristics that could result in an altered radiation environment at the earth's surface (see the discussion of stratospheric ozone destruction in Sect.6.4.5). Further, laser techniques also provide powerful means of remotely measuring meteorological conditions such as temperature, pressure, humidity, visibility and wind speed. We will discuss two active remote sensing techniques for the atmosphere - the *long-path absorption technique* and the *lidar technique*, but we will first consider a passive technique, in which lasers play an important part. The field of laser monitoring of the atmosphere is covered in several monographs and articles [10.64-70].

10.2.1 Optical Heterodyne Detection

Heterodyne detection is an important technique for low-noise signal recovery. Well-known in the radio-frequency region, it also has its counterpart in the optical regime. The principle of optical heterodyne detection is illustrated in Fig.10.15. The incoming radiation is mixed in the detector with the radiation from a local oscillator, which could be a diode laser or a CO_2 laser. Beats are generated in the detector at the difference frequency between the signal frequency ν_s and the local oscillator frequency ν_ℓ. A narrow-band electronic filter, which only transmits a fixed frequency ν_{if}, the *intermediate* frequency, selects the beat frequency $\nu_s - \nu_\ell = \nu_{if}$. ν_{if} is chosen in the radio-frequency region in which amplification can easily be performed. When the frequency of the local oscillator is swept the frequency of the recorded external signal is also swept and the spectrum will be recorded successively. The mixing of the signals can be described as

$$I = [A_s \sin(2\pi\nu_s t) + A_\ell \sin(2\pi\nu_\ell t)]^2 \qquad (10.4)$$

$$= A_s A_\ell \sin[2\pi(\nu_s - \nu_\ell)]t + \text{terms oscillating at high frequencies}.$$

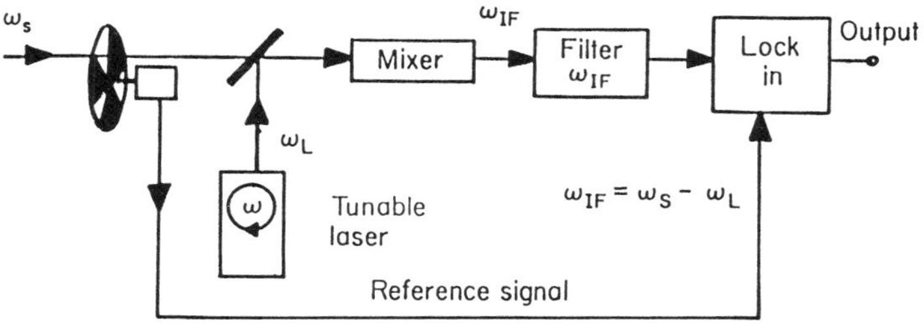

Fig.10.15. Optical heterodyne detection [10.70]

As we can see, the amplitude of the recorded signal is proportional to that of the incoming signal A_s as well as to that of the local oscillator A_ℓ. By increasing the amplitude of the local oscillator, noise-free amplification can be achieved. Heterodyne detection is frequently referred to as *coherent detection*. It is sometimes also convenient to use phase-sensitive (lock-in) detection at the frequency of a beam chopper to obtain a further increase in the signal-to-noise ratio and background rejection. An example of heterodyne detection is shown in Fig.10.16. The sun-disc is used as a radiation source and the absorption of the earth's atmosphere is monitored. The sun-disc is tracked with a heliostat. By tuning the local diode laser oscillator through the spectral region of an ozone infra-red absorption line, a signal

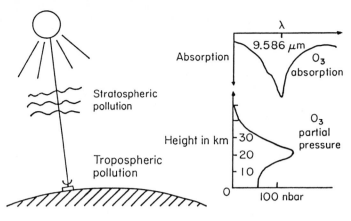

Fig.10.16. Vertical ozone profiling (Adapted from [10.71])

with a pressure-broadened component (Sect.6.1.1) from tropospheric ozone and a narrow component from the stratospheric layer is recorded. The vertical ozone concentration profile can then be calculated using a mathematical deconvolution procedure.

10.2.2 Long-Path Absorption Techniques

The principle of long-path absorption techniques is illustrated in Fig. 10.17. A laser beam is transmitted continuously into the atmosphere against a corner-cube retro-reflector (Fig. 6.21) that is placed at a distance of up to 10 km. The reflected beam is received by an optical telescope that is placed at the site of the laser and is directed towards the retro-reflector. The received light intensity is measured photo-electrically as a function of the laser wavelength. The absorption spectrum of the atmosphere between the laser and the retro-reflector is then recorded and the mean concentrations N_i of pollutant molecules can be determined using the Beer-Lambert relation

$$\ln \frac{P_0(\nu)}{P_t(\nu)} = 2R \left(\sum_{i=1}^{n} \sigma_i(\nu) N_i + K_{ext} \right). \qquad (10.5)$$

Here $P_t(\nu)$ is the received light intensity and $P_0(\nu)$ is the intensity that would have been received in the absence of atmospheric absorption. $\sigma_i(\nu)$ is

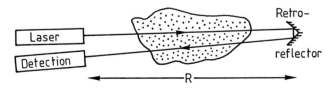

Fig.10.17. Long-path absorption measurement of atmospheric pollutants

the absorption cross section for the molecules of type i. $\sigma_i(\nu)$ normally varies strongly with the wavelength while K_{ext} represents particle extinction, which is largely wavelength independent in a small wavelength region. If several molecules absorb in the same wavelength range it is necessary to perform the measurements in a sufficiently large wavelength interval, preferably with a continuously tunable laser to allow unambiguous determination of the individual molecular species. Frequently one tries to work in a wavelength region where the gas of interest is the dominant absorber. Then the measurement can conveniently be performed by rapidly switching the laser wavelength from the line centre to a nearby off-line wavelength. Clearly, it is necessary to work in a wavelength region where the dominant atmospheric absorbers CO_2 and H_2O have a low absorption (Sect.6.4.5). Such regions can be found where the CO_2 and DF lasers emit around 10 and 4 μm, respectively. These gas lasers are normally only line tunable and accidental wavelength coincidences are utilized. Sometimes, two lasers tuned to an on-resonance and an off-resonance frequency, respectively, are used, and rapid switching between transmission from the two lasers is performed. High-pressure CO_2 lasers and diode lasers are continuously tunable. The long-path absorption technique is used mainly for monitoring gases such as C_2H_4 (ethylene), C_2H_3Cl (vinyl cloride) O_3 and CO utilizing suitably located vibrational transitions. Results from a measurement of NO across a major road are shown in Fig.10.18. Clearly, these kinds of long-path absorption measurements are closely related to those performed using classical light sources (*differential optical absorption spectroscopy*, doas, Fig. 6.73).

In a variant of the long-path absorption technique, radiation from a diode laser or a light-emitting diode is transmitted fibre-optically to re-

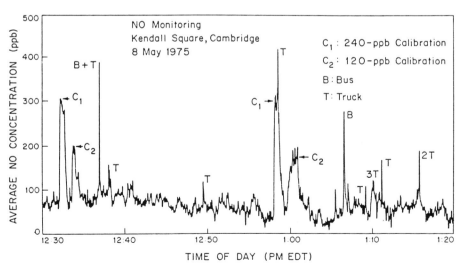

Fig.10.18. Long-path absorption measurement of the NO concentration across a major road [10.72]

motely located multi-pass absorption cells (Sect.9.2.1) and the partially absorbed beam is sent back to the measurement system also using fibre-optics. Using such a system, which operates at short IR wavelengths (1 to 2μm) at which optical fibres transmit well, many points can be monitored from a central system with laser and computer facilities [10.73].

10.2.3 Lidar Techniques

Lidar, which is an acronym for *L*ight *d*etection *a*nd *r*anging, is a measurement technique in which pulsed laser radiation is transmitted into the atmosphere and back-scattered light is detected at a certain time delay in a radar-like fashion. The principle of lidar (also called laser radar) is illustrated in Fig.10.19. Laser light that is back-scattered from a distance R arrives at the lidar receiver at a time t = 2R/c after the transmission of the pulse. The velocity of light, c, is 300 m/μs. Range-resolved information can be obtained from the time delay, and the range resolution ΔR, is given by the duration of the laser pulse t_p: $\Delta R = t_p c/2$. (The range resolution may be further impaired due to the finite electronic response of the detection system.) The intensity of the received lidar signal is given by the general lidar equation

$$P(R,\Delta R) = CWN_b(R)\sigma_b \frac{\Delta R}{R^2} \exp\left\{-2\int_0^R [\sigma(\nu)N(r) + K_{ext}(r)dr]\right\}. \quad (10.6)$$

C is a system constant, W is the transmitted pulse energy and $N_b(R)$ is the number density of scattering objects with back-scattering cross-section σ_b. The exponential factor describes the attenuation of the laser beam and the back-scattered radiation due to the presence of absorbing molecules, of concentration N(r), and absorption cross section $\sigma(\nu)$, and due to attenuating particles with wavelength-independent extinction coefficient K_{ext}. A single absorbing molecular species is assumed. The product $N_b(R)\sigma_b$ in the lidar equation determines the strength of the back-scattering which can be caused by several processes. It can be due to *fluorescence* from atoms or molecules that are resonantly excited by the laser light. At high pressures, i.e. in the troposphere, radiationless transitions due to collisions strongly quench the fluorescence light (Sect.10.1.2). On the other hand, fluorescence detection is very efficient for monitoring of mesospheric constituents. There are layers of Li, Na, K and Ca atoms in the mesosphere at a height of about 100 km. These layers of atoms, which are produced mainly through evaporation of meteorites impinging on the atmosphere, have been mapped out very successfully with ground-based fluorescence lidar systems operating with tunable dye lasers [10.74].

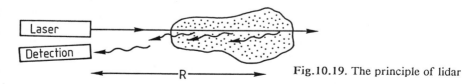

Fig.10.19. The principle of lidar

The back-scattering can also be caused by the *Raman* process. Because of the weakness of this kind of scattering, high-power laser beams are normally required even for the monitoring of major atmospheric species. Here light back-scattered with a characteristic Stokes frequency shift is detected. The technique has been used for vertical monitoring of water vapour profiles and for temperature measurements. Atmospheric visibility can also be assessed by measuring signals from N_2 molecules that are recorded with reduced intensity because of mist and fog particles. Slant visibility measurements are of great importance, e.g., at airports.

The strongest back-scattering process in the atmosphere is Mie scattering from particles (See Sect.4.5). Since this form of scattering is elastic, no information is obtained on the chemical composition of the particles. (Such information can be obtained using very high-power laser pulses which, through focusing, can induce air breakdown (laser sparks) at distances up to 100 m. The emission spectrum from the spark carries information on vaporized particles. LIBS (Laser Induced Breakdown Spectroscopy) is also a powerful laboratory technique [10.75]). The intensity of the Mie-scattered light depends on the number density, size, shape, refractive index and absorption properties of the particles (Sect.4.5). Thus, a quantitative analysis requires a calibration, whereas relative particle spatial distributions can be obtained more directly. Since particle back-scattering frequently dominates strongly over particle extinction K_{ext}, the range-resolved lidar signal directly maps out the particle distribution $N_b(r)$.

The construction of a lidar system is illustrated in Fig.10.20. A Nd:YAG laser is used, either directly or after frequency conversion in a dye laser, to produce pulses that are transmitted into the atmosphere via a planar first-surface aluminized mirror. The same mirror is used for directing back-scattered light down into a fixed Newtonian telescope. In the focal plane of the telescope there is a polished metal mirror with a small hole, which defines the field-of-view of the telescope. For a telescope with a focal length of 1 m, a 1 mm hole corresponds to a telescope field-of-view of 1 mrad, which matches typical laser beam divergences. In order to suppress background light it is essential that the telescope only observes regions from which laser photons can be backscattered. The light passing the aperture is detected by a photomultiplier tube, while all the other light is directed into a TV camera which produces a picture of the target area, except for the laser beam region which is seen as a black spot. The detected lidar signal is transferred to a transient digitizer (Sect.9.4.3a) and is read out to a computer system. The computer is used for controlling the planar mirror, the laser wavelength, etc. and also performs signal averaging and necessary processing of the lidar signals. Particle monitoring with lidar techniques is illustrated in Fig.10.21. Increased back-scattering is monitored from particle plumes while a smoothly falling $1/R^2$ intensity is obtained from the uniform background particle distribution in the atmosphere.

In order to measure the concentration of gaseous pollutants with lidar techniques, resonance absorption can be used in a similar way to the long-path-absorption method that was described above. However, in the lidar technique no retro-reflector is required and the scattering from particles is

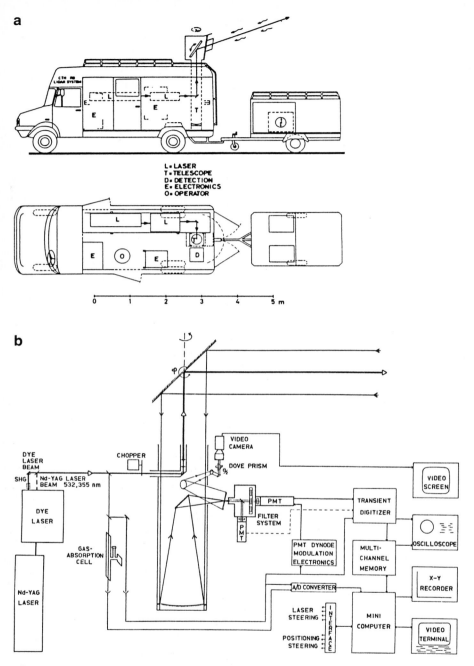

Fig.10.20a,b. Construction of a mobile lidar system. (a) General lay-out. (b) Optical and electronic systems [10.76]

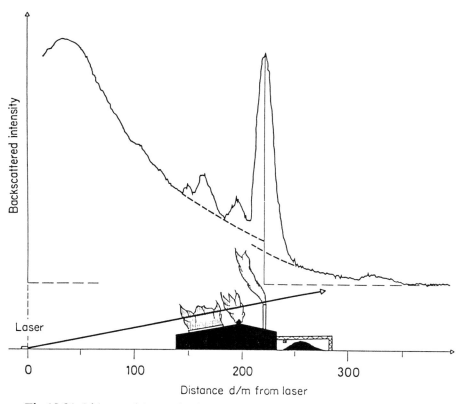

Fig.10.21. Lidar particle monitoring [10.77]

utilized to generate a reflected signal. The particles serve as a "distributed mirror". The differential absorption at close-lying wavelengths of molecules in the atmosphere is used. The method, which is called DIAL (*DIfferential Absorption Lidar*), is useful for qualitative as well as quantitative range-resolved measurements of air pollutants. The principles are illustrated in Fig.10.22. For the sake of argument we consider an atmosphere with a uniform particle distribution and with two localized clouds of absorbing gas molecules. Lidar curves are recorded at a frequency ν_1 of strong specific absorption and at a nearby reference frequency ν_2. For atmospheres with typical particle size distributions, σ_b and K_{ext} will not change in the small frequency range, since the oscillatory behaviour that is characteristic of uniform particles will be washed out. If the particle extinction K_{ext} is small the recorded curves for both frequencies will have a simple $1/R^2$ dependence for the region preceding the first gas cloud. For the non-absorbing frequency the lidar curve continues with a $1/R^2$ intensity fall-off even through the gas clouds, while intensity reductions occur in the on-resonance lidar curve at the site of the gas clouds. If no gas had been present both curves would have had the same appearance. The presence of an absorbing gas is best illustrated if the two curves are divided by each other, as illus-

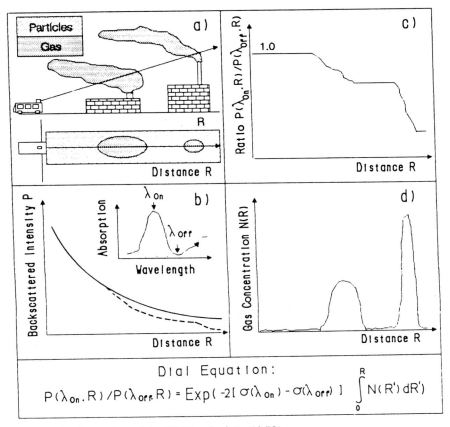

Fig. 10.22. Illustration of the DIAL principle [10.78]

trated in the figure. Mathematically, this ratio can be formed by dividing the expressions for the lidar equation (10.6) for the two frequencies ν_1 and ν_2

$$\frac{P_{\nu_1}(R,\Delta R)}{P_{\nu_2}(R,\Delta R)} = \exp\left\{-2[\sigma(\nu_1)-\sigma(\nu_2)]\int_0^R N(r)dr\right\}. \tag{10.7}$$

As can be seen, all unknown parameters such as $N_b(r)$, σ_b and $K_{ext}(r)$ are eliminated upon division and only the absorption cross-sections at the two frequencies have to be known from laboratory measurements for an evaluation of the gas concentration $N(r)$ as a function of the distance R. It can also be seen that the method does not even require a uniform particle distribution, since any bumps in the curves due to particle clouds would be cancelled out. This is important since gaseous pollutants and particles frequently occur together.

In Fig. 10.23 the absorption spectrum for SO_2 in the wavelength region around 300 nm is shown. In practical DIAL measurements of SO_2 the laser

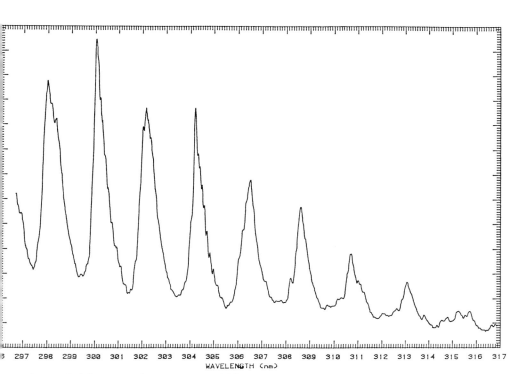

Fig.10.23. SO_2 absorption spectrum [10.79]

wavelength is changed between 300.0 and 299.3 nm every other laser pulse and the corresponding lidar curves are averaged in separate computer memories. SO_2 measurements are illustrated in Figs. 10.24a,b. Ozone can also be measured at wavelengths close to 300 nm, while NO_2 is monitored in the blue spectral region. Pollutants such as NO and Hg can be measured at short UV wavelengths, while CO, HF and hydrocarbons demand IR wavelengths. It should be noted that atmospheric back-scattering is considerably weaker at long wavelengths because of the strong wavelength dependence of Mie as well as Rayleigh scattering. In order to increase the signal-to-noise ratio of IR laser radar systems using e.g. pulsed TEA CO_2 lasers, heterodyne detection is frequently used (Sect. 10.2.1). Wind velocities can be remotely measured directly by frequency analysis of the beats between the back-scattered radiation that is Doppler shifted from moving particles and radiation from a cw laser tuned to the same frequency as the pulsed laser. Global wind field measurements as well as monitoring of molecules, especially O_3, are planned with space-borne lidar systems. Ground-based and airborne lidar systems have been used in extensive studies of ozone [10.80] as well as for monitoring of stratospheric dust particles following volcanic eruptions [10.81]. The particle content of the atmosphere is important in the consideration of the earth's radiation balance. Meteorological parameters such as humidity, temperature and pressure can be measured with DIAL techniques using weak absorption lines of abundant species such as H_2O and O_2 in the red spectral region. Tunable solid-state lasers, such as alexandrite lasers (Sect. 8.5.3), are particularly suited to such measurements [10.82].

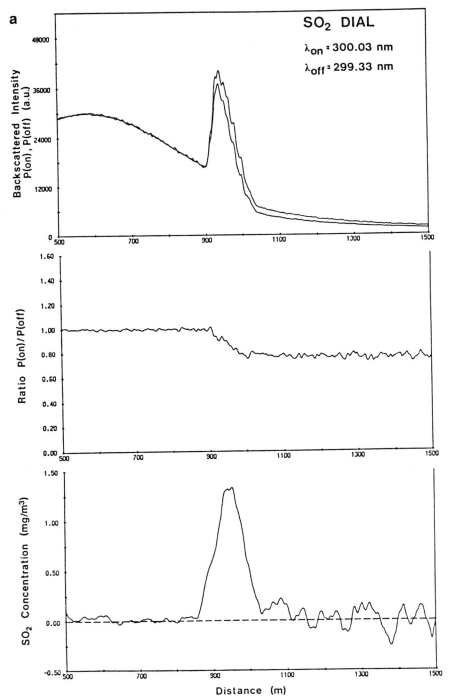

Fig.10.24a. SO_2 DIAL measurement. The R^{-2} dependence has been suppressed by using a reduced amplification at short ranges. [10.78]

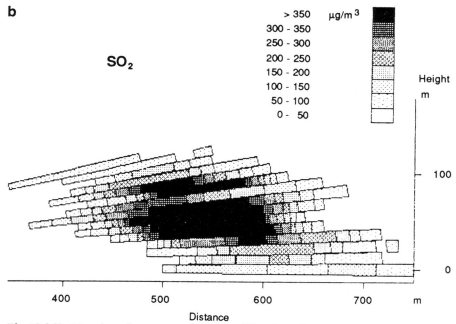

Fig.10.24b. Mapping of a cross section of an SO_2 plume from a papermill obtained by DIAL measurements. By multiplying the integrated SO_2 content by the wind velocity normal to the vertical section a total flow of 230 kg/h of SO_2 was found [10.78]

10.3 Laser-Induced Fluorescence and Raman Spectroscopy in Liquids and Solids

Laser-induced fluorescence (LIF) in liquids and solids can be used for diagnostic purposes in many contexts. Most substances have broad absorption bands in the UV region and a pulsed UV laser is frequently most useful for inducing fluorescence. The nitrogen laser ($\lambda=337$ nm) is a simple and practical source for such measurements. Other useful laser systems are tripled Nd:YAG lasers ($\lambda=355$ nm) or excimer lasers (XeCl, $\lambda=308$ nm; KrF, $\lambda=249$ nm). Liquids and solids exhibit broad fluorescence emission bands [10.83-86]. As already discussed for dyes, sharp emission features are lost because of rotational quenching and mutual interactions between the molecules. Vibrational Raman spectra, on the other hand, exhibit quite sharp lines. Raman spectroscopy is a powerful tool for studying liquids and surface layers on solids. In this section we shall describe a few applications of LIF and Raman spectroscopy of liquids and solids. We will first consider *hydrospheric remote sensing* and then discuss *monitoring of surface layers*.

10.3.1 Hydrospheric Remote Sensing

The increasing pollution of seas and inland waters calls for efficient methods of aquatic pollution monitoring. Oil spills cause drastic damage to the environment. For marine monitoring, airborne surveillance systems are the most useful ones. Allweather capability generally calls for a microwave-based system. Fluorescence can be monitored at a distance with a modified lidar set-up and an airborne laser fluorosensor can be a valuable complement to a SLAR system (Sect. 7.2) for characterizing detected oil-slicks. Clearly, it is necessary to perform laboratory measurements on the substances of interest in order to obtain the basic information necessary for a field system. A laboratory set-up for LIF studies is shown in Fig. 10.25. Either a scanning monochromator connected to a boxcar integrator is used, or better, an optical multichannel analyser (Fig. 6.38). In Fig. 10.26 LIF spectra for different substances in the aquatic environment are shown for N_2-laser excitation. A crude-oil spectrum is shown featuring a broad and strong fluorescence distribution. Oil products fluoresce quite strongly because of their aromatic hydrocarbon content, e.g. anthracene and napthalene. Refined products fluoresce even more strongly with the peak fluorescence shifted to shorter wavelenghts, while heavy residual oils, such as asphalt, fluoresce weakly and mainly towards longer wavelengths. A spectrum of sea water is also shown with a promient sharp signal due to the OH stretch Raman mode of H_2O. The broad blue fluorescence is due to organic water pollution. The spectrum also shows a fluorescence peak from a tracer dye that has been added to the water. Such dyes can be used to mark a "package" of water allowing studies of flow patterns in aquatic systems. A concentration as low as 1 μg/l is sufficient for such fluorescence tracing. Included in Fig. 10.26 is a spectrum from river water, polluted by lignine sulphonate from a pulp mill and a spectrum of water containing microscopic green algae. The prominent peak at 685 nm is due to the chlorophyll-a pigment that is present in all plants [10.89]. LIF can be used to study algal blooms, induced by eutrophication in seas and lakes. Airborne oil monitoring systems have been constructed [10.90]. Oil-slicks are detected by an increase in the blue fluorescence at the same time as the water Raman signal disappears because of full absorption in the oil layer.

An airborne lidar system operating in the blue-green transmission window of water (Fig. 6.54) can be used for measurements of water depth

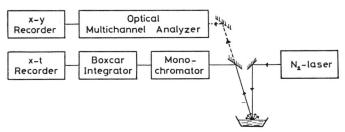

Fig. 10.25. Laboratory set-up for LIF measurements [10.87]

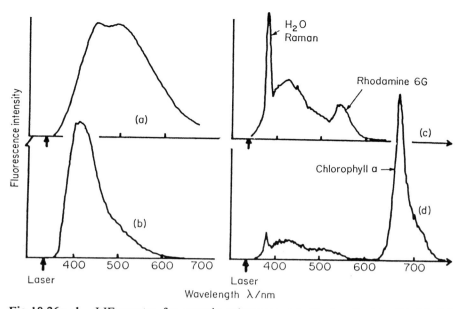

Fig.10.26a-d. LIF spectra for aquatic substances. (a) Crude oil, Abu Dhabi. (b) River water down-stream from a sulphite pulp mill. (c) Sea water containing 3 $\mu g/\ell$ Rhodamine 6G dye. The Raman peak from water is exhibited, too. (d) The green algae *Chlorella ovalis* Butcher in sea water [10.88]

[10.91]. The principle of such measurements is given in Fig.10.27. The speed of light in water is about 0.75c. A laser with a short pulse length must be used. Water is strongly attenuating and in order to enhance the weak bottom echo with respect to the strong surface echo a polarization scheme can be used. If the laser pulse is linearly polarized the surface echo remains largely polarized whereas the bottom echo is almost completely depolarized. Thus, by using a crossed linear polarizer in front of the detector a more reasonable relative strength of the two echoes is obtained. Depths down to several tens of metres can be measured by laser bathymeters of this construction. A frequency-doubled Nd:YAG laser (λ=532nm), a copper vapour laser (λ=510 and 578nm) or a HgBr laser (λ=520nm) are candidates for laser bathymeters. Space-borne green lasers are also being discussed for communication with submarines. The field of hydrospheric probing by lasers has been discussed in [10.65].

10.3.2 Monitoring of Surface Layers

We have already mentioned that oils strongly absorb UV light. Thus, LIF can be used for detecting very thin surface layers of oil [10.92,93]. Industrially, it can be of interest to be able to establish that sheet metal components are free from oil before entering a painting shop. It is also interesting to monitor the presence of surface layers, e.g. of corrosion-protective agents. Examples of spectra of this kind are given in Fig.10.28. Monitoring

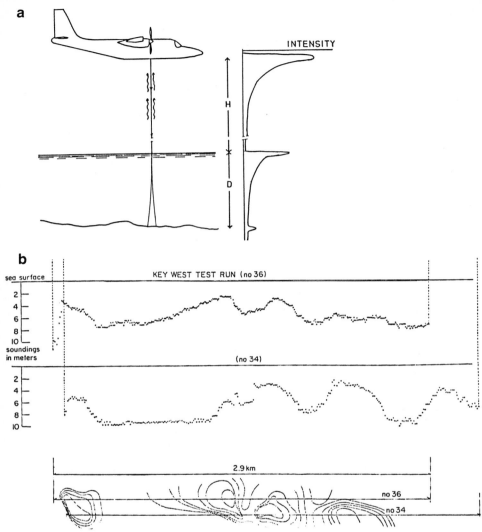

Fig.10.27. (a) Principle of airborne laser bathymetry [10.91, last Ref.], (b) Results from lidar bathymetry in shallow waters [10.91, first Ref.]

of the application of oil for rust protection during sheet metal coiling in a steel rolling mill using an industrial fluorosensor is illustrated in Fig. 10.29.

The penetration depth in organic liquids rapidly increases for longer wavelengths. By choosing a suitable excitation wavelength sensitive measurements of film thickness can be performed in different thickness ranges. The fluorescence of organic substances can also be used for sensitive detection of signals in liquid chromatography and electrophoresis [10.94]. A UV laser beam is then used for exciting fluorescence at the outlet of the column. The varying wavelength distributions for different compounds can

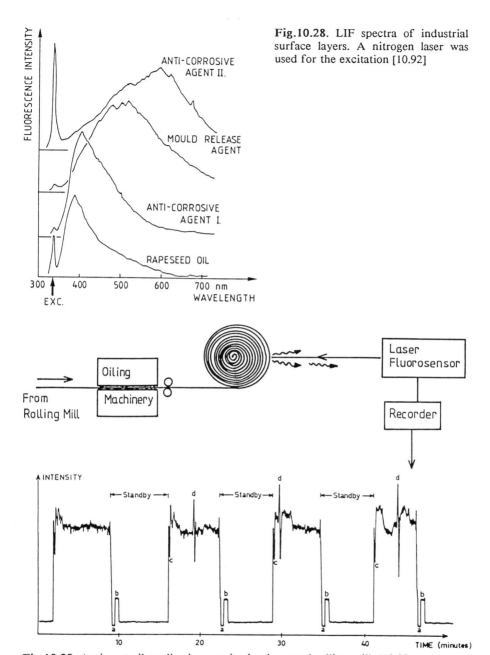

Fig.10.28. LIF spectra of industrial surface layers. A nitrogen laser was used for the excitation [10.92]

Fig.10.29. Anti-rust oil application monitoring in a steel rolling mill [10.92]

be used to further increase selectivity in complicated chromatography and electrophoresis spectra.

Certain salts of rare earths exhibit sharp fluorescence lines when excited by UV light. It has been shown that the relative strengths of different lines and also the decay time of the fluorescence at a certain fluorescence wavelength vary strongly with temperature [10.95]. Fibre thermometers with a rare-earth-salt fibre tip have been developed. By applying thin layers of such salts to surfaces it is possible to remotely measure the temperature by LIF. This technique could be particularly valable for hot rotating machine parts.

Surfaces of solids can also be investigated by Raman spectroscopy, which has a higher specificity than LIF. Thus, this method is well suited for studying chemical processes such as corrosion, electrochemical processes etc. Frequently, an Ar^+ laser is used in conjunction with a double or triple monochromator for isolating the Raman signals from the very strong elastic scattering from the surface. In Fig. 10.30 a Raman spectrum illustrating surface oxidation is shown. The Raman scattering from a surface can be strongly enhanced compared with the bulk material scattering. The enhancement is connected to periodic structures on the surface. Surface-enhanced Raman scattering has been discussed in [10.97, 98]. Surfaces can also be studied by other optical techniques. A method in which the strength of surface frequency doubling of impinging laser light is investigated, is rapidly evolving into a very powerful technique [10.99, 100]. An example of surface studies using frequency doubling is shown in Fig. 10.31. The general field of laser spectroscopy of solids was covered in [10.101, 102].

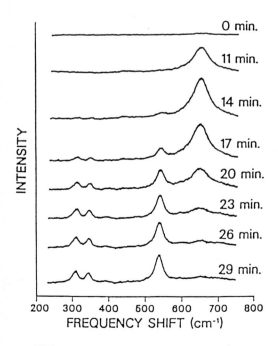

Fig. 10.30. Surface Raman spectra showing oxidation process on a surface of a Fe/Cr/Mo alloy. Initially an Fe/Cr double oxide (spinel) is formed which is successively transformed into Cr_2O_3 [10.96]

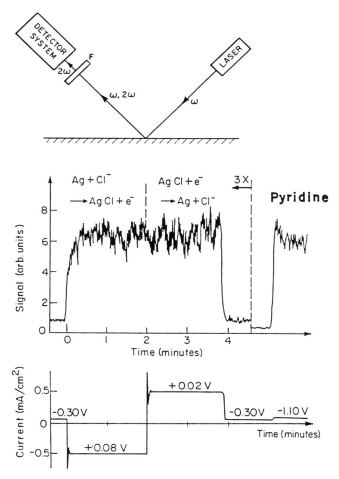

Fig.10.31. Surface monitoring by observing the generation of frequency-doubled light. The top figure depicts the experimental arrangement in this type of measurements. Below the elctrolytic build up and removal of AgCl on an Ag electrode in a cell with KCl solution is illustrated. The second harmonic signal saturates for a single monolayer. When the last monolayer is removed following voltage reversal the voltage changes and the signal disappears [10.100]

10.4 Laser-Induced Chemical Processes

Laser light can be used for inducing or controlling chemical processes in different ways. Photons of suitable energy can greatly increase the speed of certain reactions leading to a significant increase in the yield of a desired substance. In this section we will discuss *laser-induced chemical processes* in general and also consider processes in which strong isotopic selectivity is obtained leading to *laser isotope separation*. The presently very active field of laser-induced processes is covered in a number of conference proceedings and reviews [10.103-111].

10.4.1 Laser-Induced Chemistry

A primary question in connection with laser-induced chemistry has been whether it would be possible to selectively break chemical bonds in large molecules by irradiating the molecules at a frequency corresponding to the vibrational mode associated with that particular bond. If this could be accomplished the possibilities of manipulating molecules would be almost inexhaustible. However, it has been found that the energy absorbed in a certain vibrational mode is very quickly distributed over the whole molecule. Therefore, the excitation will be essentially thermal and unselective. In order to break a bond it is necessary to deposit a great deal of energy in the bond in a time less than 1 ps. This is technologically difficult. Thus, laser-induced chemistry has to rely on other processes. By exciting a molecule to a higher vibrational level it can be made to react much faster *(chemical activation)*. One example is the reaction

$$K + HCl \rightarrow KCl + H \; .$$

This reaction occurs 100 times faster if the HCl molecules are excited to v = 1 compared with v = 0. Laser radiation can also act as a catalyst in certain reactions, e.g. in the production of vinyl chloride C_2H_3Cl for the plastic industry. The process, which is basically thermal break-up of dichloroethane $C_2H_4Cl_2$ into C_2H_3Cl and HCl, can be run at considerably lower temperatures in the presence of laser light. With an XeCl excimer laser (308 nm) free Cl radicals are formed that open up new reaction paths. Only one laser photon is needed to produce 10^4 vinyl chloride molecules since a chain reaction is utilized [10.103]. Laser radiation is also very effective in many *pyrolysis reactions*. Here local heating of the reacting gases is accomplished by the laser light. As an example, the cracking of heavy hydrocarbons into lighter ones can be mentioned. In the petrochemical industry it is frequently desirable to increase the yield of ethylene C_2H_4 over methane CH_4, since ethylene is a more desirable starting material for many petrochemical processes. Laser pyrolysis results in a higher yield than that possible with thermal heating. Laser pyrolysis also has good potential in the production of ceramic materials. Using silane (SiH_4) as a starting product, powders of Si, Si_3N_4 or SiC can be produced, later to be sintered into ceramics.

Laser chemistry can also be used for *purifying* certain chemicals. The production of ultrapure silicon for the semiconductor industry and for producing solar cells is of special interest. For optical fibres the purest possible SiO_2 is desirable. A suitable starting material is silane gas. Arsine (AsH_3) and phosphine (PH_3) are typical impurities. On radiation of the natural gas mixture with an ArF excimer laser (λ=193 nm) the impurities are quickly dissociated. The technique has a good production potential and the cost for purification could be quite reasonable. Since the cost of the laser photons will always be an important factor it is likely that laser-induced chemical processes will be advantageous, especially for producing specialized and expensive chemicals, e.g. pharmaceuticals. Photo-assisted production of vitamine D and prostaglandine has also been considered.

10.4.2 Laser Isotope Separation

Using laser light of sharp frequency it is possible to induce isotope-selective processes leading to isotopic separation by utilizing optical isotopic shifts. Enriched or separated isotopes are of great interest in many contexts. Pure radio-isotopes are used in medicine. Also, stable elements, such as ^{13}C, can be used for studies of the metabolism and other biological processes. In NMR spectroscopy signals are only obtained from isotopes with a nuclear spin. For sulphur and calcium the useful isotopes ^{33}S and ^{43}Ca are naturally present only in very small concentrations. By using enriched isotopes in the building up of organic material completely new possibilities arise. Enrichment of ^{14}C from organic objects that are to be dated would mean a much higher sensitivity since the tracer element could be concentrated into a small volume allowing small detectors to be used with a resulting low background count rate. In this way the time span when dealing with the ^{14}C method could be considerably increased.

However, the greatest interest in isotope separation comes from the nuclear power industry. Production of heavy water for heavy-water reactors and separation of highly active components from the burnt-out uranium fuel are two applications. The most important aspect by far is, however, isotope separation of ^{235}U. Natural uranium contains only 0.7% ^{235}U and enrichment to about 3% is needed for use in light-water reactors. The normal separation method is diffusion of uranium hexafluoride UF_6 through porous filters. This process, which relies on the slightly different mobilities of $^{235}UF_6$ and $^{238}UF_6$, is very inefficient and the separation costs are very high and correspond to about 10% of the resulting price of the produced electricity. Several nuclear reactors are needed to provide the electricity for a typical separation plant. A technique based on high-speed gas centrifugation is now being developed as a possible alternative. At the same time, laser techniques for uranium isotope separation are being investigated in several countries. Clearly, any technique that could reduce the separation costs would be of great interest. Laser techniques can lead to a high degree of enrichment and may also be useful in processing uranium tailings from which most of the fissile material has already been removed. Laser isotope separation schemes based on free atoms or on UF_6 are possible. In both cases the presence of an isotopic shift is utilized. We will first consider free atoms. For heavy elements, such as uranium, the isotopic shift is only due to the volume effect (Sect.2.8). In Fig.10.32 the different isotopically separated components of a uranium emission line are shown for an enriched sample. The broadening in the ^{235}U line is due to unresolved hyperfine structure. The basic principle for atomic uranium separation is illustrated in Fig.10.33. A narrow-band tunable laser selectively transfers ^{235}U atoms to an excited state. Since atomic transitions can be saturated with laser pulses (Sect.9.1.2) a large fraction of the ^{235}U atoms are transferred to

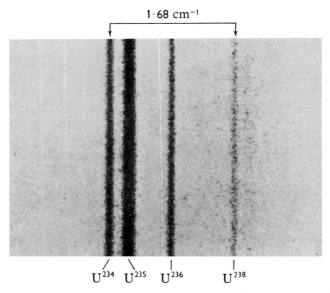

Fig.10.32. Isotope shifts in a uranium emission line. A uranium sample enriched in ^{235}U was used in the light source [10.112]

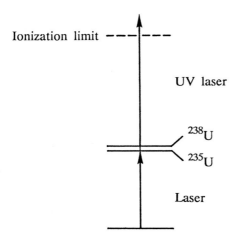

Fig.10.33. A schematic uranium isotope separation scheme

the higher state from which they can be photoionized using a second laser pulse. The photon energy for this laser is not sufficient to photo-ionize ground-state atoms. The ^{235}U ions that are formed in this way, are extracted to a collector plate using an electric field. A separation scheme with two selective excitation steps induced by dye lasers that are pumped by powerful copper vapour lasers is being explored for practical separation purposes [10.110]. A powerful excimer laser is used for the photo-ionizing step. The properties of uranium call for special evaporation techniques. It is necessary to heat the uranium to 2300 K to achieve substantial evaporation. However, molten uranium metal is extremely reactive and attacks most furnace materials. One technique which can be used to overcome this problem is local surface heating with electron guns.

Uranium isotope separation using a molecular approach is based on selective multi-photon dissocation of UF_6. The relevant vibrational isotope shift is 0.6 cm^{-1} in the primary vibrational transition at 628 cm^{-1} (16 μm). In the development of the technique, experiments on SF_6 have been very important. The conditions are much more favourable for SF_6 than for UF_6. The isotope shift is 17 cm^{-1} between ^{32}S and ^{34}S in the IR active vibrational mode that involves asymmetric stretching of two S-F bonds. The spectrum has a typical P, Q and R branch structure and the whole region of absorption for the rotational level population distribution that is obtained at room temperature is 15 cm^{-1}. Thus, the isotopic molecules are spectroscopically totally separated. Furthermore, the vibrational transition in $^{32}SF_6$ well matches the emission of a free-running pulsed CO_2 laser.

The multi-photon dissociation process is illustrated in Fig.10.34. Because of the anharmonicity of the vibrational potential a photon energy that is resonant in the first vibrational step will successively pull out of resonance higher up in the vibrational energy level ladder. However, the molecule can be excited in a multi-photon process (Sect.9.1.3c) until it is dissociated. SF_6 then disintegrates into SF_5 and F. By making the first step resonant for one isotopic molecule the probability of subsequent dissociation for this molecular species is significantly increased.

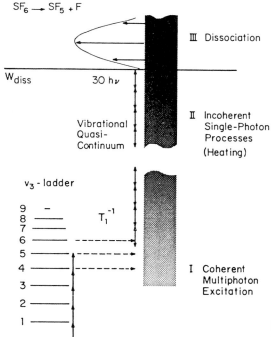

Fig.10.34. Multi-photon dissociation process [10.113]

To carry out the corresponding process for UF_6 the availability of efficient lasers in the 16 μm region is a necessity. This has resulted in considerable effort being put into producing such lasers. An efficient way of generating laser radiation at 16 μm is to Raman shift a pulsed CO_2 laser at 10.6 μm. A 16 μm beam can be produced at a 50% efficiency by stimulated Raman scattering in a multi-pass cell filled with para-H_2. Rotational transitions in this molecule with parallel nuclear spins are used. A problem with UF_6 is that the width of the sharpest feature in the spectrum (the Q branch) is 5 cm^{-1} at room temperature, which greatly exceeds the isotope shift. Thus, the selectivity is poor. The selectivity can be increased by using two laser wavelengths in the 16 μm region. By dynamic cooling of the UF_6 gas through expansion from a supersonic nozzle, the number of populated rotational levels can be drastically reduced, leading to a much better resolution of the Q branches of the two isotopic molecules. Even then, two separate 16 μm wavelengths are advantageous and further efficiency is gained by adding excimer laser UV photons for more efficient photo-dissociation of molecules that are excited in the multi-photon process. Upon UF_6 dissociation, UF_5 molecules are formed in the form of a powder that can be collected.

For the separation of ^{13}C, multi-photon dissocation of Freon-22 (CF_2HCl) or CF_3I can be used, and macroscopic quantities are being produced utilizing CO_2 TEA lasers. Heavy water (D_2O) can also be enriched with a multi-photon process utilizing CO_2 laser radiation acting on CF_3D molecules

$$CF_3D + nh\nu \rightarrow CF_2 + DF$$
$$2DF + CaO \rightarrow D_2O + CaF_2$$

The isotope shift between CF_3D and CF_3H is large and the cross-sections for multi-photon absorption differ by a factor of 6000 between the two molecules.

Formaldehyde, HCHO, has also been much studied. Using multi-photon processes this molecule can be used for the separation of hydrogen and carbon as well as oxygen.

Step-wise processes involving few photons can also be used for molecules. As a matter of fact, one of the first laser separations was performed in this way [10.114]. This experiment on NH_3 molecules is illustrated in Fig. 10.35. A mixture of $^{14}NH_3$ and $^{15}NH_3$ is kept in a cell and is irradiated by a pulsed CO_2 laser which can only excite $^{14}NH_3$ vibrationally because of the isotope shift. At the same time a strong UV source, consisting of a spark discharge in air, is fired. The UV photons dissociate the vibrationally excited molecules but thanks to suitable filtering no photons with sufficient energy to dissociate ground-state molecules are available. The following reactions occur, but only for $^{14}NH_3$

$$NH_3 + h\nu \rightarrow NH_2 + H,$$
$$NH_2 + NH_2 \rightarrow N_2H_4,$$
$$N_2H_4 + H \rightarrow N_2H_3 + H_2,$$
$$2N_2H_3 \rightarrow 2NH_3 + N_2.$$

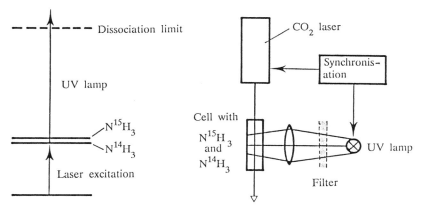

Fig.10.35. Ammonia separation scheme

Thus, isotopically enriched $^{14}N_2$ gas has been formed in the cell. Similar processes can be used for boron and chlorine separation.

For the production of pure radio-isotopes for medical use it is advantageous to induce the nuclear reactions in suitably chosen isotopically pure elements. Examples of such processes are

$$^{203}Tl \rightarrow {}^{201}Tl,$$
$$^{124}Te \rightarrow {}^{123}I,$$
$$^{68}Zn \rightarrow {}^{67}Ga$$

and

$$^{190}Os \rightarrow {}^{191}Ir.$$

The annual world demand for such isotopes may only be of the order of 100 grams, but since conventional production methods are extremely costly, laser isotope separation might here be an attractive possibility.

10.5 Spectroscopic Aspects of Lasers in Medicine

Lasers are finding increasing applications in the field of medicine. Normally, the heating effect of laser beams is utilized. Raising the temperature from 37° to 44°C causes disruptions in the enzymatic and electrical activities of the cells. At 60°C protein denaturation occurs and at 100°C the cells rupture due to the boiling of the cellular fluid. Carbonization occurs at temperatures above 200°C. The laser power densities required to attain a certain temperature rise clearly depend on the absorption properties of the tissue, which are governed by the properties of water and haemoglobin. Skin absorption is largely due to the presence of melanin. The corresponding absorption curves are included in Fig.10.36. Beams from the three most important medical laser types, the CO_2, the Nd:YAG and the Ar^+ laser

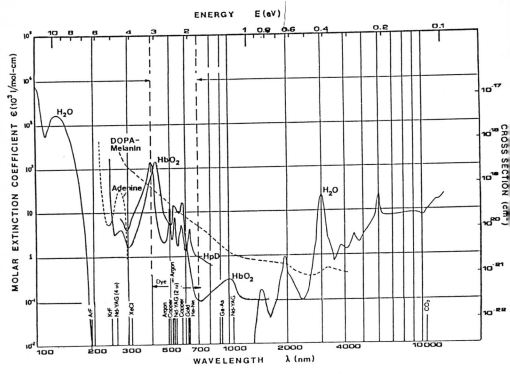

Fig.10.36. Absorption curves for tissue constituents [10.115]

have a surgical penetration depth of about 0.1, 4 and 1 mm, respectively, in normal tissue. Short-wavelength excimer lasers have a very low penetration depth and tissue removal is caused by ablation. Under pulsed irradiation the tissue is lifted off layer by layer. These lasers are now being extensively investigated for medical use, e.g., for corneal refractive surgery and opening of clogged blood vessels (angioplasty).

An important feature of laser beams in surgery is their ability to cut tissue and small vessels and at the same time coagulate the blood, which strongly reduces the need for blood transfusions ("the bloodless knife"). The thickness of the coagulated zone varies with the laser wavelength used. Other advantages are the sterile, contactless nature of the cutting and the minimum mechanical strain on the tissue. The last point is particularly important in neurosurgery. Lasers are also used to vaporize tumours. The laser radiation from Nd:YAG and Ar$^+$ lasers can conveniently be delivered through flexible optical fibres that are highly transparant to the corresponding wavelengths. However, for CO_2 no fibres have been available until very recently, when the first fibres of limited applicability appeared. Normally, CO_2 radiation is delivered through a specially designed multi-joint arm that is terminated with a handpiece.

The strong absorption of green Ar$^+$ radiation in blood can be utilized for selective coagulation of blood in small blood vessels. This is utilized in the treatment of portwine stains. This dermatological disease is caused by excessive growth of a network of tiny blood vessels close to the skin. The laser light penetrates the skin but is absorbed in the blood eliminating the cause of skin discolouration.

Ar^+ lasers are also used for photo-coagulation treatment of excessive blood vessel growth in the retina of diabetes patients. This type of laser is also used for "spot welding" of the retina in cases of retinal detachment. Because of the relatively easy optical access to the interior of the eye this organ is well suited for laser treatment. The retina consists of many layers with different absorption properties. Thus, by choosing a certain wavelength an energy dose can be delivered to a particular layer with a certain selectivity. One area in the retina has very special absorption properties because of its high content of xantophyll. The optical and spectroscopic properties of the eye largely determine the criteria for safe utilization of lasers and form the basis for legal classification of lasers. American regulations [10.116] are frequently used as a starting point in setting up national rules. The maximum permissible dose of laser light for eye exposure obviously varies strongly with wavelength. The strictest regulations apply for the wavelength region 400 to 1400 nm, in which the retina is fully exposed to the radiation. The Nd:YAG laser operating at 1.06 μm is one of the most dangerous lasers from an eye safety point of view since it penetrates into the eye, yet the presence of beams is not revealed through scattering, since the wavelength falls in the invisible region. For wavelengths below 400 nm absorption occurs in the cornea. However, high doses of UV absorbed in the cornea cause cataracts.

The most relaxed regulations pertain to the wavelength region above 1.4 μm where only the heat deposition in the cornea sets the limit. For detailed accounts of the medical use of the thermal energy in laser beams we refer the reader to [10.115, 117, 118].

Apart from the surgical application of laser beams discussed above, a new interesting field of laser utilization for tumour localization and treatment is emerging. The techniques take advantage of the special properties of certain agents, most notably hematoporphyrin derivative (HPD) [10.119-124]. After intravenous injection (typically 3 mg/kg bodyweight), the HPD molecules are distributed all over the body. However, after a few days the material is excreted from the body, but not from tumours, where it is selectively retained, as illustrated in Fig.10.37. The HPD molecules themselves have no therapeutic effect but they have two important properties that can be used in connection with the selective retention just mentioned. These properties are set on a spectroscopic foundation in Fig.10.38. The molecules have a very characteristic dual-peaked fluorescence in the red spectral region, as shown in Fig.10.38. The excitation is most efficiently performed in the Soret band peaking at 405 nm. The violet lines of a Kr^+ laser (Sect. 8.4.5) fall conveniently in this region, but Ar^+ laser lines or the N_2 laser emission can also be used. The fluorescence can be utilized for localizing tumours [10.125-130], e.g. in the lung or bladder, that would otherwise not be visible in normal endoscopic investigations with viewing devices based on fibre-optics. The other important property of HPD is that upon excita-

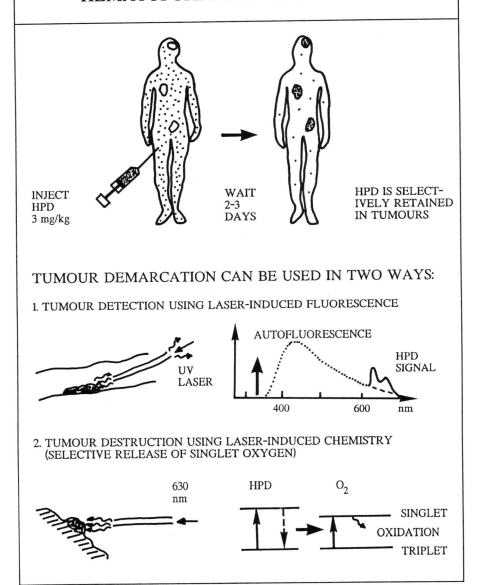

Fig.10.37. Principle of HPD-PDT and tumour fluorescence detection [10.125]

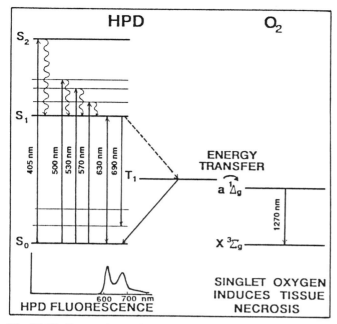

Fig. 10.38. Energy level diagrams for HPD and oxygen molecules [10.126]

tion and subsequent transfer to its triplet state, the HPD molecules can transfer their excitation energy to ground-state $X\,^3\Sigma_g$ oxygen molecules that are present in the tissue, as indicated in Figs. 10.37 and 38. The oxygen molecules are then excited from the triplet state to a singlet state, a $^1\Delta_g$. Singlet oxygen is known to strongly react with tissue through rapid oxidation and the tissue becomes necrotic. The process is thus of the laser-induced chemistry type (Sect. 10.4.1) and involves heating of the tissue only as a weak secondary effect. Since the process only occurs in the presence of HPD and HPD is accumulated in tumours, selective therapy is obtained. Again, the HPD excitation most efficiently takes place in the Soret band. However, normally the absorption peak of HPD at 630 nm is preferred since tissue transmits this radiation much better as the wavelength falls outside the region of strong haemoglobin absorption (Fig. 10.36). An effective penetration depth of about 1 cm is achieved. CW argon-ion or pulsed copper vapour laser pumping of a dye laser is normally used to generate the 630 nm radiation but the 628 nm emission from a pulsed gold vapour laser can also be utilized. The laser radiation can conveniently be delivered through a quartz optical fibre. For superficial tumours the radiation is applied directly to the surface and doses of tens of J/cm² are normally used. An example of tumour treatment is shown in Fig. 10.39, where a basal cell carcinoma was healed. The radiation can also be fibre-optically delivered through a bronchoscope or a cystoscope for lung and bladder applications, respectively. For deep-lying lesions the laser light can be transmitted fibre-optically through the lumen of a syringe needle that is inserted into the

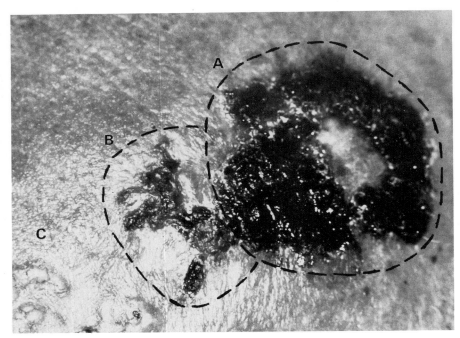

Fig.10.39. Photograph of human basal cell cancer lesions one week after HPD-PDT. The area A received a uniform light dose of 60 J/cm^2. The strong demarcatation of destructed tissue from undamaged tissue due to the selective uptake of HPD in tumour cells only is evident. The area B received only 30 J/cm^2. In area C untreated tumour parts are visible

tumour mass. A remarkable rate of success has been reported for this new type of treatment which is normally referred to as HPD-PDT (Hematoporphyrin Derivative Photo-Dynamic Therapy). New photosensitizers are being investigated. Phthalocyanines, chlorines, and bensoporphyrines have the attractive property of absorbing laser radiation at about 680 nm where the tissue penetration depth is much greater (Fig.10.36).

We will end this section by further discussing fluorescence diagnostics techniques for tissue, since these techniques also have a wide applicability in other areas. Clearly, in practical monitoring of HPD-bearing tissue one has to address the problems of how to deal with background signals that tend to obscure the specific signal. In Fig.10.40 spectra from a rat-brain tumour and surrounding tissue from an HPD-injected animal are shown. A nitrogen laser at 337 nm was used for the excitation and the fluorescence spectra were recorded with an optical multichannel analyser. In the left part of the figure a tumour spectrum is shown with the characteristic dual-peaked HPD emission sitting on the slope of the strong blue natural tissue fluorescence. To the right a spectrum for normal brain is shown. The signal intensities as defined in the spectra are also evaluated. It can be seen that the blue fluorescence intensity B decreases in the tumour tissue. The background-free HPD signal A at 630 nm clearly increases in the tumour while

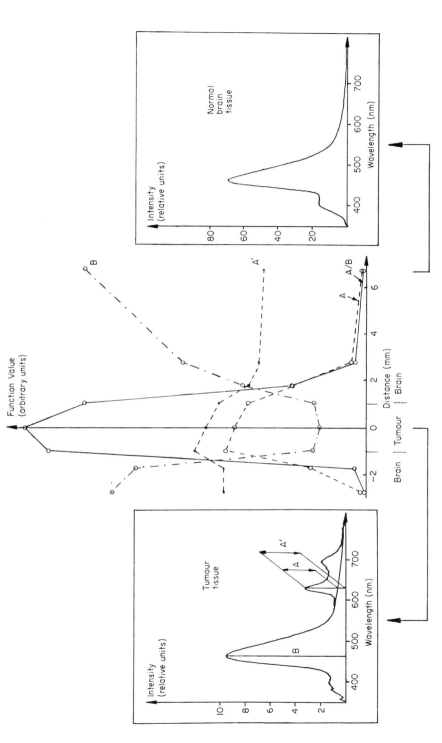

Fig.10.40. Fluorescence spectra for a rat-brain tumour and a normal brain following HPD injection. Demonstration of contrast enhancement by ratio formation [10.131]

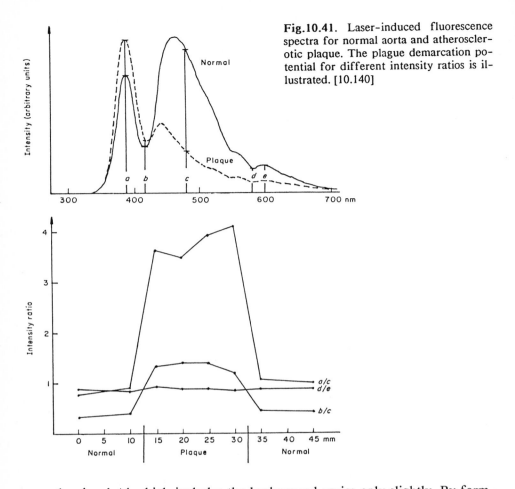

Fig.10.41. Laser-induced fluorescence spectra for normal aorta and atherosclerotic plaque. The plague demarcation potential for different intensity ratios is illustrated. [10.140]

the signal A' which includes the background varies only slightly. By forming the ratio A/B, the contrast between normal and malignant tissue is strongly enhanced. We note that monitoring a dimensionless quantity has many advantages. Temporal or spatial variations in the exciting radiation will not influence the result. The ratio is also insensitive to the surface topography, i.e. the signal is not influenced by the angles of incidence and detection but only responds to changes in the specific spectral signature. These concepts are valuable for imaging fluorescence measurement techniques that can be applied for medical as well as for industrial and remote-sensing applications [10.132].

Differences in the spectral properties of various organic substances can also be utilized for diagnostics [10.133]. Natural tissue fluorescence (autofluorescence) has been used for tumour characterization [10.134-136] as well as for caries studies in teeth [10.137,138]. An emerging application is characterization of artheroscleroris in blood vessels [10.139,140]. There are substantial spectral differences in fluorescence spectra from normal and diseased aortic wall, as illustrated in Fig.10.41, where also the demarcation

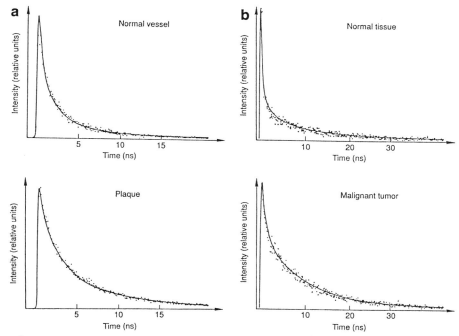

Fig. 10.42. Fluorescence-decay curves showing demarcations between normal vessel and atherosclerotic plague (*left*), and normal tissue and malignant tumor (*right*) [10.48]

potential for different intensity ratios is shown. Fluorescence diagnostics would be particularly valuable for guiding laser removal of artherosclerotic plaque [10.141, 142].

Time-resolved fluorescence spectroscopy is also a valuable tool in biological and medical research [10.143-147]. Since the lifetimes involved are normally short, picosecond spectroscopy techniques are frequently employed (Sect. 9.4). Examples of fluorescence decay curves for tissue recorded with delayed coincidence techniques employing a frequency-doubled picosecond dye laser are depicted in Fig. 10.42. The decay characteristics allow the discrimination between tumour and normal tissue, and atherosclerotic plaque and normal vessel wall, respectively. General surveys of the use of LIF for medical diagnostics can be found in [10.148, 149].

References

Chapter 2[1]

2.1 H. Haken, H.C. Wolf: *Atomic and Quantum Physics*, 2nd. ed. (Springer, Berlin, Heidelberg 1987)
2.2 B. Cagnac, J.C. Pebay-Peyroula: *Modern Atomic Physics: I Fundamental Principles, II Quantum Theory and its Applications* (Wiley, New York 1975)
2.3 H.G. Kuhn: *Atomic Spectra*, 2nd. ed. (Longmans, London 1969)
2.4 H.E. White: *Introduction to Atomic Spectroscopy* (McGraw-Hill, London 1934)
2.5 B.W. Shore, D.H. Menzel: *Principles of Atomic Spectra* (Wiley, New York 1968)
2.6 G.K. Woodgate: *Elementary Atomic Structure* (Clarendon, Oxford 1980)
2.7 M. Weissbluth: *Atoms and Molecules* (Academic, New York 1978)
2.8 J.C. Wilmott: *Atomic Physics* (Wiley, Chichester 1984)
2.9 E.U. Condon, G.H. Shortley: *The Theory of Atomic Spectra* (Cambridge Univ. Press, Cambridge 1964)
2.10 E.U. Condon, H. Odabasi: *Atomic Structure* (Cambridge Univ. Press, Cambridge 1980)
2.11 I.I. Sobelman: *Atomic Spectra and Radiative Transitions*, Springer Ser. Chem. Phys., Vol.1, (Springer, Berlin, Heidelberg 1979)
2.12 G. Herzberg: *Atomic Spectra and Atomic Structure* (Dover, New York 1944)
2.13 R.D. Cowan: *The Theory of Atomic Structure and Spectra* (Univ. California Press, Berkeley 1981)
2.14 H.A. Bethe, E.E. Salpeter: *Quantum Mechanics of One- and Two-Electron Atoms* (Springer, Berlin, Heidelberg 1957; Plenum/Rosetta, New York 1977)
2.15 C. Cohen-Tannoudji, B. Diu, F. Laloe: *Quantum Mechanics*, Vols.1 and 2 (Wiley, New York 1977)
2.16 L.I. Schiff: *Quantum Mechanics* (McGraw-Hill, New York 1968)
2.17 A. Messiah: *Quantum Mechanics* (Wiley, New York 1962)
2.18 E. Merzbacher: *Quantum Mechanics*, 2nd. ed. (Wiley, Orlando, FL 1970)
2.19 C.E. Moore: *Atomic Energy Levels as Derived from Optical Spectra*, Circular 467, National Bureau of Standards, Volumes I-III, Reissued 1971 as NSRDS-NBS 35 Vol. I 1H - 23V, Vol. II 24Cr - 41Nb, Vol. III 42Mo - 57L (NBS, Washington, DC 1971)
2.20 W.C. Martin, R. Zalubas, L. Hagan: *Atomic Energy Levels - The Rare Earth Elements*, NSRDS-NBS 60 (US Govt. Printing Office, Washington, DC 1978)
2.21 S. Bashkin, J.O. Stoner: *Atomic Energy Levels and Grotrian Diagrams*, Vol. I H I - P XV (1975), Vol. I Addenda (1978), Vol. II S I - Ti XXII (1978), Vol. III V I - Cr XXIV (1981), Vol. IV Mn I-XXV (North Holland, Amsterdam)
2.22 B. Edlén: Atomic spectra, in *Handbuch der Physik*, Vol. XXVII, ed. by S. Flügge (Springer, Berlin, Heidelberg 1964)
B. Edlén: Energy structure of highly ionized atoms. [Ref.2.59, Pt.D, p.271]
2.23 N.H. March: *Self-consistent Fields in Atoms* (Pergamon, Oxford 1978)

---

[1] Note that we make reference only to the European edition of Scientific American. The number following the slash gives the issue number. Page numbers for the American edition differ slightly.

2.24 C. Froese-Fisher: *The Hartree-Fock Method for Atoms* (Wiley, New York 1977)
2.25 C. Froese-Fisher: Multi-configuration Hartree-Fock calculations for complex atoms. [Ref.2.59, Pt.C, p.29]
2.26 B.G. Wybourne: *Spectroscopic Properties of Rare Earths* (Wiley, New York 1965)
2.27 B.R. Judd: Complex atomic spectra. Rep. Prog. Phys. **48**, 907 (1985)
2.28 M.J. Seaton: Quantum defect theory. I. General formulation. Proc. Phys. Soc. London **88**, 801 (1966); II. Illustrative one-channel and two channel problems. Proc. Phys. Soc. London **88**, 815 (1966)
2.29 U. Fano: Unified treatment of perturbed series, continous spectra and collisions. J. Opt. Soc. Am. **65**, 979 (1975)
2.30 J.A. Armstrong, P. Esherick, J.J. Wynne: Bound even parity J=0 and 2 spectra of Ca: A multichannel quantum-defect theory analysis. Phys. Rev. **A15**, 180 (1977)
P. Esherick: Bound, even parity J=0 and J=2 spectra of Sr. Phys. Rev. **A15**, 1920 (1977)
M. Aymar, O. Robaux: Multichannel quantum defect analysis of the bound even-parity J=2 spectra of neutral barium. J. Phys. **B12**, 531 (1979)
2.31 N. Ryde: *Atoms and Molecules in Electric Fields* (Almquist & Wiksell International, Stockholm 1976)
2.32 T.W. Ducas, M.G. Littman, R.R. Freeman, D. Kleppner: Stark ionization of high-lying states of sodium. Phys. Rev. Lett. **35**, 366 (1975)
2.33 T.F. Gallagher, L.M. Humphrey, R.M. Hill, S.A. Edelstein: Resolution of $[m_l]$ and $[m_j]$ levels in the electric field ionization of highly excited states of Na. Phys. Rev. Lett. **37**, 1465 (1976)
T.F. Gallagher, L.M. Humphrey, R.M. Hill, W. Cooke, S.A. Edelstein: Fine structure intervals and polarizabilities of highly excited p and d states of sodium. Phys. Rev. **A15**, 1937 (1977)
2.34 D. Kleppner, M.G. Littman, M.L. Zimmerman: Highly excited atoms. Sci. Am. **244/5**, 108 (1981)
2.35 H. Kopfermann: *Nuclear Moments* (Academic, New York 1958)
2.36 W.A. Nierenberg, I. Lindgren: In *Alpha-, Beta-, and Gamma-Ray Spectroscopy*, ed. by K. Siegbahn (North Holland, Amsterdam 1965) p.1263
2.37 G.H. Fuller, V.W. Cohen: Nuclear spins and moments. Nuclear Data Tables **A5**, 433 (1969)
2.38 E. Arimondo, M. Inguscio, P. Violino: Experimental determinations of the hyperfine structure in the alkali atoms. Rev. Mod. Phys. **49**, 31 (1977)
2.39 I. Lindgren, A. Rosén: Relativistic self-consistent field calculations with application to hyperfine interaction. Pt.I Relativistic self-consistent fields, Pt.II Relativistic theory of atomic hyperfine interaction. Case Stud. Atom. Phys. **4**, 93 (1974), Pt.III Comparison between theoretical and experimental hyperfine structure results. Case Stud. Atom. Phys. **4**, 197 (1974)
2.40 S. Büttgenbach: *Hyperfine Structure in 4d- and 5d-Shell Atoms*, Springer Tracts Mod. Phys., Vol.96 (Springer, Berlin, Heidelberg 1982)
2.41 I. Lindgren, J. Morrison: *Atomic Many-Body Theory*, 2nd ed., Springer Ser. Atoms Plasmas, Vol.3 (Springer, Berlin, Heidelberg 1987)
2.42 I. Lindgren: Effective operators in the atomic hyperfine interaction. Rep. Prog. Phys. **47**, 345 (1984)
2.43 J.R.P. Angel, P.G.H. Sandars: The hyperfine structure Stark effect, I. Theory. Proc. Roy. Soc. **A305**, 125 (1968)
2.44 A. Khadjavi, A. Lurio, W. Happer: Stark effect in the excited states of Rb, Cs, Cd, and Hg. Phys. Rev. **167**, 128 (1968)
2.45 K. Heilig, A. Steudel: Changes in mean square nuclear charge radii from optical isotope shifts. Atomic Data and Nuclear Data Tables **14**, 613 (1974)

2.46 K. Heilig, A. Steudel: New developments in classical optical spectroscopy. [Ref.2.58, Pt.A, p.263]
2.47 W.H. King: *Isotope Shifts in Atomic Spectra* (Plenum, New York 1984)
2.48 V.W. Hughes, B. Bederson, V.W. Cohen, F.M.J. Pichanick (eds.): *Atomic Physics*, New York 1968 (Plenum, New York 1969)
2.49 G.K. Woodgate, P.G.H. Sandars (eds): *Atomic Physics 2*, Oxford 1970 (Plenum, New York 1971)
2.50 S.J. Smith, G.K. Walters (eds.): *Atomic Physics 3*, Boulder, Colo. 1972 (Plenum, New York 1973)
2.51 G. zu Putlitz, E.W. Weber, A. Winnacker (eds.): *Atomic Physics 4*, Heidelberg 1974 (Plenum, New York 1975)
2.52 R. Marrus, M. Prior, H. Shugart (eds.): *Atomic Physics 5*, Berkeley 1976 (Plenum, New York 1977)
2.53 A.M. Prokhorov, R. Damburg (eds.): *Atomic Physics 6*, Riga 1978 (Plenum, New York 1979)
2.54 D. Kleppner, F.M. Pipkin (eds.): *Atomic Physics 7*, Cambridge, Mass. 1980 (Plenum, New York 1981)
2.55 I. Lindgren, A. Rosén, S. Svanberg (eds.): *Atomic Physics 8*, Göteborg 1982 (Plenum, New York 1983)
2.56 R.S. Van Dyck, Jr., E.N. Fortson (eds.): *Atomic Physics 9*, Seattle 1984 (World Scientific, Singapore 1985)
2.57 H. Narumi, I. Shimamura (eds.): *Atomic Physics 10*, Tokyo 1986 (North Holland, Amsterdam 1987)
2.58 W. Hanle, H. Kleinpoppen (eds.): *Progress in Atomic Spectroscopy*, Pts.A and B (Plenum, New York 1978 and 1979)
2.59 H.J. Beyer, H. Kleinpoppen (eds.): *Progress in Atomic Spectroscopy*, Pts.C and D (Plenum, New York 1984 and 1987)

Chapter 3

3.1 C.E. Banwell: *Fundamentals of Molecular Spectroscopy* (McGraw-Hill, London 1983)
3.2 G.B. Barrow: *The Structure of Molecules* (Benjamin, New York 1963)
3.3 J.I. Steinfeld: *Molecules and Radiation*, 2nd. ed. (MIT Press, Cambridge, Mass. 1985)
3.4 G.W. King: *Spectroscopy and Molecular Structure* (Holt, Rinehart and Winston, New York 1964)
3.5 M. Weissbluth: *Atoms And Molecules* (Academic, New York 1978)
3.6 W.G. Richards, P.R. Scott: *Structure and Spectra of Molecules* (Wiley, Chichester 1985)
3.7 D.A. McQuarrie: *Quantum Chemistry* (Oxford Univ. Press, Oxford 1983) Chap.10
3.8 A.C. Hurley: *Introduction to the Electron Theory of Small Molecules* (Academic, London 1976)
3.9 R. McWeeny, B.T. Pickup: Quantum theory of molecular electronic structure. Rep. Prog. Phys. 43, 1065 (1980)
3.10 A. Hinchliffe: *Ab Initio Determination of Molecular Electronic Structure* (Hilger, Bristol 1987)
3.11 G. Herzberg: *Molecular Spectra and Molecular Structure. I The Spectra of Diatomic Molecules* (Van Nostrand, Princeton 1963)
3.12 G. Herzberg: *Molecular Spectra and Molecular Structure. II Infrared and Raman Spectra of Polyatomic Molecules* (Van Nostrand, Princeton 1965)

3.13 G. Herzberg: *Molecular Spectra and Molecular Structure. III Electronic Spectra and Electronic Structure of Polyatomic Molecules* (Van Nostrand, Princeton 1966)

3.14 K.P. Huber, G. Herzberg: *Molecular Spectra and Molecular Structure. IV Constants of Diatomic Molecules* (Van Nostrand Reinhold, New York 1979)
H. Lefebre-Brion, R.W. Field: *Perturbations in the Spectra of Diatomic Molecules* (Academic, Orlando 1986)

3.15 L.M. Sverdlov, M.A. Kovner, E.P. Krainov: *Vibrational Spectra of Polyatomic Molecules* (Israel Program for Scientific Translations, Jerusalem 1974)

3.16 R.F. Hout, Jr.: *Pictorial Approach to Molecular Structure and Reactivity* (Wiley, New York 1984)

3.17 R.W.B. Pearse, A.G. Gaydon: *The Identification of Molecular Spectra* (Chapman and Hall, London 1950)

Chapter 4

4.1 D.J. Jackson: *Classical Electrodynamics*, 2nd. ed. (Wiley, New York 1975)

4.2 P. Lorraine, D. Corson: *Electromagnetic Fields and Waves*, 2nd. ed. (Freeman, New York 1970)

4.3 F.A. Hopf, G.I. Stegeman: *Applied Classical Electrodynamics*, Vol.1: Linear Optics, Vol.2: Non-Linear Optics (Wiley, New York 1986)

4.4 W. Heitler: *The Quantum Theory of Radiation*, 2nd. ed. (Oxford Univ. Press, Oxford 1944)

4.5 R. Loudon: *The Quantum Theory of Light*, 2nd. ed. (Clarendon, Oxford 1983)

4.6 F.H.M. Faisal: *Theory of Multiphoton Processes* (Plenum, New York 1987)

4.7 H.C. van de Hulst: *Light Scattering by Small Particles* (Wiley, New York 1957)
H.C. van de Hulst: *Multiple Light Scattering*, Vols.1 and 2 (Academic, Orlando 1980)

4.8 B. Chu: *Laser Light Scattering* (Academic, New York 1974)

4.9 C.F. Bohren, D.R. Huffman: *Absorption and Scattering of Light by Small Particles* (Wiley, New York 1983)

4.10 A. Anderson (ed.): *The Raman Effect, Principles*, and *The Raman Effect, Applications* (Dekker, New York 1971 and 1973)

4.11 D.A. Long: *Raman Spectroscopy* (McGraw-Hill, New York 1977)

4.12 A. Einstein: On the quantum theory of radiation. Phys. Zs. **18**, 124 (1917)

4.13 T.F. Gallagher, W.E. Cooke: Interaction of blackbody radiation with atoms. Phys. Rev. Lett. **42**, 835 (1979)

4.14 J. Farley, W.H. Wing: Accurate calculation of dynamic Stark shifts and depopulation rates of Rydberg energy levels induced by blackbody radiation. Hydrogen, helium and alkali-metal atoms. Phys. Rev. **A23**, 2397 (1981)

4.15 D. Kleppner: Turning off the vacuum, Phys. Rev. Lett. **47**, 233 (1981)
R.G. Hulet, E.S. Hilfer, D. Kleppner: Inhibited spontaneous emission by a Rydberg atom. Phys. Rev. Lett. **55**, 2137 (1985)
W. Jhe, A. Anderson, E.A. Hinds, O. Meschede, L. Moi, S. Haroche: Suppression of spontaneous decay at optical frequencies: Test of vaccuum-field anisotropy in confined space. Phys. Rev. Lett. **58**, 666 (1987)

4.16 S. Haroche, J.M. Raimond: *Advances in Atomic and Molecular Physics* **20**, 350 (Academic, New York 1985)
J.A.C. Gallas, G. Leuchs, H. Walther, H. Figger: ibid. p.414.

4.17 D.F. Walls: Evidence for the quantum nature of light. Nature **280**, 451 (1979)

4.18 R. Loudon: Non-classical effects in the statistical properties of light. Rep. Prog. Phys. **43**, 913 (1980)

4.19 D.F. Walls: Squeezed states of light. Nature **306**, 141 (1983)
G. Leuchs: Squeezing the quantum fluctuations of light. Contemp. Phys. **29**, 299 (1988)
R.E. Slusher, B. Yurke: Squeezed light. Sci. Am. **258**/5, 50 (1988)
4.20 R.E. Slusher, L.W. Hollberg, B. Yurke, J.C. Mertz, J.F. Valley: Observation of squeezed states generated by four-wave mixing in an optical cavity. Phys. Rev. Lett. **55**, 2409 (1985)
L.-A. Wu, H. J. Kimble, J.L. Hall, H. Wu: Generation of squeezed states by parametric down conversion. Phys. Rev. Lett. **57**, 2520 (1986)
4.21 H.J. Kimble, D.F. Walls (eds.): Squeezed states of the electromagnetic field. J. Opt. Soc. Am. **4**, No.10 (p.1450) (1987) (Feature Issue)
4.22 J.D. Harvey, D.F. Walls (eds.): *Quantum Optics IV*, Springer Proc. Phys., Vol.12 (Springer, Berlin, Heidelberg 1986)
J.D. Harvey, D.F. Walls (eds.): *Quantum Optics V*, Springer Proc. Phys., Vol.41 (Springer, Berlin, Heidelberg 1989)
4.23 S. Sarkar (ed.): *Nonlinear Phenomena and Chaos* (Hilger, Bristol 1986)
4.24 E.R. Pike, S. Sarkar (eds.): *Frontiers in Quantum Optics* (Hilger, Bristol 1986)
E.R. Pike, H. Walther (eds.): *Photons and Quantum Fluctuations* (Hilger, Bristol 1988)
4.25 P. Meystre, M. Sargent III: *Elements of Quantum Optics* (Springer, Berlin, Heidelberg 1990)
4.26 G. Herzberg: *Molecular Spectra and Molecular Structure I, The Spectra of Diatomic Molecules* (van Nostrand, Princeton, NJ 1963)
4.27 H. Inaba: In *Laser Monitoring of the Atmosphere*, ed. by E.D. Hinkley, Topics Appl. Phys., Vol.14 (Springer, Berlin, Heidelberg 1976)
4.28 C.K. Sloan: J. Phys. Chem. **59**, 834 (1955)
4.29 A. D'Alessio, A. Di Lorenzo, A.F. Serafim, F. Beretta, S. Masi, C. Venitozzi: Soot formation in methane-oxygen flames. 15 th Int'l Symp. on Combustion (The Combustion Institute, Pittsburg 1975) p.1427
A. D'Alessio, A. Di Lorenzo, A. Borghese, F. Beretta, S. Masi: Study of the soot nucleation zone of rich methane-oxygen combustion. 16 th Int'l Symp. on Combustion (The Combustion Institute, Pittsburg 1977) p.695
4.30 H.M. Nussenzveig: The theory of the rainbow. Sci. Am. **236**/4, 116 (1977)
4.31 D.K. Lynch: Atmospheric haloes. Sci. Am. **238**/4, 144 (1978)
A.B. Fraser, W.H. Mach: Mirages. Sci. Am. **234**/1, 102 (1976)
W. Tape: The topology of mirages. Sci. Am **252**/6, 100 (1985)
D.J.K. O'Connel: The green flash. Sci. Am. **202**/1, 112 (1960)
H.C. Bryant, N. Jarmie: The glory. Sci. Am. **231**/1, 60 (1974)
J. Walker (ed.): *Light from the Sky*, Reprint collection (Scientific American) (Freeman, San Francisco 1980)
4.32 I. Shimamura, K. Takayanagi (eds.): *Electron-Molecule Collisions* (Plenum, New York 1984)
4.33 G.F. Drukarev: *Collisions of Electrons with Atoms and Molecules* (Plenum, New York 1987)
4.34 F. Brouillard (ed.): *Atomic Processes in Electron-Ion and Ion-Ion Collisions*, NATO ASI Series, Vol.145 (Plenum, New York 1987)
4.35 J.M. Bowman (ed.): *Molecular Collision Dynamics*, Topics Current Phys., Vol.33 (Springer, Berlin, Heidelberg 1983)
4.36 U. Fano, A.R.P. Rau: *Atomic Collisions and Spectra* (Academic, New York 1986)
4.37 P. Schattschneider: *Fundamentals of Inelastic Electron Scattering* (Springer, Wien 1986)

Chapter 5

5.1 B. Crasemann (ed.): *Atomic Inner-Shell Physics* (Plenum, New York 1985)
5.2 G.L. Clark: *Applied X-Rays* (McGraw-Hill, New York 1955)
5.3 B. Agarwal: *X-Ray Spectroscopy*, 2nd edn., Springer Ser. Opt. Sci., Vol.15 (Springer, Berlin, Heidelberg 1990)
5.4 R. Jenkins, R.W. Gould, D. Gedcke: *Quantitative X-Ray Spectrometry* (Dekker, New York 1981)
5.5 S.A.E. Johansson: Proton-induced X-ray emission spectrometry - state of the art. Fresenius Z. Anal. Chem. 324, 635 (1986)
5.6 S. Johansson (ed.): Proc. 1st Int'l PIXE Conf., Nucl. Instrum. Methods 142, 1 (1977); Proc. 2nd Int'l PIXE Conf., Nucl. Instrum. Methods 181, 1 (1981)
B. Martin (ed.): Proc. 3rd Int'l PIXE Conf., Nucl. Instrum. Methods 231, 1 (1984)
H. van Rinsvelt, S. Bauman, J.W. Nelson, J.W. Winchester (eds.): Proc. 4th Int'l PIXE Conf., Nucl. Instr. Methods B22, 1 (1987)
5.7 R.W. Shaw: Air pollution by particles. Sci. Am. 257/2, 84 (1987)
T.G. Dzubiey (ed.): *X-Ray Fluorescence Analysis of Environmental Samples* (Ann Arbor Science Publ., Ann Arbor 1977)
5.8 J. Nordgren, L. Selander, L. Pettersson, C. Nordling, K. Siegbahn, H. Ågren: Core state vibrational excitations and symmetry breaking in the CK and OK emission spectra of CO_2. J. Chem. Phys. 76, 3928 (1982)
5.9 B. Crasemann (ed.): *Proc. Conf. on X-ray and Atomic Inner Shell Physics*, AIP Conf. Proc. 94 (American Inst. Phys., New York 1982), especially article by C. Nordling
5.10 A. Meisel, G. Leonhardt, R. Szargon: *X-Ray Spectra and Chemical Bonding*, Springer Ser. Chem. Phys., Vol.37 (Springer, Berlin, Heidelberg 1989)
5.11 J. Nordgren: Ultra-soft X-ray emission spectroscopy - a progress report, in Proc. 14th Int'l Conf. on X-Ray and Inner Shell Processes. J. Physique C 9, Suppl.12 (1987)
5.12 J. Nordgren, C. Nordling: Ultrasoft X-ray emission from atoms and molecules. Comments. At. Mol. Phys. 13, 229 (1983)
5.13 J. Nordgren, H. Ågren: Interpretation of ultra-soft X-ray emission spectra. Comments. At. Mol. Phys. 14, 203 (1984)
5.14 J. Berkowitz: *Photoabsorption, Photoionization, and Photoelectron Spectroscopy* (Academic, New York 1979)
5.15 E.A. Stern: Structure determination by X-ray absorption. Contemp. Phys. 19, 289 (1978)
5.16 A. Bianconi, L. Incoccia, S. Stipchich (eds.): *EXAFS and Near Edge Structure*, Springer Ser. Chem. Phys., Vol.27 (Springer, Berlin, Heidelberg 1983)
5.17 B.K. Teo: *EXAFS: Basic Principles and Data Analysis, Inorganic Chemistry Concepts*, Vol.9 (Springer, Berlin, Heidelberg 1986)
5.18 P.A. Lee, P.H. Citrin, B.M. Kincaid: Extended X-ray absorption fine structure - its strengths and limitations as a structural tool. Rev. Mod. Phys. 53, 769 (1981)
5.19 B. Lindberg, R. Maripuu, K. Siegbahn, R. Larsson, C.G. Gölander, J.C. Eriksson: ESCA studies of heparinized and related surfaces. J. Coll. Int. Sci. 95, 308 (1983)
5.20 K. Siegbahn et al. (eds.): *ESCA: Atomic, Molecular and Solid State Structure Studied by Means of Electron Spectroscopy* (Almqvist and Wiksell, Uppsala 1967)
5.21 H. Siegbahn, L. Karlsson: Photoelectron Spectroscopy, in *Handbuch der Physik*, Vol.31, ed. by E. Mehlhorn (Springer, Berlin, Heidelberg 1982)

5.22 C.R. Brundle, A.D. Baker (eds.): *Electron Spectroscopy: Theory, Techniques and Applications*, Vols.1-5 (Academic, New York 1977-84)
5.23 R.E. Ballard: *Photoelectron Spectroscopy and Molecular Orbital Theory* (Hilger, London 1978)
5.24 P. Weightman: X-ray excited Auger and photoelectron spectroscopy. Rep. Prog. Phys. **45**, 753 (1982)
5.25 K. Siegbahn: Some current problems in electron spectroscopy, in *Atomic Physics 8*, ed. by I. Lindgren, A. Rosén, S. Svanberg (Plenum, New York 1983) p.243
5.26 K. Siegbahn: Photoelectron spectroscopy: retrospects and prospects. Phil. Trans. Roy. Soc. Lond. A **318**, 3 (1986)
5.27 N.V. Smith, F.J. Himpsel: Photoelectron spectroscopy, in *Handbook of Synchrotron Radiation*, ed. by E.E. Koch (North-Holland, Amsterdam 1983)
5.28 J.W. Rabalais: *Principles of Ultraviolet Photoelectron Spectroscopy* (Wiley-Interscience, New York 1977)
5.29 S. Svensson, N. Mårtensson, E. Basilier, P.Å. Malmquist, U. Gelius, K. Siegbahn: Core and valence orbitals in solid and gaseous mercury by means of ESCA. J. Electr. Spectrosc. Rel. Phenom. **9**, 51 (1976)
5.30 K. Siegbahn: Electron spectroscopy and molecular structure. J. Pure Appl. Chem. **48**, 77 (1976)
5.31 H. Siegbahn: Electron spectroscopy for chemical analysis of liquids and solutions. J. Phys. Chem. **89**, 897 (1985)
5.32 L.-G. Petersson, S.-E. Karlsson: Clean and oxygen exposed potassium studied by photoelectron spectroscopy. Phys. Scri. **16**, 425 (1977)
5.33 R.H. Williams: Electron spectroscopy of surfaces. Contemp. Phys. **19**, 389 (1978) (1978)
5.34 D. Briggs, M.P. Seah (eds.): *Practical Surface Analysis by Auger and X-Ray Photoelectron Spectroscopy* (Wiley, Chichester 1983)
5.35 K. Siegbahn: Electron spectroscopy for solids, surfaces, liquids and free molecules, in *Molecular Spectroscopy*, ed. by A.R. West (Heyden, London 1977) Chap.15

Chapter 6

6.1 *MIT Wavelength Tables*, ed. by G.R. Harrison (Wiley, New York 1969)
6.2 F.M. Phelps III: *MIT Wavelength Tables*, Vol.2: *Wavelengths by Element* (MIT Press, Cambridge, Mass. 1982)
6.3 R.L. Kelly, L.J. Palumbo: Atomic and ionic emission lines below 2000 Angstroms, hydrogen through krypton. NRL Report 7599 (Naval Research Laboratory, Washington, DC 1973)
R.L. Kelly: Atomic and ionic spectral lines below 2000 Angstroms, hydrogen through krypton. J. Phys. Chem. Ref. Data, Suppl. No.1 to Vol.16 (1987)
6.4 J. Reader, C.H. Corliss, W.L. Wiese, G.A. Martin: *Wavelengths and Transition Probabilities for Atoms and Atomic Ions*. NSRDS-NBS 68 (US Govt. Prtg. Off., Washington, DC 1980)
6.5 A.R. Striganov, N.S. Sventitskii: *Tables of Spectral Lines in Neutral and Ionized Atoms* (IFI/Plenum, New York 1968)
6.6 A. Thorne: *Spectrophysics* (Chapman & Hall, London 1974)
B.P. Straugan, S. Walker: *Spectroscopy*, Vols. 1, 2, and 3 (Chapman & Hall, London 1976)
6.7 P.F.A. Klinkenberg: In *Methods of Experimental Physics*, Vol.13, Spectroscopy (Academic, New York 1976) p.253
6.8 P. Bousquet: *Spectroscopy and its Instrumentation* (Hilger, London 1971)

6.9 J. Kuba, L. Kucera, F. Plzak, M. Dvorak, J. Mraz: *Coincidence Tables for Atomic Spectroscopy* (Elsvier, Amsterdam 1965)

6.10 R. Beck, W. Englisch, K. Gürs: *Tables of Laser Lines in Gases and Vapors*, 3rd. ed., Springer Ser. Opt. Sci., Vol.2 (Springer, Berlin, Heidelberg 1980)

6.11 A.C.G. Mitchell, M.W. Zemansky: *Resonance Radiation and Excited Atoms* (Cambridge Univ. Press, Cambridge 1961)

6.12 I.I. Sobelmann, L.A. Vainshtein, E.A. Yukov: *Excitation of Atoms and Broadening of Spectral Lines*, Springer Ser. Chem. Phys., Vol.7 (Springer, Berlin, Heidelberg 1981)

6.13 B. Wende (ed.): *Spectral Line Shapes* (Conf. Proc.) (de Gruyter, Berlin 1981)

6.14 K. Burnett (ed.): *Spectral Line Shapes* (Conf. Proc.) (de Gruyter, Berlin 1983)

6.15 C.H. Corliss, W.R. Bozman: *Experimental Transition Probabilities for Spectral Lines of Seventy Elements*, NBS Monograph 53 (National Bureau of Standards, Wash., DC 1962)

6.16 C.H. Corliss, J.L. Tech: Revised lifetimes of energy levels in neutral iron. J. Res. Nat. Bur. Stand. Sect. A **80**, 787 (1976)

6.17 C. de Michelis, M. Mattioli: Spectroscopy and impurity behaviour in fusion plasmas. Rep. Prog. Phys. **47**, 1233 (1984)

6.18 E.T. Kennedy: Plasmas and intense laser light. Contemp. Phys. **25**, 31 (1984)

6.19 T.P. Hughes: *Plasmas and Laser Light* (Wiley, New York 1975)

6.20 G. Bekefi (ed.): *Principles of Laser Plasmas* (Wiley, New York 1976)

6.21 K. Laqua: Analytical Spectroscopy using Laser Atomizers, in *Analytical Laser Spectroscopy*, ed. by N. Omenetto (Wiley, New York 1979)

6.22 R.J. Rosner, R. Feder, A. Ng, F. Adams, P. Celliers, R.J. Speer: Nondestructive single-shot soft X-ray lithography and contact microscopy using a laser-produced plasma source. Appl. Spectr. **26**, 4313 (1987)

6.23 G. Schmahl, D. Rudolph (eds.): *X-Ray Microscopy*, Springer Ser. Opt. Sci., Vol.43 (Springer, Berlin, Heidelberg 1984)
D. Sayre, M. Howells, J. Kirz, H. Rarback (eds.): *X-Ray Microscopy II*, Springer Ser. Opt. Sci., Vol.56 (Springer, Berlin, Heidelberg 1988)
A.G. Michette: X-ray microscopy. Rep. Prog. Phys. **51**, 1525 (1988)

6.24 N.G. Basov, Yu. A. Zakharenkov, N.N. Zorev, G.V. Sklizkov, A.A. Rupasov, A.S. Shikanov: *Heating and Compression of Thermonuclear Targets by Laser Beams* (Cambridge Univ. Press, Cambridge 1986)

6.25 R.D. Cowan: Progress in the spectroscopy of highly ionized atoms and its use in plasma diagnostics. Phys. Scr. **24**, 615 (1981)
H.W. Drawin: Atomic physics and thermonuclear fusion research. Phys. Scr. **24**, 622 (1981)
R.C. Isler: Impurities in Tokamaks. Nuclear Fusion **24**, 1599 (1984)
E. Källne, J. Källne: X-ray spectroscopy in fusion research. Phys. Scr. **T17**, 152 (1987)

6.26 S. Bashkin: Optical spectroscopy with van de Graaff accelerators. Nucl. Instr. Meth. **28**, 88 (1964)

6.27 L. Kay: A van de Graaff beam as a source of atomic emission spectra. Phys. Lett. **5**, 36 (1963)

6.28 J.O. Stoner, J.A. Leavitt: Reduction in Doppler broadening of spectral lines in fast-beam spectroscopy. Appl. Phys. Lett. **18**, 477 (1971)

6.29 R. Hutton, L. Engström, E. Träbert: Nucl. Instrum. Meth. in Phys. Res. B **31**, 294 (1988)

6.30 L.J. Curtis, H.J. Berry, J. Bromander: A meanlife measurement of the 3d ^{2}D resonance doublet in SiII by a technique which exactly accounts for cascading. Phys. Lett. **34 A**, 169 (1971)
L.J. Curtis: In [6.37]

6.31 L. Engström: CANDY, a computer program to perform ANDC analysis of cascade corrected decay curves. Nucl. Instr. Meth. **202**, 369 (1982)

6.32 I. Martinson, A. Gaupp: Atomic physics with ion accelerators - beam-foil spectroscopy. Phys. Rep. 15, 113 (1974)
6.33 H.G. Berry, L.J. Curtis, D.G. Ellis, R.M. Schectman: Hyperfine quantum beats in oriented ^{14}N IV. Phys. Rev. Lett. 35, 274 (1975)
6.34 U. Fano, J.H. Macek: Impact excitation and polarization of the emitted light. Rev. Mod. Phys. 45, 553 (1973)
6.35 W. Wittmann, K. Tillmann, H.J. Andrä, P. Dobberstein: Fine-structure measurement of ^{4}He by zero-field quantum beats. Z. Physik 257, 279 (1972)
6.36 O. Poulsen, J.L. Subtil: Hyperfine structure measurement in Be III. J. Phys. B 7, 31 (1974)
6.37 S. Bashkin (ed.): *Beam-Foil Spectroscopy*, Topics Current Phys., Vol.1 (Springer, Berlin, Heidelberg 1976)
6.38 I.A. Sellin, D.J. Pegg (eds.): *Beam-Foil Spectroscopy*, Vols.1,2 (Plenum, New York 1976)
6.39 S. Bashkin (ed.): Beam-Foil Spectroscopy, Proc. 3rd Int'l Conf.. Nucl. Instr. Meth. 110 (1973).
6.40 Proc. Int. Conf. on Fast Ion Beam Spectroscopy, Proc. Colloque No. 1. J. Physique 40 (1978)
6.41 E.J. Knystautas, R. Drouin (eds.): Proc. 6th Int'l Conf. on Fast Ion Beam Spectroscopy. Nucl. Instr. Meth. 202 (1982)
6.42 J. D. Silver, N.J. Peacock (eds): The Physics of Highly Ionized Atoms. Nucl. Instr. Meth. in Phys. Res. B9, 359-787 (1985)
6.43 H.J. Andrä: Fast Beam (Beam-Foil) Spectroscopy, in *Progress in Atomic Spectroscopy*, Pt.B, ed. by W. Hanle, H. Kleinpoppen (Plenum, New York 1979) p.829
6.44 D.J. Pegg: In *Methods of Experimental Physics*, Vol.17, ed. by P. Richard (Academic, New York 1980) p.529
6.45 I. Martinson: Recent progress in the studies of atomic spectra and transition probabilities by beam-foil spectroscopy. Nucl. Instr. Meth. 202, 1 (1982)
I. Martinson: Beam-Foil Spectroscopy, in *Treatise on Heavy-Ion Science*, Vol.5, ed. by D.A Bromley (Plenum, New York 1985)
I. Martinson: The spectroscopy of highly ionized atoms. Rep. Prog. Phys. 52, 157 (1989)
6.46 C.L. Cocke: Beam-Foil Spectroscopy, in *Methods of Experimental Physics*, Vol.13 (Academic, New York 1976)
6.47 H.G. Berry: Beam-foil spectroscopy. Rep. Progr. Phys. 40, 155 (1977)
6.48 H.G. Berry, M. Mass: Beam-foil spectroscopy. Ann. Rev. Nucl. Part. Sci. 32, 1 (1982)
6.49 J. Schwinger: On the classical radiation of accelerated electrons. Phys. Rev. 75, 1912 (1949)
R.P. Madden, K. Codling: Phys. Rev. Lett. 10, 516 (1963)
K. Codling: Applications of synchrotron radiation, Rep. Progr. Phys. 36, 541 (1973)
See also O.J. Jackson: *Classical Electrodynamics*, 2nd. ed. (Wiley, New York 1975)
6.50 D.H. Tomboulian, P.L. Hartman: Spectral and angular distribution of ultraviolet radiation from the 300 MeV Cornell synchrotron. Phys. Rev. 102, 1423 (1956)
6.51 E. Matthias, R.A. Rosenberg, E.D. Poliakoff, M.G. White, S.-T. Lee, D.A. Shirley: Time resolved VUV spectroscopy using synchrotron radiation: Fluorescent lifetimes of atomic Kr and Xe. Chem. Phys. Lett. 52, 239 (1977).
6.52 T. Möller, G. Zimmerer: Time-resolved spectroscopy with synchrotron radiation in the vacuum ultraviolet. Phys. Scr. T17, 177 (1987)
R. Rigler, O. Kristensen, J. Roslund, P. Thyberg, K. Oba, M. Eriksson: Molecular structures and dynamics: Beamline for time resolved spectroscopy at the MAX synchrotron in Lund. Phys. Scr. T17, 204 (1987)

6.53 H. Motz: Undulators and free-electron lasers. Contemp. Phys. **20**, 547 (1979)
H.P. Freund, R.K. Parker: Free-electron lasers. Sci. Am. **260**/4, 56 (1989)
6.54 V.L. Granatstein, C.W. Robertson (eds.): Third special issue on free electron lasers. IEEE J. QE-**21**, 804-1113 (1985)
6.55 J.M.J. Madey, A. Renieri (eds.): *Free Electron Lasers* (Conf. Proc.) (North Holland, Amsterdam 1985)
6.56 F.C. Marshall: *Free Electron Lasers* (Macmillan, New York 1985)
6.57 J.M. Ortega, Y. Lapierre, B. Girard, M. Billardon, P. Elleaume, C. Baziin, M. Bergher, M. Velghe, Y. Petroff: Ultraviolet coherent generation from an optical klystron. IEEE J. QE-**21**, 909 (1985)
S. Werin, M. Eriksson, J. Larsson, A. Persson, S. Svanberg: First results in coherent harmonic generation using the undulator at the MAX-Lab electron storage ring. Nucl. Instr. Meth. Phys. Res. A **290**, 589 (1990)
6.58 C. Joshi, T. Katsouleas (eds.): *Laser Accelerators of Particles*. AIP Conf. Proc. **130** (Am. Inst. Physics, New York 1985)
6.59 K. Siegbahn: Electron spectroscopy for solids, surfaces, liquids and free molecules, in *Molecular Spectroscopy* (Heyden & Son, London 1983) Chap.15, p.227
6.60 C. Kunz (ed.): *Synchrotron Radiation. Techniques and Applications*, Topics Current Phys., Vol.10 (Springer, Berlin, Heidelberg 1979).
6.61 H. Winich, S. Doniach (eds.): *Synchrotron Radiation Research* (Plenum, New York 1980)
6.62 E.E. Koch (ed.): *Handbook on Synchrotron Radiation*, Vols.1-3, (North Holland, Amsterdam 1983, 1986, 1987)
6.63 E.J. Ansaldo: Uses of synchrotron radiation. Contemp. Phys. **18**, 527 (1977)
W. Jitschin: Inner-Shell Spectroscopy with Hard Synchrotron Radiation, in *Progress in Atomic Spectroscopy*, Pt.D, ed. by H.J. Beyer, H. Kleinpoppen (Plenum, New York 1987) p.295
6.64 H.H. Malitson: The solar energy spectrum. Sky and Telescope **29**/4, 162 (1965)
6.65 W.K. Pratt: *Laser Communication Systems* (Wiley, New York 1969)
6.66 S.P. Davis: *Diffraction Grating Spectrometers* (Holt, Rinehard, Winston, New York 1970)
R.A. Sawyer: *Experimental Spectroscopy* (Dover, New York 1963)
6.67 D.A. Skoog, D.M. West: *Principles of Instrumental Analysis* (Holt-Saunders, Philadelphia 1980)
6.68 M.C. Hutley: *Diffraction Gratings* (Academic, London 1982)
Handbook of Diffraction Gratings, Ruled and Holographic (Jobin-Yvon Optical Systems, 20 Highland Ave., Metuchen, NJ 1970)
6.69 H. Walther: Das Kernquadrupolmoment des ^{55}Mn. Z. Physik **170**, 507 (1962)
6.70 J.M. Vaughan: *The Fabry-Pérot Interferometer* (Hilger, Bristol 1989)
6.71 W. Demtröder: *Laser Spectroscopy*, 2nd prt., Springer Ser. Chem. Phys., Vol.5 (Springer, Berlin, Heidelberg 1982)
W. Demtröder: *Laser Spetroskopie*, 2. Aufl. (Springer, Berlin, Heidelberg 1990)
6.72 R.W. Ramirez: *The FFT: Fundamentals and Concepts* (Prentice Hall, Englewood Cliffs, NJ 1985)
H.J. Nussbaumer: *Fast Fourier Transform and Convolution Algorithms*, 2nd. ed., Springer Ser. Inf. Sci., Vol.2 (Springer, Berlin, Heidelberg 1982)
6.73 G. Guelachvili: High accuracy Doppler-limited 10^6 samples Fourier transform spectroscopy. Appl. Opt. **17**, 1322 (1978)
6.74 S. Tolansky: *An Introduction to Interferometry* (Longmans, London 1973)
6.75 W.H. Steel: *Interferometry*, 2nd ed. (Cambridge Univ. Press, Cambridge 1983)
6.76 P. Hariharan: *Optical Interferometry* (Academic, New York 1986)
6.77 R.J. Bell: *Introductory Fourier Transform Spectroscopy* (Academic, New York 1972)
6.78 *The Optical Industry & Systems Purchasing Directory*, 26th edn. (Laurin Publ. Co., Pittsfield, MA 1980) p.B-114

6.79 R.J. Keyes (ed.): *Optical and Infrared Detectors*, 2nd. ed., Topics Appl. Phys., Vol.19 (Springer, Berlin, Heidelberg 1980)
6.80 R.H. Kingston: *Detection of Optical and Infrared Radiation*, 2nd Pr., Springer Ser. Opt. Sci., Vol.10 (Springer, Berlin, Heidelberg 1979)
6.81 R.W. Boyd: *Radiometry and the Detection of Optical Radiation* (Wiley, New York 1983)
6.82 H.H. Melchior: Demodulation and photodetection techniques, in *Laser Handbook*, Vol.1, ed. by T. Arecchi, E.O. Schulz-Dubois (North-Holland, Amsterdam 1972) Chap.7
6.83 E.L. Dereniak, D.G. Crowe: *Optical Radiation Detectors* (Wiley, New York 1984)
6.84 M. Lampton: The microchannel image intensifier. Sci. Am. **245**/5, 46 (1981)
6.85 Proc. Topical Meeting on Quantum-Limited Imaging and Image Processing (Optical Society of America, Washington, DC 1986)
6.86 *The Photonics Design & Application Handbook* (Laurin Publ. Comp., Pittsfield, MA 1990)
6.87 G.R. Fowles: *Introduction to Modern Optics* (Holt, Rinehart and Winston, New York 1968)
6.88 J. Strong: *Procedures in Experimental Physics* (Prentice Hall, New York 1945)
6.89 M. Kasha: Transmission filters for the ultraviolet. J. Opt. Soc. Am. **38**, 929 (1948)
6.90 K. Bennett, R.L. Byer: Computer controllable wedge-plate optical variable attenuator. Appl. Opt. **19**, 2408 (1980)
6.91 B. Edlén: The refractive index of air. Metrologia **2**, 71 (1966)
6.92 R. Revelle: Carbon dioxide and world climate. Sci. Am. **247**/5, 33 (1982)
R.A. Houghton, G.W. Woodwell: Global climatic change. Sci. Am. **260**/4, 18 (1989)
S.H. Schneider: The changing climate. Sci. Am. **261**/3, 38 (1989)
B.J. Mason: The greenhouse effect. Contemp. Phys. **30**, 417 (1989)
6.93 J.C. Farman, B.G. Gardiner, J.D. Shanklin: Large losses of total ozon in Antarctica reveal seasonal ClO_x/NO_2 interaction. Nature **315**, 207 (1985)
6.94 R.S.. Stolarski: The Antarctic ozone hole. Sci. Am. **258**/1, 30 (1988)
6.95 J.H. Seinfeld: *Atmospheric Chemistry and Physics of Air Pollution* (Wiley, New York 1986)
6.96 R.P. Wayne: *Chemistry of Atmospheres* (Clarendon, Oxford 1985)
6.97 T.E. Graedel, D.T. Hawkins, L.D. Claxton: *Atmospheric Chemical Compounds: Sources, Ocurrence, Bioassay* (Academic, Orlando 1986)
6.98 B.A. Thrush: The chemistry of the stratosphere. Rep. Prog. Phys. **51**, 1341 (1988)
T.H. Graedel, P.J. Crutzen: The changing atmosphere. Sci. Am. **261**/3, 28 (1989)
6.99 S.L. Valley (ed.): *Handbook of Geophysics and Space Environments* (McGraw-Hill, New York 1965)
6.100 M. Vergez-Deloncle: Absorption des radiations infrarouges par les gas atmospheriques. J. Physique **25**, 773 (1964)
6.101 Hudson and Hudson (1975), quoted in [6.106]
6.102 L.S. Rotman et al.: The HITRAN Database: 1986 edition, Appl. Opt. **26**, 4058 (1987)
6.103 B.A. Thompson, P. Harteck, R.R. Reeves, Jr.: Ultraviolet absorption coefficients of CO_2, CO, O_2, H_2O, N_2O, NH_3, NO, SO_2, and CH_4 between 1850 and 4000 Å. J. Geophys. Res. **68**, 6431 (1963)
6.104 W. Eppers: Atmospheric Transmission, in *Handbook of Lasers with Selected Data on Optical Technology*, ed. by R.J. Pressley (CRC Press, Cleveland 1977)
6.105 N.G. Jerlov: *Optical Oceanography* (Elsevier, Amsterdam 1968)
T. Stefanick: The nonacoustic detection of submarines. Sci. Am. **258**/3, 25 (1988)

6.106 R.M. Measures: *Laser Remote Sensing* (Wiley-Interscience, New York 1984)
6.107 D.B. Northam, M.A. Guerra, M.E. Mock, I. Itzkan, C. Deradourian: High repetition rate frequency-doubled Nd:YAG laser for airborne bathymetry. Appl. Opt. **20**, 968 (1981)
6.108 B. Welz: *Atomic Absorption Spectroscopy* (VCH, Weinheim 1985)
6.109 C.Th.J. Alkemade, R. Herrmann: *Fundamentals of Analytical Flame Spectroscopy* (Hilger, Bristol 1979)
6.110 D.A. Skoog: *Principles of Instrumental Analysis*, 3rd ed. (Saunders, Philadelphia 1985)
D.A. Skoog, M.D. West: *Fundamentals of Analytical Chemistry*, 4th ed. (Saunders, Philadelphia 1986)
6.111 G.D. Christian, J.E. O'Reilly (eds.): *Instrumental Analysis*, 2nd ed. (Allyn and Bacon, Boston 1986)
6.112 H.H. Willard, L.L. Merritt, Jr., J.A. Dean, F.A. Settle, Jr.: *Instrumental Methods of Analysis*, 6th ed. (Wadsworth, Belmont, Calif. 1981)
6.113 J.U. White: Long optical paths of large aperture. J. Opt. Soc. Am. **32**, 285 (1942)
J.U. White: Very long paths in air. J. Opt. Soc. Am. **66**, 411 (1976)
6.114 H. Edner, A. Sunesson, S. Svanberg, L. Unéus, S. Wallin: Differential optical absorption spectroscopy system used for atmospheric mercury monitoring. Appl. Opt. **25**, 403 (1986)
6.115 J.E. Stewart: *Infrared Spectroscopy* (Marcel Dekker, New York 1970)
6.116 H.A. Szymanski: *Interpreted Infrared Spectra*, Vols.1-3 (Plenum, New York 1964-67)
6.117 S. Hüfner: *Optical Spectra of Transparent Rare Earth Compounds* (Academic, New York 1978)
6.118 A.P.B. Lever: *Inorganic Electronic Spectroscopy*, 2nd ed. (Elsvier, Amsterdam 1984)
6.119 H.A. Szymanski, R.E. Erickson: *Infrared Band Handbook*, Vols.1, 2 (IFI/Plenum, New York 1970)
6.120 IUPAP Tables of Wavenumbers for the Calibration of Infrared Spectrometers (Butterworths, London 1961) p.560
6.121 R.J. Pressley (ed.): *Handbook of Lasers* (with Selected Data on Optical Technology (CRC Press, Cleveland, Ohio 1971) p.407
6.122 H.A. Szymanski (ed.): *Raman Spectroscopy* (Plenum, New York 1967)
6.123 A. Weber (ed.): *Raman Spectroscopy of Gases and Liquids* (Springer, Berlin, Heidelberg 1979)
6.124 D.P. Strommen, K. Nakamoto: *Laboratory Raman Spectroscopy* (Wiley, New York 1984)
6.125 M.M. Sushchinskii: *Raman Spectra of Molecules and Crystals* (Israel Progr. for Sci. Transl., Jerusalem 1972)
6.126 H. Bergström, Lund Institute of Technology (unpublished)
6.127 G.L. Eesley: *Coherent Raman Spectroscopy* (Pergamon, Oxford 1981)
6.128 S. Svanberg: Lasers as probes for air and sea. Contemp. Phys. **21**, 541 (1980)
6.129 E. Schanda: *Physical Fundamentals of Remote Sensing* (Springer, Berlin, Heidelberg 1986)
6.130 S. Svanberg: Fundamentals of atmospheric spectroscopy, in *Surveillance of Electromagnetic Pollution and Resources by Electromagnetic Waves*, ed. by T. Lund (Reidel, Dordrecht 1978)
6.131 E.J. McCartney: *Absorption and Emission by Gases: Physical Processes* (Wiley, New York 1983)
6.132 C.B. Ludwig, M. Griggs, W. Malkmus, E.R. Bartle: Measurements of air pollutants from satellites 1: Feasibility considerations. Appl. Opt. **13**, 1494 (1974)
6.133 U. Platt, D. Perner, H.W. Pätz: Simultaneous measurement of atmospheric CH_2O, O_3, and NO_2 by differential optical absorption. J. Geophys. Res. **84**, 6329 (1979)

6.134 U. Platt, D. Perner: Measurements of atmospheric trace gases by long path differential UV/visible absorption spectroscopy, in *Optical and Laser Remote Sensing*, ed. by D.K. Killinger, A. Mooradian, Springer Ser. Opt. Sci., Vol.39 (Springer, Berlin, Heidelberg 1983).
6.135 P.V. Johnston, R.L. McKenzie: Long-path absorption measurements of tropospheric NO_2 in rural New Zealand. Geophys. Lett. 11, 69 (1984)
6.136 M.M. Millan, R.M. Hoff: Dispersive correlation spectroscopy: a study of mask optimization procedures. Appl. Opt. 16, 1609 (1977)
D.M. Hamilton, H.R. Varey, M.M. Millan: Atmos. Env. 12, 127 (1978)
6.137 J.A. Hodgeson, W.A. McClenney, P.L. Hanst: Science 182, 248 (1973)
T.V. Ward, H.H. Zwick: Gas cell correlation spectrometer: GASPEC. Appl. Opt. 14, 2896 (1975)
H.S. Lee, H.H. Zwick: Gas filter correlation instrument for the remote sensing of gas leaks. Rev. Sci. Instr. 56, 1812 (1985)
6.138 S.C. Cox (ed.): The Multispectral Imaging Sciences Working Group: Final Report, NASA Conf. Publ. No 2260 (NASA, Washington, DC 1983)
Earth Observing Systems Reports, Vol. IIc, High Resolution Imaging Spectrometry (NASA, Washington, DC 1986)
G. Vane (ed.): *Imaging spectroscopy II*. Proc. Soc. Photo. Opt. Instrum. Eng. 834 (1987)
6.139 P.N. Slater: *Remote Sensing: Optics and Optical Systems* (Addison Wesley, Reading, Mass. 1980)
6.140 A.F.H. Goetz, J. Wellman, W. Barnes: Optical remote sensing of the Earth. Proc. IEEE 73 (June 1985)
6.141 H.S. Chen: *Space Remote Sensing Systems* (Academic, Orlando 1985)
6.142 T.A. Croft: Nighttime images of the Earth from space. Sci. Am. 239/1, 68 (1978)
6.143 A. Dalgarno, D. Layzer (eds): *Spectroscopy of Astrophysical Plasmas* (Cambridge Univ. Press, Cambridge 1987)
6.144 G.B. Rybicki, A.P. Lightman: *Radiative Processes in Astrophysics* (Wiley, New York 1979)
6.145 D.F. Gray: *The Observation and Analysis of Stellar Photospheres* (Wiley, New York 1976)
6.146 R.H. Baker: *Astronomy* (van Nostrand, Princeton, NJ 1964)
6.147 D.J. Schroeder: *Astronomical Optics* (Academic, San Diego 1987)
6.148 B. Aschenbach: X-ray telescopes. Rep. Prog. Phys. 48, 579 (1985)
6.149 R.F. Griffin: *A Photometric Atlas of the Spectrum of Arcturus* (Cambridge Phil. Soc., Cambridge 1968)
6.150 J.M. Beckers, C.A. Bridges, L.B. Gilliam: A high resolution atlas of the solar irradiance from 380-700 nm. Sacramento Peak Observatory (1983)
6.151 D. Dravins: In KOSMOS 1980. Swedish Phys. Soc., Stockholm (1980)
6.152 B. Edlén: Z. Astrophysik 22, 30 (1942)
B. Edlén: Forbidden lines in hot plasmas. Phys. Scr. T 8, 5 (1984)
6.153 R. Giacconi: The Einstein X-ray observatory. Sci. Am. 242/2, 70 (1980)
A. Vidal-Madjar, Th. Encrenaz, R. Ferlet, J.C. Henoux, R. Lallement, G. Vaudair: Galactic ultraviolet astronomy. Rep. Progr. Phys. 50, 65 (1987)
6.154 H.J. Habing, G. Neugebauer: The infrared sky. Sci. Am. 251/5, 42 (1984)
6.155 J.B. Bahcall, L. Spitzer, Jr.: The space telescope. Sci. Am. 247/1, 38 (1982)
6.156 D.W. Weedman: *Quasar Astronomy* (Cambridge Univ. Press, Cambridge 1986)
6.157 P.S. Osmer: Quasars as probes of the distant and early universe. Sci. Am 246/2, 96 (1982)
6.158 P. Murdin: The supernova in the Large Magellanic Cloud. Contemp. Phys. 28, 441 (1987)
W. Hillebrandt, P. Höflich: The supernova 1987A in the Large Magellanic Cloud. Rep. Progr. Phys. 52, 1421 (1989)
S. Woosley, T. Weaver: The great supernova of 1987. Sci Am. 261/2, 24 (1989)

6.159 R. Fosburg: The spectrum of supernova 1987A. ESO Messenger **47**, 32 (1987)
P. Andreani, R. Ferlet, R. Vidal-Madjar: ESO Messenger **47**, 33 (1987)
6.160 P. Connes, G. Michel: Astronomical Fourier spectrometer. Appl. Opt. **14**, 2067 (1975)
6.161 L.A. Soderblom, T.V. Johnson: The moons of Saturn. Sci. Am. **246**/1, 72 (1982)
6.162 T. Oen: Titan. Sci. Am. **246**/2, 76 (1982)
6.163 R.P. Laeser, W.I. McLaughlin, D.M. Wolff: Engineering Voyager 2's encounter with Uranus. Sci. Am. **255**/5, 34 (1986)
A.P. Ingersoll: Uranus. Sci. Am. **256**/1, 30 (1987)
T.J. Johnson, R.H. Brown, L.A. Soderblom: The moons of Uranus. Sci. Am. **256**/4, 40 (1987)
J. Kinoshita: Neptune. Sci. Am. **261**/5, 60 (1989)
J.N. Cuzzi, L.W. Esposito: The rings of Uranus. Sci. Am. **257**/1, 42 (1987)
6.164 Sky and Telescope **73**, No.3 (1987) (Feature issue)
Nature **321**, No.6067 (1987) (Feature issue)
H. Balsiger, H. Fechtig, J. Geiss: A close look at Halley's comet. Sci. Am. **259**/3, 62 (1988)
6.165 C. Arpigny, F. Dossin, J. Manfroid, P. Magain, A.C. Danks, D.L. Lambert, C. Sterken: Spectroscopy, photometry and direct filter imagery of comet P/Halley. ESO Messenger **45**, 10 (1986)

Chapter 7

7.1   N. Ramsey: *Molecular Beams* (Clarendon, Oxford 1956, Paperback 1985)
J.M. Pendlebury, K.F. Smith: Molecular beams. Contemp. Phys. **28**, 3 (1987)
7.2   A.N. Nesmeyanov: *Vapor Pressure of the Elements* (Academic, New York 1963)
7.3   R.E. Honig, D.A. Kramer: RCA Rev. **30**, 285 (1969)
7.4   O. Stern, W. Gerlach: Der experimentelle Nachweis der Richtungsquantelung im Magnetfeld. Das magnetische Moment des Silberatoms. Z. Physik **9**, 349, 353 (1922)
7.5   I.I. Rabi, J.R. Zacharias, S. Millman, P. Kusch: A new method of measuring nuclear magnetic moment. Phys. Rev. **53**, 318 (1938)
7.6   N.F. Ramsey: A new molecular beam resonance method. Phys. Rev. **76**, 996 (1949); A molecular beam resonance method with separated oscillating fields. Phys. Rev. **78**, 695 (1950); Phase shifts in the molecular beam method of separated oscillating fields. Phys. Rev. **84**, 506 (1951)
7.7   C. Ekström, I. Lindgren: Atomic beam experiments at the ISOLDE facility at CERN, in *Atomic Physics 5*, ed. by R. Marrus, M. Prior, H. Shugart (Plenum, New York 1977) p.201
7.8   W.J. Childs: Case Studies Atomic Phys. **3**, 215 (1973)
7.9   S. Büttgenbach, G. Meisel, S. Penselin, K.H. Schneider: A new method for the production of atomic beams of highly refractory elements and first atomic beam magnetic resonances in Ta^{181}. Z. Physik **230**, 329 (1970)
H. Rubinsztein, I. Lindgren, L. Lindström, H. Riedl, A. Rosén: Atomic beam measurements on refractory elements. Nucl. Instr. Meth. **119**, 269 (1974).
7.10 U. Brinkmann, J. Goschler, A. Steudel, H. Walther: Experimente mit Erdalkaliatomen in Metastabilen Zuständen. Z. Physik **228**, 427 (1969)
S. Garpman, G. Lidö, S. Rydberg, S. Svanberg: Lifetimes of some highly excited levels in the Pb-I spectrum measured by the Hanle method. Z. Physik **241**, 217 (1971)
7.11 S. Penselin: Recent developments and results of the atomic beam magnetic resonance method, in *Progress in Atomic Spectroscopy*, Pt.A, ed. by W. Hanle, H. Kleinpoppen (Plenum, New York 1979) p.463

7.12 A. Kastler: Quelques suggestions concernant la production optique et la détetection optique d'une inégalité de population des niveaux de quantification spatiale des atomes. Application a l'expérience de Stern et Gerlach et a la résonance magnétique. J. Phys. Radium 11, 255 (1950); Méthodes optiques d'etude de la resonance magnétique. Physica 17, 191 (1951); Optical methods of atomic orientation and of magnetic resonance. J. Opt. Soc. Am. 47, 460 (1957)

7.13 H.J. Besch, U. Köpf, E.W. Otten: Optical pumping of shortlived beta emitters. Phys. Lett. 25B, 120 (1967)
E.W. Otten: Hyperfine and isotope shift measurements far from stability by optical pumping, in *Atomic Physics 5*, ed. by R. Marrus, M. Prior, H. Shugart (Plenum, New York 1977) p.239

7.14 J. Bonn, G. Huber, H.J. Kluge, U. Köpf, L. Kugler, E.W. Otten, J. Rodrigues: Orientation of short-lived mercury isotopes by means of optical pumping detected by β and γ radiation, in *Atomic Physics 3*, ed. by S.J. Smith, G.K. Walters (Plenum, New York 1973) p.471

7.15 R. Bernheim: *Optical Pumping* (Benjamin, New York 1965)

7.16 W. Happer: Optical pumping. Rev. Mod. Phys. 44, 169 (1972)

7.17 G.W. Series: Thirty years of optical pumping. Contemp. Phys. 22, 487 (1981)

7.18 M. Arditi, T.R. Carver: Optical detection of zero field hyperfine structure in Na^{23}. Phys. Rev. 109, 1012 (1958); Frequency shift of the zero field hyperfine splitting of Cs^{133} produced by various buffer gases. Phys. Rev. 112, 449 (1958)

7.19 H.M. Goldenberg, D. Kleppner, N.F. Ramsey: Atomic hydrogen maser. Phys. Rev. Lett. 5, 361 (1960)
D. Kleppner, H.M. Goldenberg, N.F. Ramsey: Properties of the hydrogen maser. Appl. Opt. 1, 55 (1962)
S.B. Crampton, D. Kleppner, N.F. Ramsey: Hyperfine structure of ground state atomic hydrogen. Phys. Rev. Lett. 11, 338 (1963)

7.20 P. Karpaschoff: *Frequency and Time* (Academic, London 1978)
H. Hellwig: Atomic frequency standards. Proc. IEEE 63, 212 (1974)
J. Vanier, C. Audoin: *The Quantum Physics of Atomic Frequency Standards* (Hilger, Bristol 1989)

7.21 F.L. Walls: Frequency standards based on atomic hydrogen. Proc. IEEE 74, 142 (1986)
D.J. Wineland: Frequency standards based on stored ions. Proc. IEEE 74, 147 (1986)

7.22 A. Kastler, J. Brossel: La détection de la résonance magnétique des niveaux excités: L'effet de dépolarisation des radiations de résonance optique et de fluorescence. Comp. Rend. 229, 1213 (1949)

7.23 J. Brossel, F. Bitter: A new "double resonance" method for investigating atomic energy levels. Application to Hg $^3P_1$. Phys. Rev. 86, 308 (1952)

7.24 G. Belin, I. Lindgren, I. Holmgren, S. Svanberg: Hyperfine interaction, Zeeman and Stark effects for excited states in potassium. Phys. Scr. 12, 287 (1975)

7.25 W. Hanle: Über magnetische Beeinflussung der Polarisation der Resonanzfluoreszenz. Z. Physik 30, 93 (1924); Erg. Ex. Naturwiss. 4, 214 (1925)

7.26 F.D. Colgrove, P.A. Franken, R.R. Lewis, R.H. Sands: Novel method of spectroscopy with applications to precision fine structure measurements. Phys. Rev. Lett. 3, 420 (1959)

7.27 G. Breit: Quantum theory of dispersion (continued). Pts.VI and VII. Rev. Mod. Phys. 5, 91 (1933)

7.28 P. Franken: Interference effects in the resonance fluorescence of "crossed" excited states. Phys. Rev. 121, 508 (1961)

7.29 T.G. Eck, L.L. Foldy, H. Wiedner: Observation of "anticrossings" in optical resonance fluorescence. Phys. Rev. Lett. 10, 239 (1963)
H. Wiedner, T.G. Eck: "Anticrossing" signals in resonance fluorescence. Phys. Rev. 153, 103 (1967)

H.J. Beyer, H. Kleinpoppen: Anticrossing spectroscopy, in *Progress in Atomic Spectroscopy*, Pt.A, ed. by W. Hanle, H. Kleinpoppen (Plenum, New York 1979) p.607

7.30 G. Belin, S. Svanberg: Electronic g_J factors, natural lifetimes and electric quadrupole interaction in the np $^2P_{3/2}$ series of the RbI spectrum. Phys. Scr. **4**, 269 (1971)

7.31 R. Gupta, S. Chang, C. Tai, W. Happer: Cascade radio-frequency spectroscopy of excited S and D states of rubidium; anomalous D-state hyperfine structure. Phys. Rev. Lett. **29**, 695 (1972)
R. Gupta, W. Happer, L. Lam, S. Svanberg: Hyperfine structure measurements of excited S states of the stable isotopes of potassium, rubidium and cesium by cascade radio-frequency spectroscopy. Phys. Rev. **A8**, 2792 (1973)

7.32 M.E. Rose, R.L. Carovillano: Coherence effects in resonance fluorescence. Phys. Rev. **122**, 1185 (1961)

7.33 G. zu Putlitz: Double resonance and level-crossing spectroscopy, in *Atomic Physics*, ed. by V.W. Hughes, B. Bederson, V.W. Cohen, F.M.J. Pichanick (Plenum, New York 1969)

7.34 B. Budick: In *Advances in Atomic and Molecular Physics*, ed. by R.D. Bates, I. Esterman (Academic, New York 1967)

7.35 W. Happer, R. Gupta: Perturbed fluorescence spectroscopy. in *Progress in Atomic Spectroscopy*, Pt.A, ed. by W. Hanle, H. Kleinpoppen (Plenum, New York 1979) p.391

7.36 E. Arimondo, M. Inguscio, P. Violino: Experimental determinations of the hyperfine structure in the alkali atoms. Rev. Mod. Phys. **49**, 31 (1977)

7.37 P.R. Johnson, R. Pearson, Jr.: *Methods in Experimental Physics*, **13**, 102 (Academic, New York 1976)

7.38 C.P. Slichter: *Principles of Magnetic Resonance*, 3rd. ed., Springer Ser. Solid-State Sci., Vol.1 (Springer, Berlin, Heidelberg 1990)

7.39 D.A. Skoog, D.M. West: *Principles of Instrumental Analysis* (Saunders, Philadelphia 1980)

7.40 R. Brewer, E.L. Hahn: Atomic memory. Sci. Am. **251**/6, 42 (1984)

7.41 H. Gunther: *NMR Spectroscopy - An Introduction* (Wiley, Chichester 1985)

7.42 D.A.R. Williams: *Nuclear Magnetic Resonance Spectroscopy* (Wiley, Chichester 1986)

7.43 W. Kemp: *NMR in Chemistry* (McMillan, London 1986)

7.44 I.L. Pykett: NMR imaging in medicine. Sci. Am. **246**/5, 54 (1982)

7.45 D.R. Bailes, D.J. Bryant: NMR imaging. Contemp. Phys. **25**, 441 (1984)

7.46 R.S. MacKay: *Medical Images and Displays: Comparison of Nuclear Magnetic Resonance, Ultrasound, X-Rays and Other Modalities* (Wiley, New York 1984)

7.47 R.S. Alger: *Electron Paramagnetic Resonance* (Wiley, New York 1968)

7.48 J.E. Wertz: *Electron Spin Resonance: Elementary Theory and Practical Applications* (Chapman and Hall, New York 1986)

7.49 J.J. Davies: Optically detected magnetic resonance and its applications. Contemp. Phys. **17**, 275 (1976)

7.50 C.H. Townes, A.L. Schawlow: *Microwave Spectroscopy* (Dover, New York 1975)

7.51 H.W. Kroto: *Molecular Rotation Spectra* (Wiley, London 1975)

7.52 W. Gordy, R.L. Cook: *Microwave Molecular Spectra*, 3rd. ed., Techniques of Chemistry, Vol.XVIII (Wiley, New York 1984)

7.53 T. Lund (ed.): Surveillance of environmental pollution and resources by electromagnetic waves. NATO Adv. St. Inst. Ser. (Reidel, Dordrecht 1978)

7.54 E. Schanda: *Physical Fundamentals of Remote Sensing* (Springer, Berlin, Heidelberg 1986)

7.55 D.T. Gjessing: Remote surveillance by electromagnetic waves for air - water - land (Ann Arbor Science, Ann Arbor 1978)

7.56 K.A. Browning: Uses of radar in metrology. Contemp. Phys. **27**, 499 (1986)
7.57 E. Schanda: Microwave radiometry applications to remote sensing, in [7.53]
7.58 E.P.W. Attema: The radar signature of natural surfaces and its application in active microwave remote sensing, in [7.53]
7.59 Ch. Elachi: Radar images of the Earth from space. Sci. Am. **247**/6, 46 (1982)
7.60 O.E.H. Rydbeck: Interstellar molecules, in *Kosmos 1974*, ed. by N.R. Nilsson (Swedish Phys. Soc., Stokholm 1975)
7.61 W.M. Irvine, P.F. Goldsmith, Å. Hjalmarsson: Chemical abundances in molecular clouds, in *Interstellar Processes*, ed. by D.J. Hollenback, H.A. Thronson Jr. (Reidel, Dordrecht 1987)
7.62 K. Rohlfs: *Tools of Radio Astronomy* (Springer, Berlin, Heidelberg 1986)
7.63 A.S. Webster, M.S. Longair: Millimetre and sub-millimetre astronomy. Contemp. Phys. **25**, 519 (1984)
7.64 A.C.S. Readhead: Radio astronomy and very long baseline interferometry. Sci. Am. **246**/6, 38 (1982)
7.65 A.R. Thompson, J. Moran, G.W. Swenson, Jr.: *Interferometry and Synthesis in Radio Astronomy* (Wiley, New York 1986)
7.66 M. Elitzur: Physical characteristics of astronomical masers. Rev. Mod. Phys. **54**, 1225 (1982)
    D.F. Dickinson: Cosmic masers. Sci. Am. **238**/6, 68 (1978)
7.67 P. Morrison, J. Billingham, J. Wolfe: *The Search for Extraterrestial Intelligence* (prepared by NASA) (Dover, New York 1979; Academic, New York 1986)

Chapter 8

8.1 A. Yariv: *Introduction to Quantum Electronics*, 2nd. ed. (Holt, Rinehart and Winston, New York 1976)
8.2 A. Yariv: *Quantum Electronics*, 3rd edn. (Wiley, New York 1989)
8.3 M. Sargent III, M.O. Scully, W.E. Lamb, Jr.: *Laser Physics* (Addison Wesley, London 1974)
8.4 O. Svelto: *Principles of Lasers*, 3rd edn. (Plenum, New York 1989)
8.5 A.E. Siegman: *Lasers* (University Science Books, Mill Valley, Calif. 1986)
8.6 H. Haken: *Laser Theory* (Springer, Berlin, Heidelberg 1983)
8.7 K. Shimoda: *Introduction to Laser Physics*, 2nd. ed., Springer Ser. Opt. Sci., Vol.44 (Springer, Berlin, Heidelberg 1984)
8.8 M. Young: *Optics and Lasers*, 3rd. ed., Springer Ser. Opt. Sci., Vol.5 (Springer, Berlin, Heidelberg 1986)
8.9 M.J. Weber (ed.): *CRC Handbook of Laser Science and Technology*, Vols.1 and 2 (CRC Press, Boca Raton, FL 1982)
8.10 *Laser Handbook*, Vols.1 and 2, ed. by F.T. Arecchi, E.O. Schulz-Dubois (1972); Vol.3, ed. by M.L. Stitch (1979); Vol.4, ed. by M.L. Stitch, M. Bass (1985); Vol.5, ed. by M. Bass, M.L. Stitch (1986) (North-Holland, Amsterdam)
8.11 T.H. Maiman: Stimulated optical radiation in ruby. Nature **187**, 493 (1960)
8.12 A.L. Schawlow, C.H. Townes: Infrared and optical masers. Phys. Rev. **112**, 1940 (1958)
8.13 J.P. Gordon, H.J. Zeiger, Ch.H. Townes: The maser - new type of microwave amplifier, frequency standard and spectrometer. Phys. Rev. **99**, 1264 (1955)
8.14 C.H. Townes: In *Nobel Lectures in Physics* (Elsevier, Amsterdam 1972) Vol.4
8.15 N.G. Basov: In *Nobel Lectures in Physics* (Elsevier, Amsterdam 1972) Vol.4
8.16 A.M. Prokhorov: In *Nobel Lectures in Physics* (Elsevier, Amsterdam 1972) Vol.4

8.17 D.L. Matthews, P.L. Hagelstein, M.D. Rosen, M.J. Eckart, N.H. Ceglio, A.U. Hazi, H. Medicki, B.J. MacGowan, J.E. Trebes, B.L. Witten, E.M. Campbell, C.W. Hatcher, A.H. Hawryluk, R.L. Kaufmann, L.D. Pleasance, G. Rambach, J.H. Scoefield, G. Stone, T.A. Weaver: Demonstration of a soft X-ray amplifier. Phys. Rev. Lett. **54**, 110 (1985); J. Opt. Soc. Am. **4**, 575 (1987)

8.18 S. Suchewer, C.H. Skinner, M. Milchberg, C. Keane, D. Voorhees: Amplification of stimulated X-ray emission in a confined plasma column. Phys. Rev. Lett. **55**, 1753 (1985)

8.19 D.L. Matthews, R.R. Freeman (eds.): The generation of coherent XUV and soft X-ray radiation. J. Opt. Soc. Am. **B4**, 529-618 (1987) (feature issue)
D.L. Matthews, M.D. Rosen: Soft X-ray lasers. Sci. Am. **256**/6, 60 (1988)

8.20 H. Kogelnik, T. Li: Laser beams and resonators. Proc. IEEE **54**, 1312 (1966)
H.K.V. Lotsch: The confocal resonator system. Optik **30**, 1, 181, 217, 563 (1969/70)

8.21 G.D. Boyd, J.P. Gordon: Bell Syst. Tech. J. **40**, 489 (1961)
G.D. Boyd, H. Kogelnik: Bell Syst. Tech. J. **41**, 1347 (1962)

8.22 S. Svanberg: Lasers as probes for air and sea. Contemp. Phys. **21**, 541 (1980)

8.23 W. Koechner: *Solid-State Laser Engineering*, 2nd edn., Springer Ser. Opt. Sci., Vol.1 (Springer, Berlin, Heidelberg 1988)

8.24 D.C. Brown: *High-Peak-Power Nd:Glass Laser Systems*, Springer Ser. Opt. Sci., Vol.25 (Springer, Berlin, Heidelberg 1981)

8.25 A.A. Kaminskii: *Laser Crystals*, 2nd edn., Springer Ser. Opt. Sci., Vol.14 (Springer, Berlin, Heidelberg 1990)

8.26 A.F. Gibson: Lasers for compression and fusion. Contemp. Phys. **23**, 285 (1982)

8.27 R.S. Craxton, R.L. McCrory, J.M. Sources: Progress in laser fusion. Sci. Am. **255**/2, 60 (1986)

8.28 N.G. Basov, Yu. A. Zakharenkov, N.N. Zorev, G.V. Sklizkov, A.A. Rupasov, A.S. Shikanov: *Heating and Compression of Thermonuclear Targets by Laser Beams* (Cambridge Univ. Press, Cambridge 1986)

8.29 J.E. Eggleston, T.J. Kane, K. Kuhn, J. Unternahrer, R.L. Byer: The slab geometry laser. IEEE J. **QE-20**, 289 (1984)

8.30 D. Findlay, D.W. Goodwin: The neodymium in YAG laser, in *Advances in Quantum Electronics*, ed. by D.W. Goodwin (Academic, London 1970) Vol.1

8.31 B. Zhou, T.J. Kane, G.J. Dixon, R.L. Byer: Efficient, frequency-stable laser-diode-pumped Nd:YAG laser. Opt. Lett. **10**, 62 (1985)
A. Owyoung, G.R. Hadley, P. Esherick: Gain switching of a monolithic single-frequency laser-diode-excited Nd:YAG laser. Opt. Lett. **10**, 484 (1985)
W.R. Trutna, D.K. Donald, M. Nazarathy: Unidirectional diode-laser-pumped Nd:YAG ring laser with a small magnetic field. Opt. Lett. **12**, 248 (1987)

8.32 R.L. Byer, G.J. Dixon, T.J. Kane, W. Kozlovsky, B. Zhou: Frequency-doubled, laser diode pumped, miniature Nd:YAG oscillator - progress toward an all solid state sub-kilohertz linewidth coherent source, in *Laser Spectroscopy VII*, ed. by T.W. Hänsch, Y.R. Shen, Springer Ser. Opt. Sci., Vol.49 (Springer, Berlin, Heidelberg 1985) p.350

8.33 C.K. Rhodes (ed.): *Excimer Lasers*, 2nd. ed., Topics Appl. Phys., Vol.30 (Springer, Berlin, Heidelberg 1984)

8.34 M.H.R. Hutchinson: Excimers and excimer lasers. Appl. Phys. **21**, 15 (1980)

8.35 C.K. Rhodes, H. Egger, H. Plummer (eds.): *Excimer Lasers*, Conf. Proc. Series No.100 (Am. Inst. Phys., New York 1983)

8.36 A. Javan, W.R. Bennet, Jr., D.R. Herriott: Population inversion and continous optical maser oscillation in a gas discharge containing a He-Ne mixture. Phys. Rev. Lett. **6**, 48 (1961)

8.37 W.T. Silfvast, J.J. Macklin, O.R. Wood II: High-gain inner-shell photoionization laser in Cd vapor pumped by soft X-ray radiation from a laser produced plasma source. Opt. Lett. **8**, 551 (1983)

W.T. Silfvast, O.R. Wood II: Photoionization lasers pumped by broadband soft-X-ray flux from laser-produced plasmas. J. Opt. Soc. Am. **4**, 609 (1987)

8.38 R.A. Lacy, A.C. Nilsson, R.L. Byer, W.T. Silfvast, O.R. Wood II, S. Svanberg: Photoionization-pumped gain at 185 nm in a laser-ablated indium plasma. J. Opt. Soc. Am. B **6**, 1209 (1989)

8.39 H.C. Kapteyn, R.W. Lee, R.W. Falcone: Observation of a short-wavelength laser pumped by Auger decay. Phys. Rev. Lett. **57**, 2939 (1986)
M.H. Sher, J.J. Macklin, J.F. Young, S.E. Harris: Saturation of the XeIII 109-nm laser using traveling-wave laser-produced-plasma excitation. Opt. Lett. **12**, 891 (1987)

8.40 C.C. Davis, T.A. King: Gaseous ion lasers, in *Advances in Quantum Electronics*, ed. by D.W. Goodwin (Academic, London 1975) Vol.3

8.41 S.D. Smith, R.B. Dennis, R.G. Harrison: The spin-flip Raman laser. Prog. Quant. Electr. **5**, 205 (1977)

8.42 M.J. Colles, C.R. Pigeon: Tunable lasers. Rep. Prog. Phys. **38**, 329 (1975)

8.43 A. Mooradian: Tunable infrared lasers. Rep. Prog. Phys. **42**, 1533 (1979)

8.44 J. White, L. Mollenauer (eds.): *Tunable Lasers*, Topics Appl. Phys., Vol.59 (Springer, Berlin, Heidelberg 1986)

8.45 P.P. Sorokin, J.R. Lankard: Stimulated emission observed from an organic dye, chloro-aluminium phthalocyanine. IBM J. Res. Dev. **10**, 306 (1966)

8.46 F.P. Schäfer, W. Smidt, J. Volze: Organic dye solution laser. Appl. Phys. Lett. **9**, 306 (1966)

8.47 B.H. Soffer, B.B. McFarland: Continously tunable narrow-band organic dye laser. Appl. Phys. Lett. **10**, 266 (1967)

8.48 T.W. Hänsch: Repetitively pulsed tunable dye laser for high resolution spectroscopy. Appl. Opt. **11**, 895 (1972)

8.49 M.G. Littman, H.J. Metcalf: Spectrally narrow pulsed dye laser without beam expander. Appl. Opt. **25**, 375 (1978)
I. Shoshan, U. Oppenheim: The use of a diffraction grating as a beam expander in a dye laser cavity. Opt. Commun. **25**, 375 (1978)

8.50 M.G. Littman: Single-mode operation of grazing incidence pulsed dye laser. Opt. Lett. **3**, 138 (1978)
H.S. Saikan: Nitrogen laser pumped single mode dye laser. Appl. Phys. **17**, 41 (1978)
M.G. Littman: Single mode pulsed tunable dye laser. Appl. Opt. **23**, 4465 (1984)

8.51 R. Wallenstein, T.W. Hänsch: Powerful dye laser oscillator-amplifier system for high-resolution spectroscopy. Opt. Commun. **14**, 353 (1975)
R. Wallenstein, H. Zacharias: High-power narrowband pulsed dye laser oscillator-amplifier system. Opt. Commun. **32**, 429 (1980)

8.52 O.G. Peterson, S.A. Tuccio, B.B. Snavely: CW Operation of an organic dye solution laser. Appl. Phys. Lett. **17**, 245 (1970)

8.53 J. Evans: The birefringent filter. J. Opt. Soc. Am. *39*, 229 (1949)

8.54 F.P. Schäfer (ed.): *Dye Lasers*, 3rd edn., Topics Appl. Phys., Vol.1 (Springer, Berlin, Heidelberg 1990)

8.55 T.F. Johnston: Tunable dye lasers, in *Encyclopedia of Physical Science and Technology*, Vol.14 (Academic, New York 1987)
T.F. Johnston, R.H. Brady, W. Proffitt: Powerful single-frequency ring dye laser spanning the visible spectrum. Appl. Opt. **21**, 2307 (1982)

8.56 M. Maeda: *Laser Dyes* (Academic, Orlando 1984)
K. Brackman: Lambdachrome Laser Dyes (Lambda Physik, Göttingen 1986)

8.57 L.F. Mollenauer: Tunable lasers, in [Ref.8.44, Chap.6]
L.F. Mollenauer: In [Ref.8.10, Vol.4, Chap.2]

8.58 R.L. Byer (ed.): Special issue on tunable solid state lasers. IEEE J. QE-21, 1567-1636 (1985)
B. Henderson, G.F. Imbusch: Optical processes in tunable transition-metal-ion lasers. Contemp. Phys. **29**, 235 (1988)

8.59 P. Hammerling, A.B. Budgor, A. Pinto (eds.): *Tunable Solid-State Lasers*, Springer Ser. Opt. Sci., Vol.47 (Springer, Berlin, Heidelberg 1985)
8.60 A.B. Budgor, L. Esterowitz, L.G. DeShazer (eds): *Tunable Solid-State Lasers II*, Springer Ser. Opt. Sci., Vol.52 (Springer, Berlin, Heidelberg 1986)
8.61 D.C. Tyle: Carbon dioxide lasers, in *Advances in Quantum Electronics*, ed. by D.W. Goodwin (Academic, New York 1970) Vol.1
8.62 W.J. Witteman: *The CO_2 Laser*, Springer Ser. Opt. Sci., Vol.53 (Springer, Berlin, Heidelberg 1987)
8.63 F. O'Neill, W.T. Whitney: A high-power tunable laser for the 9-12.5 μm spectral range. Appl. Phys. Lett. **31**, 271 (1977)
8.64 R. Beck, W. Englisch, K. Gürs: *Table of Laser Lines in Gases and Vapors*, 3rd. edn., Springer Ser. Opt. Sci., Vol.2 (Springer, Berlin, Heidelberg 1978)
8.65 R.M. Measures: *Laser Remote Sensing: Fundamentals and Applications* (Wiley, New York 1984)
8.66 H. Kressel, J.K. Butler: *Semiconductor Lasers and Heterojunction LEDs* (Academic, New York 1977)
8.67 H.C. Lasey, M.B. Panisch: *Heterostructure Lasers I and II* (Academic, New York 1978)
8.68 J.C. Camparo: The diode laser in atomic physics. Contemp. Phys. **26**, 443 (1985)
8.69 E.D. Hinkley, K.W. Nill, F.A. Blum: Infrared spectroscopy with tunable lasers, in *Laser Spectroscopy of Atoms and Molecules*, ed. by H. Walther, Topics Appl. Phys., Vol.2 (Springer, Berlin, Heidelberg 1976)
8.70 R. Lang: Recent progress in semiconductor lasers, in *Laser Spectroscopy VIII*, ed. by W. Persson, S. Svanberg, Springer Ser. Opt. Sci., Vol.55 (Springer, Berlin, Heidelberg 1987) p. 434
8.71 T.F. Johnston, Jr., T.J. Johnston: Angle matched doubling in $LiIO_3$ intracavity to a ring dye laser, in *Laser Spectroscopy VI*, ed. by H.P. Weber, W. Lüthy, Springer Ser. Opt. Sci., Vol.40 (Springer, Berlin, Heidelberg 1983) p. 417
8.72 B. Couillaud, L.A. Bloomfield, T.W. Hänsch: Generation of continous-wave radiation near 243 nm by sum frequency mixing in an external ring cavity. Opt. Lett. **8**, 259 (1983)
8.73 A.S. Pine: IR spectroscopy via difference-frequency generation, in *Laser Spectroscopy III*, ed. by J.L. Hall, J.L. Carlsten, Springer Ser. Opt. Ser., Vol.7 (Springer, Berlin, Heidelberg 1977) p.376
8.74 S. Singe: In *Handbook of Laser Science and Technology*, ed. by M.J. Weber (CRC Press, Boca Raton, FL 1986) Vol.3
8.75 D.S. Chemla, J. Zyss (eds): *Nonlinear Optical Properties of Organic Molecules and Crystals*, Vols.1 and 2 (Academic, Orlando 1987)
8.76 R.C. Eckardt, Y.X. Fan, M.M. Fejer, W.J. Kozlovsky, C.N. Nabors, R.L. Byer, R.K. Route, R.S. Feigelson: Recent developments in nonlinear optical materials, in *Laser Spectroscopy VIII*, ed. by W. Persson, S. Svanberg, Springer Ser. Opt. Sci., Vol.55 (Springer, Berlin, Heidelberg 1987) p.426
8.77 S.E. Harris: Tunable optical parametric oscillators. Proc. IEEE **57**, 2096 (1969)
8.78 Y.X. Fan, R.L. Byer: Progress in optical parametric oscillators. SPIE **461**, 27 (1984)
8.79 P.P. Sorokin, J.A. Armstrong, R.W. Dreyfus, R.T. Hodgeson, J.R. Lankard, L.H. Manganaro, J.J. Wynne: Generation of vacuum ultraviolet radiation by nonlinear mixing in atomic and ionic vapors, in *Laser Spectroscopy*, ed. by S. Haroche, J.C. Pebay-Peyroula, T.W. Hänsch, S.E. Harris, Lecture Notes Phys., Vol.43 (Springer, Berlin, Heidelberg 1975) p.46
J.F. Rentjes: *Nonlinear Optical Parametric Processes in Liquids and Gases* (Academic, New York 1984)
W. Jamroz, B.P. Stoicheff: Generation of tunable coherent vacuum-ultraviolet radiation. *Progress in Optics* **XX**, 325 (North-Holland, Amsterdam 1983)

8.80 B. Ya. Zel'dovich, N.F. Pilipetsky, V.V. Shkunov: *Principles of Phase Conjugation*, Springer Ser. Opt. Sci., Vol.42 (Springer, Berlin, Heidelberg 1985)
V.V. Shkunov, B. Ya. Zel'dovich: Optical phase conjugation. Sci. Am. 253/6, 40 (1985)
D.M. Pepper: Applications of optical phase conjugation. Sci. Am. 254/1, 56 (1986)
8.81 H.J. Eichler, P. Gunther, D.W. Pohl: *Laser Induced Dynamic Gratings*, Springer Ser. Opt. Sci., Vol.50 (Springer, Berlin, Heidelberg 1986)
8.82 C.R. Vidal: Coherent VUV sources for high-resolution spectroscopy. Appl. Opt. 19, 3897 (1980).
8.83 R. Hilbig, G. Hilber, A. Lago, B. Wolff, R. Wallenstein: Tunable coherent VUV radiation generated by nonlinear optical frequency conversion in gases. Comments At. Mol. Phys. 18, 157 (1986)
R. Hilbig, G. Hilber, A. Timmermann, R. Wallenstein: Generation of coherent tunable VUV radiation, in *Laser Spectroscopy VI*, ed. by H.P. Weber, W. Lüthy, Springer Ser. Opt. Sci., Vol.40 (Springer, Heidelberg, Berlin 1983) p.387
G. Hilber, A. Lago, R. Wallenstein: Generation and application of coherent tunable VUV radiation at 60 to 200 nm, in *Laser Spectroscopy VIII*, ed. by W. Persson, S. Svanberg, Springer Ser. Opt. Sci., Vol.55 (Springer, Berlin, Heidelberg 1987) p.446
8.84 T.J. McIlrath, R.R. Freeman (eds): *Laser Techniques for Extreme Ultraviolet Spectroscopy*, Conf. Proc. Series, No.90 (Am. Inst. Phys., New York 1982)
8.85 S.E. Harris, T.B. Lucatorto (eds.): *Laser Techniques in the Extreme Ultraviolet*, Conf. Proc. Series, No.119 (American Inst. of Physics, New York 1984)
8.86 D.T. Attwood, J. Bokor (eds.): *Short Wavelength Coherent Radiation: Generation and Application*, Conf. Proc. Series, No.147 (Am. Inst. Phys., New York 1986)
8.87 R.W. Falcone, J. Kirz (eds.): *Short Wavelength Coherent Radiation: Generation and Applications* (Opt. Soc. Am., Washington, DC 1988)
C. Yamanaka (ed.): *Short-Wavelength Lasers*, Springer Proc. Phys., Vol.30 (Springer, Berlin, Heidelberg 1988)
8.88 V. Wilke, W. Smidt: Tunable coherent radiation source covering a spectral range from 185 - 880 nm. Appl. Phys. 18, 177 (1979). See also Appl. Phys. 18, 235 (1979)
8.89 J. Paisner, S. Hargrove: A tunable laser system for UV, visible, and IR regions, in Energy and Technology Review (Lawrence Livermore Nat'l Lab., Livermore 1979)
A.P. Hickman, J.A. Paisner, W.K. Bishel: Theory of multiwave propagation and frequency conversion in a Raman medium. Phys. Rev. A33, 1788 (1986)
8.90 F. Moya, S.A.J. Druet, J.P. Taran: Rotation-vibration spectroscopy of gases by coherent anti-Stokes Raman scattering: Application to concentration and temperature measurements, in *Laser Spectroscopy*, ed. by S. Haroche, J.C. Pebay Peyroula, T.W. Hänsch, S.E. Harris (Springer, Berlin, Heidelberg 1975)
8.91 M. Aldén, H. Edner, S. Svanberg: Coherent anti-Stokes Raman spectroscopy (CARS) applied to combustion probing. Phys. Scripta 27, 29 (1983)
8.92 A.C. Eckbreth: BOXCARS: Crossed beam phase-matched CARS generation in gases. Appl. Phys. Lett. 32, 421 (1978)
8.93 N. Bloembergen: *Nonlinear Optics*, 3rd Pr. (Benjamin, New York 1977)
8.94 Y.R. Shen: *The Principles of Nonlinear Optics* (Wiley, New York 1984)
8.95 M. Schubert, B. Wilhelmi: *Nonlinear Optics and Quantum Electronics, Theoretical Concepts* (Wiley, New York 1986)
8.96 V.S. Letokhov, V.P. Chebotayev: *Nonlinear Laser Spectroscopy*, Springer Ser. Opt. Sci., Vol.4 (Springer, Berlin, Heidelberg 1977)
8.97 M.D. Levenson, S. Kano: *Introduction to Nonlinear Spectroscopy*, 2nd. edn. (Academic, New York 1988)

Chapter 9

9.1 W. Demtröder: *Laser Spectroscopy*, 2nd. printing, Springer Ser. Chem. Phys., Vol.5 (Springer, Berlin, Heidelberg 1982)
9.2 A. Corney: *Atomic and Laser Spectroscopy* (Clarendon, Oxford 1977)
9.3 L.J. Radziemski, R.W. Solarz, J.A. Paisner (eds): *Laser Spectroscopy and its Applications* (Dekker, New York 1987)
9.4 V.S. Letokhov, V.P. Chebotayev: *Nonlinear Laser Spectroscopy*, Springer Ser. Opt. Sciences, Vol.4 (Springer, Berlin, Heidelberg 1977)
9.5 M.D. Levenson, S. Kano: *Introduction to Nonlinear Spectroscopy*, rev. ed. (Academic, New York 1988)
9.6 Y.R. Shen: *The Principles of Nonlinear Optics* (Wiley, New York 1984)
9.7 M. Schubert, B. Wilhelmi: *Nonlinear Optics and Quantum Electronics, Theoretical Concepts* (Wiley, New York 1986)
9.8 S. Stenholm: *Foundations of Laser Spectroscopy* (Wiley, New York (1984)
9.9 A.L. Schawlow: Spectroscopy in a new light. Rev. Mod. Phys. **54**, 697 (1982)
9.10 N. Bloembergen: Nonlinear optics and spectroscopy. Rev. Mod. Phys. **54**, 685 (1982)
9.11 G.W. Series: Laser spectroscopy. Contemp. Phys. **25**, 3 (1984)
9.12 B. Couillaud, A. Ducasse: New methods in high-resolution laser spectroscopy, in *Progress in Atomic Spectroscopy*, ed. by H.J. Beyer, H. Kleinpoppen (Plenum, New York 1984) Pt.C, p.57
9.13 R.C. Thompson: High-resolution laser spectroscopy of atomic systems. Rep. Prog. Phys. **48**, 531 (1985)
9.14 H. Walther (ed.): *Laser Spectroscopy of Atoms and Molecules*, Topics Appl. Phys., Vol.2 (Springer, Berlin, Heidelberg 1976)
9.15 K. Shimoda (ed.): *High-Resolution Laser Spectroscopy*, Topics Appl. Phys., Vol.13 (Springer, Berlin, Heidelberg 1976)
9.16 Y. Prior, A. Ben-Reuven, M. Rosenbluh: *Methods of Laser Spectroscopy* (Plenum, New York 1986)
R.A. Smith (ed.): *Very High Resolution Spectroscopy* (Academic, London 1976)
9.17 A. Mooradian, T. Jaeger, P. Stokseth (eds.): *Tunable Lasers and Applications*, Springer Ser. Opt. Sci., Vol.3 (Springer, Berlin, Heidelberg 1976)
9.18 M.D. Levenson, W.H. Yen (eds.): *Lasers, Spectroscopy and New Ideas. A Tribute to A.L. Schawlow*, Springer Ser. Opt. Sci., Vol.54 (Springer, Berlin, Heidelberg 1987)
9.19 R.G. Brewer, A. Mooradian (eds.): *Laser Spectroscopy*, Proc. 1st. Int'l Conf., Vail 1973 (Academic, New York 1974)
9.20 S. Haroche, J.C. Pebay-Peyroula, T.W. Hänsch, S.E. Harris (eds.): *Laser Spectroscopy*, Proc. 2nd. Int'l Conf., Megeve 1975, Lecture Notes Phys., Vol.43 (Springer, Berlin, Heidelberg 1975)
9.21 J.L. Hall, J.L. Carlsten (eds.): *Laser Spectroscopy III*, Proc. 3rd. Int'l Conf., Jackson Lake 1977, Springer Ser. Opt. Sci., Vol.7 (Springer, Berlin, Heidelberg 1977)
9.22 H. Walther, K.W. Rothe (eds.): *Laser Spectroscopy IV*, Proc. 4th Int'l Conf., Rottach-Egern 1979, Springer Ser. Opt. Sci., Vol.21 (Springer, Berlin, Heidelberg 1979)
9.23 A.R.W. McKellar, T. Oka, B.P. Stoicheff (eds.): *Laser Spectroscopy V*, Proc. 5th Int'l Conf., Jasper 1981, Springer Ser. Opt. Sci., Vol.30 (Springer, Berlin, Heidelberg 1981)
9.24 H.P. Weber, W. Lüthy (eds.).: *Laser Spectroscopy VI*, Proc. 6th Int'l Conf., Interlaken 1983, Springer Ser. Opt. Sci., Vol.40 (Springer, Berlin, Heidelberg 1983)
9.25 T.W. Hänsch, Y.R. Shen (eds.): *Laser Spectroscopy VII*, Proc. 7th Int'l Conf., Maui 1985, Springer Ser. Opt. Sci., Vol.49 (Springer, Berlin, Heidelberg 1985)

9.26 W. Persson, S. Svanberg (eds.): *Laser Spectroscopy VIII*, Proc. 8th Int'l Conf., Åre 1987, Springer Ser. Opt. Sci., Vol.55 (Springer, Berlin, Heidelberg 1987)
9.27 M.S. Feld, J.E. Thomas, A. Mooradian (eds.): *Laser Spectroscopy IX* (Academic, Boston 1989)
9.28 C.J. Latimer: Recent experiments involving highly excited atoms. Contemp. Phys. 20, 631 (1979)
9.29 D. Kleppner: The spectroscopy of highly excited atoms, in *Progress in Atomic Spectroscopy*, ed. by W. Hanle, H. Kleinpoppen (Plenum, New York 1979) Pt.B, p.713
D. Kleppner, M.G. Littman, M.L. Zimmerman: Highly excited atoms. Sci. Am. 244, 130 (1981)
9.30 R.F. Stebbings, F.B. Dunning (eds): *Rydberg States of Atoms and Molecules* (Cambridge Univ. Press, Cambridge 1983)
9.31 T.F. Gallagher: Rydberg atoms. Rep. Progr. Phys. 51, 143 (1988)
9.32 K.C. Harvey, B.P. Stoicheff: Fine structure of the n ^{2}D series in rubidium near the ionization limit. Phys. Rev. Lett. 38, 537 (1977)
9.33 K. Niemax: Spectroscopy using thermionic diode detectors, Appl. Phys. B38, 147 (1985)
9.34 T.W. Ducas, M.G. Littman, R.R. Freeman, D. Kleppner: Stark ionization of high-lying states of sodium. Phys. Rev. Lett. 35, 366 (1975)
9.35 T.F. Gallagher, L.M. Humphrey, R.M. Hill, S.A. Edelstein: Resolution of [m_l] and [m_j] levels in the electric field ionization of highly excited d-states of Na. Phys. Rev. Lett. 37, 1465 (1976)
T.F. Gallagher, L.M. Humphrey, R.M. Hill, W. Cooke, S.A. Edelstein: Fine structure intervals and polarizabilities of highly excited p and d states of sodium. Phys. Rev. A15, 1937 (1977)
9.36 F.V. Kowalski, R.T. Hawkins, A.L. Schawlow: Digital wavemeter for cw lasers. J. Opt. Soc. Am. 66, 965 (1976).
9.37 J.L. Hall, S.A. Lee: Interferometric real time display of cw dye laser wavelengths with sub-Doppler accuracy. Appl. Phys. Lett. 29 367 (1976)
9.38 A. Fischer, K. Kullmer, W. Demtröder: Computer-controlled Fabry-Pérot wavemeter. Opt. Commun. 39, 277 (1981)
9.39 L.S. Lee, A.L. Schawlow: Multi-wedge wavemeter for pulsed lasers, Opt. Lett. 6, 610 (1981)
9.40 P. Juncar, J. Pinard: A new method for frequency calibration and laser control. Opt. Commun 14, 438 (1975)
9.41 T.W. Hänsch, J.J. Snyder: Wavemeters, *Dye Lasers*, 3rd edn., ed. F.P. Schäfer, Topics Appl. Phys., Vol.1 (Springer, Berlin, Heidelberg 1990)
9.42 R. Castell, W. Demtröder, A. Fischer, R. Kullmer, H. Weickenmeier, K. Wickert: The accuracy of laser wavelength meters. Appl Phys. B38, 1 (1985)
9.43 M. Herscher: The spherical mirror Fabry-Perot interferometer. Appl. Opt. 7, 951 (1968)
9.44 J.U. White: Long optical paths of large aperture. J. Opt. Soc. Am. 32, 285 (1942)
9.45 J.U. White: Very long paths in air. J. Opt. Soc. Am. 66, 411 (1976)
9.46 G. Yale Eastman: The heat pipe. Sci. Am. 218, 38 (1968)
9.47 C.R. Vidal, J. Cooper: Heat pipe oven. A new well defined metal vapor device for spectroscopic measurements. J. Appl. Phys. 40, 3370 (1969)
9.48 H.-L. Chen: Applications of laser absorption spectroscopy, in [Ref.9.3, p.261]
9.49 T.W. Hänsch, A.L. Schawlow, P. Toschek: Ultrasensitive response of a cw dye laser to selective extinction. IEEE J. QE-8, 802 (1972)
9.50 T.H. Harris: Laser intracavity-enhanced spectroscopy, in [Ref.9.76, p.343]
V.M. Baev, T.P. Belikova, E.A. Sviridenkov, A.F. Suchkov: JETP 74, 21 (1978)
9.51 D.J. Bradley, P. Ewart, J.V. Nicholas, J.R.D. Shaw: Excited state absorption spectroscopy of alkaline earths selectively pumped by tunable dye lasers. I. Barium arc spectra. J. Phys. B6, 1594 (1973)

9.52 J.R. Rubbmark, S.A. Borgström, K. Bockasten: Absorption spectroscopy of laser-excited barium. J. Phys. B10, 421 (1977)
9.53 M.E. Kaminsky, R.T. Hawkins, F.V. Kowalski, A.L. Schawlow: Identification of absorption lines by modulated lower-level population: Spectrum of Na. Phys. Rev. Lett. 36, 671 (1976)
9.54 A.L. Schawlow: Simplifying spectra by laser level labeling. Phys. Scr. 25, 333 (1982)
9.55 R. Teets, R. Feinberg, T.W. Hansch, A.L. Schawlow: Simplification of spectra by polarization labeling. Phys. Rev. Lett. 37, 683 (1976)
9.56 P. Esherick: Bound, even-parity J=0 and J=2 states of Sr. Phys. Rev. A15, 1920 (1977)
9.57 J.E.M. Goldsmith, J.E. Lawler: Optogalvanic spectroscopy. Contemp. Phys. 22, 235 (1981)
9.58 C.J. Sansonetti, K.-H. Weber: Reference lines for dye-laser wavenumber calibration in the optogalvanic spectra of uranium and thorium. J. Opt. Soc. Am. 131, 361 (1984)
9.59 O. Axner, I. Lindgren, I. Magnusson, H. Rubinsztein-Dunlop: Trace element determination in flames by laser-enhanced ionization spectrometry. Anal. Chem. 57, 773 (1985)
9.60 J.E.M. Goldsmith: Recent advances in flame diagnostics using fluorescence and ionization techniques, in [Ref.9.26, p.337]
9.61 J.A. Paisner, R.W. Solarz: Resonance photoionization spectroscopy, in [Ref.9.3, p.175]
9.62 P. Camus (ed.): Optogalvanic Spectroscopy and its Applications. J. Physique Coll. C7, Suppl. no 11, Tome 44 (1983)
9.63 P. Hannaford: Spectroscopy with sputtered atoms. Contemp. Phys. 24, 251 (1983)
9.64 K.C. Smith, P.K. Schenck: Optogalvanic spectroscopy of a neon discharge. Chem. Phys. Lett. 55, 466 (1978)
9.65 V.S. Letokhov: *Laser Photoionization Spectroscopy* (Academic Press, Orlando 1987)
9.66 J.C. Travis, G.C. Turk, J.R. DeVoe, P.K. Schenck, C.A. van Dijk: Progr. Anal. Atom. Spectr. 7, 199 (1984)
9.67 I. Magnusson, O. Axner, I. Lindgren, H. Rubinsztein-Dunlop: Laser-enhanced ionization detection of trace elements in a graphite furnace. Appl. Spectr. 40, 968 (1986)
O. Axner, I. Magnusson, J. Petersson, S. Sjöström: Investigation of the multi-element capability of laser-enhanced ionization spectrometry in flames for analysis of trace elements in water solution. Appl. Spectr. 41, 19 (1987)
9.68 N. Omenetto: The impact of several atomic and molecular laser spectroscopic techniques for chemical analysis. in H. Medin, S. Svanberg (eds.): *Laser Technology in Chemistry*, Special issue, Appl. Phys. B46, No. 3 (1988)
9.69 G.S. Hurst, M.G. Payne (eds.): *Resonance Ionization Spectroscopy and its Applications* 1984, Conf. Series Number 71 (The Institute of Physics, Bristol 1984)
G.S. Hurst, C. Grey Morgan (eds.): *Resonance Ionization Spectroscopy*, Conf. Series Number 84 (The Institute of Physics, Bristol 1987)
9.70 G.S. Hurst, M.G. Payne (eds.): *Principles and Applications of Resonance Ionization Spectroscopy* (Adam Hilger, Bristol 1988)
9.71 C.H. Chen, G.S. Hurst, M.G. Payne: Resonance ionization spectroscopy: Inert gas detection. in H.J. Beyer, H. Kleinpoppen (eds.): *Progress in Atomic Spectroscopy*, Pt.C (Plenum, New York 1984) p. 115;
9.72 G.S. Hurst, M.G. Payne, S.D. Kramer, C.H. Chen, R.C. Phillips, S.L. Allman, G.D. Alton, J.W.T. Dabbs, Rd. Willis, B.E. Lehman: Method for counting noble gas atoms with isotopic selectivity. Rep. Prog. Phys. 48, 1333 (1985)
V.S. Letokhov: Detecting individual atoms and molecules with lasers. Sci. Am. 259/3, 44 (1988)

9.73 J.A. Gelbwachs (ed.): *Laser Spectroscopy for Detection*. Proc. SPIE Vol. 286 (SPIE, Washington 1981)
9.74 R.A. Keller: *Laser Based Ultrasensitive Spectroscopy and Detection*. Proc. SPIE Vol. 426 (SPIE, Washington 1983)
9.75 J.J. Snyder, R.A. Keller (eds.): *Ultrasensitive Laser Spectroscopy* Special issue, J. Opt. Soc. Am. B2, No 9 (1985)
9.76 D. Kliger (ed.): *Ultrasensitive Laser Spectroscopy* (Academic Press, New York 1983)
9.77 N. Omenetto (ed.): *Analytical Laser Spectroscopy* (Wiley, New York 1979)
9.78 E.H. Piepmeier (ed.): *Analytical Applications of Lasers* (Wiley, New York 1986)
9.79 V.S. Letokhov (ed.): *Laser Analytical Spectrochemistry* (Hilger, Bristol 1986)
9.80 S. Svanberg: Fundamentals of atmospheric spectroscopy, in T. Lund (ed.): *Surveillance of Environmental Pollution and Resources by Electromagnetic Waves* (D. Reidel, Dordrecht 1978)
9.81 L.B. Kreutzer: Laser optoacoustic spectroscopy. A new technique of gas analysis. Anal. Chem. 46, 239A (1974)
9.82 A. Rosencwaig: *Photoacoustics and Photoacoustic Spectroscopy* (Wiley, New York 1980)
9.83 V. Letokhov, V. Zhaorov: *Laser Opto-Acoustic Spectroscopy*, Springer Ser. Opt. Sci., Vol. 37 (Springer, Berlin, Heidelberg 1986)
9.84 A.C. Tam: Applications of photoacoustic sensing techniques. Rev. Mod. Phys. 58, 381 (1986)
9.85 P. Hess, J. Pelzl (eds.): *Photoacoustics and Photothermal Phenomena*, Springer Ser. Opt. Sci., Vol. 58 (Springer, Berlin, Heidelberg 1988)
9.86 C.K.N. Patel, A.C. Tam: Pulsed optoacoustic spectroscopy of condensed matter. Rev. Mod. Phys. 53, 517 (1981)
9.87 S. Svanberg, P. Tsekeris, W. Happer: Hyperfine structure studies of highly excited D and F levels in alkali atoms using a CW dye laser. Phys. Rev. Lett. 30, 817 (1973)
S. Svanberg, P. Tsekeris: Hyperfine-structure investigation of highly excited ^{2}D levels in ^{87}Rb and ^{133}Cs using a cw tunable laser in a two-step excitation scheme. Rev. A11, 1125 (1975)
9.88 G. Belin, I. Lindgren, L. Holmgren, S. Svanberg: Hyperfine interaction, Zeeman and Stark effects for excited states in potassium. Phys. Scr. 12, 287 (1975)
9.89 G. Belin, L. Holmgren, S. Svanberg: Hyperfine interaction, Zeeman and Stark effects for excited states in rubidium. Phys. Scr. 13, 351 (1976)
9.90 G. Belin, L. Holmgren, S. Svanberg: Hyperfine interaction, Zeeman and Stark effects for excited states in cesium. Phys. Scr. 14, 39 (1976)
9.91 S. Svanberg: Measurement and calculation of excited alkali hyperfine and Stark parameters, in [Ref.9.21, p. 183]
9.92 R. Neumann, F. Träger, G. zu Putlitz: Laser-microwave spectroscopy, in H.K. Beyer, H. Kleinpoppen (eds.): *Progress in Atomic Spectroscopy*, Part D (Plenum, New York 1987) p. 1
T.F. Gallagher: Radiofrequency spectroscopy of Rydberg atoms, in *Progress in Atomic Spectroscopy*, Pt.D, ed. by H.K. Beyer, H. Kleinpoppen (Plenum, New York 1987) p.12
9.93 K. Fredriksson, S. Svanberg: Precision determination of the fine structure of the 4d state in sodium using level crossing spectroscopy. Phys. Lett. 53A, 61 (1975)
9.94 E. Matthias, R.A. Rosenberg, E.D. Poliakoff, M.G. White, S.-T. Lee, D.A. Shirley: Time resolved VUV spectroscopy using synchrotron radiation: Fluorescent lifetimes of atomic Kr and Xe. Chem. Phys. Lett. 52, 239 (1977)
T. Möller, G. Zimmerer: Time-resolved spectroscopy with synchrotron radiation in the vacuum ultraviolet. Phys. Scr. T17, 177 (1987)

R. Rigler, O. Kristensen, R. Roslund, P. Thyberg, K. Oba, M. Eriksson: Molecular structure and dynamics: Beamline for time resolved spectroscopy at the MAX synchrotron in Lund. Phys. Scr. T17, 204 (1987)

9.95 G.H.C. New: The generation of ultrashort laser pulses. Rep. Prog. Phys. 46, 877 (1983)

S. de Silvestri, P. Laporta, V. Magni: Generation and applications of femtosecond laser pulses. Europhys. News 17 (9), 105 (1986).

9.96 T.F. Johnston, Jr: Tunable dye lasers, in *Encyclopedia of Physical Science and Technology*, Vol.14 (Academic, New York 1987)

9.97 B. Couillaud, V. Fossati-Bellani: Modelocked lasers and ultrashort pulses. Lasers & Appl. (January 1985) pp.79-83 and (February 1985) pp.91-94

9.98 W.H. Know, R.L. Fork, M.C. Downer, R.H. Stolen, C.V. Shank, J.A. Valdamis: Optical pulse compression to 8 fs at a 5 kHz repetition rate. Appl. Phys. Lett. 46, 1120 (1985)

R.L. Fork, C.H. Brito Cruz, P.C. Becker, C.V. Shank: Compression of optical pulses to six femtoseconds by using cubic phase compression. Opt. Lett. 12, 483 (1987)

9.99 P.B. Coates: The correction for photon "pile-up" in the measurement of radiative lifetimes. J. Phys. E1, 878 (1968)

9.100 M. Gustavsson, H. Lundberg, L. Nilsson, S. Svanberg: Lifetime measurements for excited states of rare-earth atoms using pulse-modulation of a cw dye laser beam. J. Opt. Soc. Am. 69, 984 (1979)

J. Carlsson: Accurate time-resolved laser spectroscopy on sodium and bismuth atoms. Z. Phys. D 9, 147 (1988)

9.101 K. Bhatia, P. Grafström, C. Levinson, H. Lundberg, L. Nilsson, S. Svanberg: Natural radiative lifetimes in the perturbed 6snd $^1D_2$ sequence of barium. Z. Physik A 303, 1 (1981)

T.F. Gallagher, W. Sandner, K.A. Safinya: Probing configuration interaction of the Ba 5d7d $^1D_2$ state using radiofrequency spectroscopy and lifetime measurements. Phys. Rev. A 23, 2969 (1981)

M. Aymar, R.-J. Champeau, C. Delsart, J.C. Keller: Lifetimes of Rydberg levels in the perturbed 6snd $^{1,3}D_2$ series of barium I. J. Phys. B 14, 4489 (1981)

9.102 S. Svanberg: Perturbations in Rydberg sequences probed by lifetime, Zeeman-effect and hyperfine structure measurements, in [Ref.9.23, p. 301]

9.103 S. Letzring: Buying and using a streak camera. Lasers & Appl. (March 1983) p.49

9.104 S.L. Shapiro (ed.): *Ultrashort Light Pulses*, 2nd Ed., Topics Appl. Phys., Vol. 18 (Springer, Berlin, Heidelberg 1984)

W. Kaiser (ed.): *Ultrashort Laser Pulses and Applications*, Springer Ser. Opt. Sci., Vol.60 (Springer, Berlin, Heidelberg 1988)

9.105 C.V. Shank, E. Ippen, S.L. Shapiro: *Picosecond Phenomena*, Springer Ser. Chem. Phys., Vol.4 (Springer, Berlin, Heidelberg 1978)

9.106 R.M. Hochstrasser, W. Kaiser, C.V. Shank: *Picosecond Phenomena II*, Springer Ser. Chem. Phys., Vol.14 (Springer, Berlin, Heidelberg 1980)

9.107 K. Eisenthal, R.M. Hochstrasser, W. Kaiser, A Lauberau (eds.): *Picosecond Phenomena III*, Springer Ser. Chem. Phys., Vol.23 (Springer, Berlin, Heidelberg 1982)

9.108 D. Auston, K. Eisenthal (eds.): *Ultrafast Phenomena IV*, Springer Ser. Chem. Phys., Vol. 38 (Springer, Berlin, Heidelberg 1984)

9.109 A. Siegman, G. Fleming (eds.): *Ultrafast Phenomena V*, Springer Ser. Chem. Phys., Vol. 46 (Springer, Berlin, Heidelberg 1986)

9.110 C. De Michelis, M. Mattioli: Spectroscopy and impurity behaviour in fusion plasmas. Rep. Prog. Phys. 47, 1233 (1984)

R.C. Isler: Impurities in Tokomaks. Nuclear Fusion 24, 1599 (1984)

9.111 R.E. Imhof, F.H. Read: Measurements of lifetimes of atoms, molecules and ions. Rep. Prog. Phys. 40, 1 (1977)
9.112 P. Erman: Time resolved spectroscopy of small molecules, in Specialists Periodical Reports, *Molecular Spectroscopy*, Vol. 6, Ch. 5 (The Chemical Society, London 1979) p. 174
9.113 J.N. Dodd, G.W. Series: Time resolved fluorescence spectroscopy, in *Progress in Atomic Spectroscopy*, Pt.A, ed. by W. Hanle, H. Kleinpoppen (Plenum, New York 1978) p.639
W.L. Wiese: Atomic transition probabilities and lifetimes, in *Progress in Atomic Spectroscopy*, Pt.B, ed. by W. Hanle, H. Kleinpoppen (Plenum, New York 1979) p.1101
9.114 M.C.E. Huber, R.J. Sandeman: The measurement of oscillator strengths. Rep. Prog. Phys. 49, 397 (1986)
9.115 O. Poulsen, J.L. Hall: Spectroscopic investigation in ^{209}Bi I using tunable-cw-dye-laser spectroscopy. Phys. Rev. A 18, 1089 (1978)
9.116 S. Svanberg: Natural radiative lifetimes of some excited Bi I levels belonging to the $6p^2 7s$ and $6p^2 6d$ configurations measured by the Hanle method. Phys. Scr. 5, 73 (1972)
9.117 H.J. Andrä, A. Gaupp, W. Wittmann: New method for precision lifetime measurements by laser excitation of fast-moving atoms. Phys. Rev. Lett. 31, 501 (1973)
9.118 P. Erman, J. Brzozowski, B. Sigfridsson: Gas excitations using high-frequency deflected electron beams: A convenient method for determinations of atomic and molecular lifetimes. Nucl. Instr. Methods 110, 471 (1973)
9.119 P. Erman: High resolution measurements of atomic and molecular lifetimes using the high-frequency deflection technique. Phys. Scr. 11, 65 (1975)
P. Erman: Astrophysical applications of time-resolved molecular spectroscopy. Phys. Scr. 20, 575 (1979); Studies of perturbations using time resolved techniques. Phys. Scr. 25, 365 (1982)
9.120 J. Brzozowski, P. Bunker, N. Elander, P. Erman: Predissociation effects in the A, B, and C states of CN and the interstellar formation rate of CH via inverse predissociation. Astrophys. J. 207, 414 (1976)
9.121 J.K. Link: Measurement of the radiative lifetimes of the first excited states of Na, K, Rb, and Cs by means of the phase-shift method. J. Opt. Soc. Am. 56, 1195 (1966)
P.T. Cunningham, J.K. Link: Measurement of lifetimes of excited states of Na, Tl, In, Ga, Cu, Ag, Pb, and Bi by the phase-shift method. J. Opt. Soc. Am. 57, 1000 (1967)
L. Armstrong, Jr., S. Ferneuille: Theoretical analysis of the phase shift measurement of lifetimes using monochromatic light. J. Phys. B 8, 546 (1975)
9.122 C.H. Corliss, W.R. Bozman: Experimental transition probabilities for spectral lines of seventy elements. NBS Monograph 53 (National Bureau of Standards, Wash., D.C. 1962)
9.123 W. Marlow: Hakenmethode. Appl Opt. 6, 1715 (1967)
9.124 N.P. Penkin: Experimental determination of electronic transition probabilities and the lifetimes of the excited atomic and ionic states, in *Atomic Physics 6*, ed. by R. Damburg (Plenum, New York 1979) p.33
9.125 W.A. van Wijngaarden, K.D. Bonin, W. Happer: Inverse hook method for measuring oscillator strengths for transitions between excited atomic states. Hyperf. Interact. 38, 471 (1987)
9.126 S. Svanberg: Atomic spectroscopy by resonance scattering. Phil. Trans. Roy. Soc. (London) A293, 215 (1979)
9.127 J.N. Dodd, G.W. Series: Time-resolved fluorescence spectroscopy, in *Progress in Atomic Spectroscopy*, Pt.A, ed. by W. Hanle, H. Kleinpoppen (Plenum, New York 1978) p.639

9.128 S. Haroche: Quantum beats and time-resolved fluorescence spectroscopy, in [Ref.9.15, p.253]

9.129 P. Grundevik, H. Lundberg, A.-M. Martensson, K. Nyström, S. Svanberg: Hyperfine-structure study in the P sequence of ^{23}Na using quantum-beat spectroscopy. J. Phys. B 12, 2645 (1979)

9.130 G. Jönsson, C. Levinson, I. Lindgren, A. Persson, C.G. Wahlström: Experimental and theoretical studies of the $4s^2 np\ ^2P$ sequence in neutral gallium. Z. Phys. A 322, 351 (1985)

9.131 J. Bengtsson, J. Larsson, S. Svanberg, C.-G. Wahlström: Hyperfine-structure study of the $3d^{10} 5p\ ^2P_{3/2}$ level of neutral copper using pulsed level-crossing spectroscopy at short wavelengths. Phys. Rev. A 41, 233 (1990)
J. Bengtsson, J. Larsson, S. Svanberg, C.G. Wahlström: High-resolution pulsed laser spectroscopy in the UV/VUV spectral region. [Ref.9.27, p.86]
S. Svanberg: High-resolution laser spectroscopy in the UV/VUV spectral region, in *Applied Laser Spectroscopy*, ed. by M. Inguscio, W. Demtröder (Plenum, New York 1990)

9.132 D.P. O'Brien, P. Meystre, H. Walther: Subnatural linewidths in atomic spectroscopy, in *Advanced Atomic and Molecular Physics*, Vol.21, ed. by D.R. Bates, B. Bederson (Academic, Orlando 1985)
H. Figger, H. Walther: Optical resolution beyond the natural linewidth: A level-crossing experiment on the $3\ ^2P_{3/2}$ level of sodium using a tunable dye laser. Z. Phys. 267, 1 (1974)

9.133 P. Schenk, R.C. Hilborn, H. Metcalf: Time-resolved fluorescence from Ba and Ca excited by a pulsed tunable dye laser. Phys. Rev. Lett. 31, 189 (1974)

9.134 K. Fredriksson, H. Lundberg, S. Svanberg: Fine- and hyperfine structure investigation in the 5 D - n F series of cesium. Phys. Rev. A21, 241 (1980)

9.135 K. Fredriksson, L. Nilsson, S. Svanberg: Stark interaction for alkali atoms (unpublished report, 1980)

9.136 M.A. Zaki Ewiss, W. Hogervorst, W. Vassen, B.H. Post: The Stark effect in the 6snf Rydberg series of barium. Z. Phys. A 322, 211 (1985)

9.137 F. Touchard, J.M. Serre, S. Büttgenbach, P. Guimbal, R. Klapisch, M. de Saint Simon, C. Thibault, H.T. Duong, P. Juncar, S. Liberman, J. Pinard, J.-L. Vialle: Electric quadrupole moments and isotopic shifts of radioactive sodium isotopes. Phys. Rev C 25, 2756 (1982)

9.138 S. Liberman: High resolution laser spectroscopy of radioactive atoms, in [Ref.9.26, p.162]

9.139 H.T. Duong, P. Juncar, S. Liberman et al.: First observation of the blue optical lines of francium. Europhys. Lett. 3, 175 (1987)
S.V. Andreev, V.S. Letokhov, V.I. Mishin: Laser resonance photoionization spectroscopy of Rydberg levels in Fr. Phys. Rev. Lett. 59, 1274 (1987)

9.140 W. Ertmer, B. Hofer: Zero-field hyperfine structure measurements of metastable states $3d^2 4s\ ^4F_{3/2,9/2}$ of ^{45}Sc using laser fluorescence atomic-beam-magnetic-resonance technique. Z. Phys. A 276, 9 (1976)
S.D. Rosner, R.A. Holt, T.D. Gaily: Measurement of the zero-field hyperfine structure of a single vibration-rotation level of Na_2 by a laser-fluorescence molecular-beam-resonance technique. Phys. Rev. Lett. 35, 785 (1975)

9.141 P. Grundevik, M. Gustavsson, I. Lindgren, G. Olsson, L. Robertsson, A. Rosén, S. Svanberg: Precision method for hyperfine structure studies in low-abundance isotopes: The quadrupole moment of ^{43}Ca. Phys. Rev. Lett. 42, 1528 (1979)

9.142 W.H. Wing, G.A. Ruff, W.E. Lamb, J.J. Spezeski: Observation of the infrared spectrum of the hydrogen molecular ion HD^+. Phys. Rev. Lett. 36, 1488 (1976)

9.143 S.L. Kaufmann: High resolution laser spectroscopy in fast beams. Opt. Commun. 17, 309 (1976)

9.144 E.W. Otten: Hyperfine and isotope shift measurements, in *Atomic Physics 5*, ed. by R. Marrus, M. Prior, H. Shugart (Plenum, New York 1977)

9.145 P. Jaquinot, R. Klapisch: Hyperfine spectroscopy of radioactive atoms. Rep. Prog. Phys. **42**, 773 (1979)
9.146 R. Neugart, S.L. Kaufman, W. Klempt, G. Moruzzi: High resolution spectroscopy in fast beams, in [Ref.9.21, p.446]
R. Neugart: Collinear fast-beam laser spectroscopy, in *Progress in Atomic Spectroscopy*, Pt.D, ed. by H.K. Beyer, H. Kleinpoppen (Plenum, New York 1987) p.75
9.147 H.J. Kluge: Optical spectroscopy of shortlived isotopes, in *Progress in Atomic Spectroscopy*, Pt.B, ed. by W. Hanle, H. Kleinpoppen (Plenum, New York 1979) p.727
H.J. Kluge: Hyperf. Interac. **24-26**, 69 (1985)
9.148 H.H Stroke: Isotopic shifts, in I. Lindgren, A. Rosen, S. Svanberg (eds.): *Atomic Physics 8* (Plenum, New York 1983) p. 509
9.149 S. Svanberg: Laser spectroscopy applied to the study of hyperfine interactions. Hyperf. Interact. **15/16**, 111 (1983)
9.150 P.H. Lee, M.L. Skolnick: Saturated neon absorption inside a 6328Å laser. Appl. Phys. Lett. **10**, 303 (1967)
9.151 W.E. Lamb, Jr.: Theory of the optical laser. Phys. Rev. **134**, A1429 (1964)
9.152 W.R. Rowley, B.W. Jolliffe, K.C. Schotton, A.J. Wallard, P.T. Woods: Laser wavelength measurements and the speed of light. Opt. Quant. Electr. **8**, 1 (1976)
J.L. Hall: Stabilized lasers and precision measurements. Science **202**, 147 (1978)
9.153 K.M. Evenson, D.A. Jennings, F.R. Peterson, J.S. Wells: Laser frequency measurements: A review, limitations, extension to 197 THz (1.5 mm), in [Ref. 9.21, p.57]
D.A. Jennings, F.R. Petersen, K.M. Evenson: Direct frequency measurement of the 260 THz (1.15μ) ^{20}Ne laser: And beyond, in [Ref.9.22, p. 31]
9.154 D.A. Jennings, C.R. Pollock, F.R. Petersen, R.E. Drullinger, K.M. Evenson, J.S. Wells: Direct frequency measurement of the I_2 stabilized He-Ne 473 THz (633 nm) laser. Opt. Lett. **8** 136 (1983)
9.155 R.G. DeVoe, R.G. Brewer: Laser frequency division and frequency stabilization. Phys. Rev. **A30**, 2827 (1984)
R.G. DeVoe, C. Fabre, K. Jungmann, J. Hoffnagle, R.G. Brewer: Precision optical-frequency difference measurements. Phys. Rev. **A37**, 1802 (1988)
9.156 T. Wilkie: Time to remeasure the metre. New Scientist (Oct. 27, 1983)
9.157 Documents concerning the new definition of the metre. Metrologia **19**, 163 (1984)
9.158 M.D. Levenson, A.L. Schawlow: Hyperfine interactions in molecular iodine. Phys. Rev. A **6**, 10 (1972)
9.159 T.W. Hänsch, I.S. Shahin, A.L. Schawlow: High resolution saturation spectroscopy of the sodium D line with a pulsed tunable dye laser. Phys. Rev. Lett. **27**, 707 (1971)
9.160 C. Bordé: Spectroscopie d'absorption saturée de diverses molécules au moyen des lasers á gas carbonique et á prooxyde d'azote. C. R. Acad. Sci. B **271**, 371 (1970).
9.161 S. Svanberg, G.-Y. Yan, T.P. Duffey, A.L. Schawlow: High-contrast Doppler-free transmission spectroscopy. Opt. Lett. **11**, 138 (1986)
S. Svanberg, G.-Y. Yan, T. P. Duffey, W.-M. Du, T.W. Hänsch, A.L. Schawlow: Saturation spectroscopy for optically thick atomic samples. J. Opt. Soc. Am. B **4**, 462 (1987)
9.162 C. Wieman, T.W. Hänsch: Doppler-free laser polarization spectroscopy. Phys. Rev. Lett. **36**, 1170 (1976)
9.163 T.W. Hänsch, I.S. Shahin, A.L. Schawlow: Optical resolution of the Lamb shift in atomic hydrogen. Nature **235**, 56 (1972)
T.W. Hänsch, M.H. Nayfeh, S.A. Lee, S.M. Curry, I.S. Shahin: Precision measurement of the Rydberg constant by laser saturation spectroscopy of the Balmer-α line in hydrogen and deuterium. Phys. Rev. Lett. **32**, 1336 (1974)

9.164 J.E.M. Goldsmith, E.W. Weber, T.W. Hänsch: New measurement of the Rydberg constant using polarization spectroscopy of H. Phys. Rev. Lett. **41**, 1525 (1978)
9.165 P. Zhao, W. Lichten, J.C. Bergquist, H.P. Layer: Remeasurement of the Rydberg constant. Phys. Rev. **A34**, 5138 (1986)
9.166 P. Zhao, W. Lichten, H. Layer, J. Bergquist: New value for the Rydberg constant from the hydrogen Balmer-β transition. Phys. Rev. Lett. **58**, 1293 (1987)
9.167 F. Biraben, J.C. Garreau, L. Julien: Determination of the Rydberg constant by Doppler-free two-photon spectroscopy of hydrogen Rydberg states. Europhys. Lett. **2**, 925 (1986); and in [Ref.9.26, p. 8]
9.168 T.W. Hänsch, A.L. Schawlow, G.W. Series: The spectrum of atomic hydrogen. Sci. Am. 3 240/3 72 (1979)
G.W. Series (ed.): *The Spectrum of Atomic Hydrogen: Advances* (World Scientific, Singapore 1988)
G.F. Bassani, M. Inguscio, T.W. Hänsch (eds.): *The Hydrogen Atom* (Springer, Berlin, Heidelberg 1989)
9.169 A.I. Ferguson, J.M. Tolchard: Laser spectroscopy of atomic hydrogen. Contemp. Phys. **28**, 383 (1987)
9.170 H.R. Schlossberg, A. Javan: Saturation behaviour of a Doppler-broadened transition involving levels with closely spaced structure. Phys. Rev. **150**, 267 (1966)
9.171 T.W. Hänsch, P. Toschek: Theory of a three-level gas laser amplifier. Z. Physik **236**, 213 (1970)
9.172 M.A. Bouchiat, L. Pottier: An atomic preference between left and right. Sci. Am. **250**/6, 76 (1984)
M.-A. Bouchiat, L. Pottier: Optical experiments and weak interactions. Nature **234**, 1203 (1986).
9.173 E.D. Commins: Parity violation in atoms, in [Ref.9.26, p.43]
9.174 T.P. Emmons, E.N. Fortson: Parity conservation in atoms, in *Progress in Atomic Spectroscopy*, Pt.D, ed. by H.K. Beyer, H. Kleinpoppen (Plenum, New York 1987) p.237
9.175 F.V. Kowalski, W.T. Hill, A.L. Schawlow: Saturated interference spectroscopy. Opt. Lett. **2**, 112 (1978)
9.176 R. Schieder: Interferometric nonlinear spectroscopy. Opt. Commun. **26**, 113 (1978)
9.177 M.S. Sorem, A.L. Schawlow: Saturation spectroscopy in molecular iodine by intermodulated fluorescence. Opt. Commun. **5**, 148 (1972)
9.178 J.E. Lawler, A.I. Ferguson, J.E.M. Goldsmith, D.J. Jackson, A.L. Schawlow: Doppler-free intermodulated optogalvanic spectroscopy. Phys. Rev. Lett. **42**, 1046 (1979)
9.179 D.R. Lyons, A.L. Schawlow, G.-Y. Yan: Doppler-free radiofrequency optogalvanic spectroscopy. Opt. Commun. **38**, 35 (1981)
9.180 E.E. Marinero, M. Stuke: Doppler-free optoacoustic spectroscopy. Opt. Commun. **30**, 349 (1979)
9.181 T.P. Duffey, D. Kammen, A.L. Schawlow, S. Svanberg, H.-R. Xia, G.-G. Xiao, G.-Y. Yan: Laser spectroscopy using beam overlap modulation. Opt. Lett. **10**, 597 (1986)
9.182 T.W. Hänsch, D.R. Lyons, A.L. Schawlow, A. Siegel, Z.-Y. Wang, G.-Y. Yan: Polarization intermodulated excitation (POLINEX) spectroscopy of helium and neon. Opt. Commun. **37**, 87 (1981)
9.183 G.C. Bjorklund: Frequency modulation spectroscopy: A new method for measuring weak absorptions and dispersions. Opt. Lett. **5**, 15 (1980)
M. Gehrtz, G.C. Bjorklund, E.A. Whittaker: Quantum-limited laser frequency-modulation spectroscopy. J. Opt. Soc. Am. **B2**, 1510 (1985)
9.184 N.H. Tran, R Kachru, P. Pillet, H.B. van Linden van dem Heuvell, T.F. Gallagher, J.P. Watjen: Frequency-modulation spectroscopy with a pulsed dye laser: Experimental investigations of sensitivity and useful features. Appl. Opt. **23**, 1353 (1984); Appl. Opt. **25**, 510 (1986)

N.H. Tran, T.F. Gallagher, J.P. Watjen, G.R. Janik, C.B. Carlisle: High efficiency resonant cavity microwave optical modulator. Appl. Opt. 24, 4282 (1984)
9.185 J. Bialas, R. Blatt, W. Neuhauser, P. Toschek: Ultrasensitive detection of light absorption by a few ions. Opt. Commun. 59, 27 (1986)
9.186 D.J. Wineland, W.M. Itano, J.C. Bergquist: Absorption spectroscopy at the limit: Detection of a single atom. Opt. Lett. 12, 389 (1987)
9.187 M. Goeppert-Mayer: Über Elementarakte mit zwei Quantensprüngen. Ann. Phys. 9, 273 (1931)
9.188 L.S. Vasilenko, V.P. Chebotayev, A.V. Shishaev: Line shape of a two-photon absorption in a standing-wave field in a gas. JETP Lett. 12, 113 (1970)
9.189 F. Biraben, B. Cagnac, G. Grynberg: Experimental evidence of two-photon transition without Doppler broadening. Phys. Rev. Lett. 32, 643 (1974)
9.190 G. Grynberg, B. Cagnac: Doppler-free multiphoton spectroscopy. Rep. Prog. Phys. 40, 791 (1977)
N. Bloembergen, M.D. Levenson: Dopplerfree two-photon absorption spectroscopy. Phys. Rev. Lett. 31, 645 (1974)
9.191 T.W. Hänsch, K.C. Harvey, G. Meisel, A.L. Schawlow: Two-photon spectroscopy of Na 3s-4d without Doppler broadening using a cw dye laser. Opt. Commun. 11, 50 (1974)
9.192 H. Rinneberg, J. Neukammer, G. Jönsson, H. Hieronymus, A. König, K. Vietzke: High-n Rydberg atoms and external fields. Phys. Rev. Lett. 55, 382 (1985)
J. Neukammer, H. Rinneberg, K. Vietzke, A. König, H. Hieronymus, M. Kohl, H.-J. Grabka: Spectroscopy of Rydberg atoms at n $\simeq$ 500: Observation of quasi-Landau resonances in low magnetic fields. Phys. Rev. Lett. 59, 2947 (1987)
9.193 B.P. Stoicheff, E. Wineberger: Doppler-free two-photon absorption spectrum of rubidium. Can. J. Phys. 57, 2143 (1979)
9.194 C.-J. Lorenzen, K. Niemax, L.R. Pendrill: Precise measurements of ^{39}K nS and nD energy levels with an evacuated wavemeter. Opt. Commun. 39, 370 (1981)
9.195 K.-H. Weber, C.J. Sansonetti: Accurate energies of nS, nP, nF and nG levels of neutral cesium. Phys. Rev. A35, 4650 (1987)
9.196 M.J. Seaton: Quantum defect theory. Rep. Progr. Phys. 46, 167 (1983)
9.197 E. Matthias, H. Rinneberg, R. Beigang, A. Timmermann, J. Neukammer, K. Lücke: Hyperfine structure and isotope shifts in alkaline earth atoms, in *Atomic Physics 8*, ed. by I. Lindgren, A. Rosén, S. Svanberg (Plenum, New York 1983) p.543
9.198 H. Rinneberg: Rydberg series of two-electron systems studied by hyperfine interactions, in *Progress in Atomic Spectroscopy*, Pt.D, ed. by H.K. Beyer, H. Kleinpoppen (Plenum, New York 1987) p.157
9.199 M. Aymar: Rydberg series of alkaline-earth atoms Ca through Ba. The interplay of laser spectroscopy and multi-channel quantum defect theory. Phys. Repts. 110, 163 (1984)
9.200 W. Hogervorst: Laser spectroscopy of Rydberg states of two-electron atoms. Comments At. Mol. Phys. 13, 69 (1983)
9.201 C. Wieman, T.W. Hänsch: Precision measurement of the 1S Lamb shift and the 1S - 2S isotope shift of hydrogen and deuterium. Phys. Rev. A22, 192 (1980)
9.202 E.A. Hildum, U. Boesl, D.H. McIntyre, R.G. Beausoleil, T.W. Hänsch: Measurements of the 1s-2s frequency in atomic hydrogen. Phys. Rev. Lett. 56, 576 (1986)
9.203 R.G. Beausoleil, D.H. McIntyre, C.J. Foot, E.A. Hildum, B. Couillaud, T.W. Hänsch: Continous wave measurement of the 1S Lamb shift in atomic hydrogen. Phys. Rev. A35, 4878 (1987); *ibid* A 39, 4591 (1989)
C. Zimmermann, R. Kallenback, T.W. Hänsch: High-resolution spectroscopy of the hydrogen 1s-2s transition in an atomic beam, Phys. Rev. Lett. 65, 571 (1990)

9.204 J. Cariou, P. Luc: *Atlas du Spectre d'Absorption de la Molécule de Tellure* (Laboratoire Aime' Cotton, CNRS II, Orsay 1980)
9.205 S. Gersternkorn, P. Luc: *Atlas du Spectre d'Absorption de la Molécule d'Iode 14800 - 20000 cm^{-1}* (Editions du CNRS, Paris 1978)
9.206 C.J. Foot, B. Couillaud, R.G. Beausoleil, T.W. Hänsch: Continous two-photon spectroscopy of hydrogen 1s - 2s. Phys. Rev. Lett. **54**, 1913 (1985)
9.207 M.G. Boshier, P.E.G. Baird, C.J. Foot, E.A. Hinds, M.D. Plimmer, D.N. Stacey, J.B. Swan, D.A. Tate, D.M. Warrington, G.K. Woodgate: Precision cw laser spectroscopy of hydrogen and deuterium, in [Ref.9.26, p.18]
9.208 T.W. Hänsch, R.G. Beausoleil, C.J. Foot, E.A. Hildum, D.H. McIntyre: The hydrogen atom in a new light, in *Methods in Laser Spectroscopy*, ed. by Y. Prior (Plenum, New York 1986)
9.209 P. Zhao, W. Lichten, H.P. Layer, J.C. Bergquist: Absolute wavelength measurements and fundamental atomic physics, in [Ref.9.26, p. 12]
9.210 M.M. Salour, C. Cohen-Tannoudji: Observation of Ramsey's interference fringes in the profile of Doppler-free two-photon resonances. Phys. Rev. Lett. **38**, 757 (1977)
9.211 R. Teets, J. Eckstein, T.W. Hänsch: Coherent two-photon excitation by multiple light pulses. Phys. Rev. Lett. **38**, 760 (1977)
9.212 J.C. Bergquist, S.A. Lee, J.L. Hall: Saturated absorption with spatially separated laser fields: Observation of optical "Ramsey" fringes. Phys. Rev. Lett. **38**, 159 (1977); and in [Ref.9.21, p. 142]
9.213 R.G. Beausoleil, T.W. Hänsch: Two-photon optical Ramsey spectroscopy of freely falling atoms. Opt. Lett. **10**, 547 (1985)
R.G. Beausoleil, T.W. Hänsch: Ultra-high-resolution two-photon optical Ramsey spectroscopy of an atomic fountain. Phys. Rev. A **33**, 1661 (1986)
M.A. Kasevich, E. Riis, S. Chu, R.G. DeVoe: RF spectroscopy in an atomic fountain. Phys. Rev. Lett. **63**, 612 (1989)
9.214 J. Prodan, A. Migdal, W.D. Phillips, I. So, H. Metcalf, J. Dalibard: Stopping atoms with laser light. Phys. Rev. Lett. **54**, 992 (1985)
W. Phillips, J. Prodan, H. Metcalf: Laser cooling and electromagnetic trapping of neutral atoms. J. Opt. Soc. Am. **B2**, 1751 (1985)
H. Metcalf, W.D. Phillips: Laser cooling of atomic beams: Comments. At. Mol. Phys. **16**, 79 (1985).
9.215 W. Ertmer, R. Blatt, J.L. Hall, M. Zhu: Laser manipulation of atomic velocities: Demonstration of stopped atoms and velocity reversal. Phys. Rev. Lett. **54**, 996 (1985)
9.216 T.W. Hänsch, A.L. Schawlow: Cooling of gases by laser radiation. Opt. Commun. **13**, 68 (1975)
9.217 W. Neuhauser, M. Hohenstatt, P. Toschek, H. Dehmelt: Localized visible Ba$^+$ mono-ion oscillator. Phys. Rev. A**22**, 1137 (1980)
9.218 F. Diedrich, E. Peik, J.M. Chen, W. Quint, H. Walther: Observation of a phase transition of stored laser-cooled ions. Phys. Rev. Lett. **59**, 2931 (1987); and Phys. Blätter **44**, 12 (1988); Nature **334**, 309 (1988)
D.J. Wineland, J.C. Bergquist, Wayne M. Itano, J.J. Bollinger, C.H. Manney: Atomic-ion Coulomb cluster in an ion trap. Phys. Rev. Lett. **59**, 2935 (1987)
J. Hoffnagle, R.V. DeVoe, L. Reyna, R.G. Brewer: Order-chaos transition of two trapped ions. Phys. Rev. Lett. **61**, 255 (1988)
R.G. Brewer, J. Hoffnagle, R.G. DeVoe, L. Reyna, W. Henshaw: Collision-induced two-ion chaos. Nature **344**, 305 (1990)
9.219 A.L. Migdall, J.V. Prodan, W.D. Phillips, T.H. Bergeman, H.J. Metcalf: First observation of magnetically trapped neutral atoms. Phys. Rev. Lett. **54**, 2596 (1985)
V.S. Bagnato, G.P. Lafyatis, A.G. Martin, E.L. Raab, R.N. Ahmad-Bitar, D.E. Pritchard: Continous stopping and trapping of neutral atoms. Phys. Rev. Lett. **58**, 2194 (1987)

9.220 S. Chu, L. Hollberg, J.E. Bjorkholm, A. Cable, A. Ashkin: Three-dimensional viscous confinement and cooling of atoms by resonance radiation pressure. Phys. Rev. Lett. 55, 48 (1985)
S. Chu, J.E. Bjorkholm, A. Ashkin, A. Cable: Experimental observation of optically trapped atoms. Phys. Rev. Lett. 57, 314 (1986)
S. Chu, M.G. Prentiss, A.E. Cable, J.E. Bjorkholm: Laser cooling and trapping of atoms, in [Ref.9.26, p.58]
P.D. Lett, R.N. Walls, Ch. I. Westbrook, W.D. Phillips, P.L. Gould, H.J. Metcalf: Observation of atoms laser cooled below the Doppler limit. Phys. Rev. Lett. 61, 169 (1988)
9.221 A. Bárány, A. Kerek, M. Larsson, S. Mannervik, L.-O. Norlin (eds.): Workshop and Symposium on the Physics of Low-Energy Stored and Trapped Particles. Phys. Scripta T 22, 1 (1988)
9.222 P. Meystre, S. Stenholm (eds.): The Mechanical Effects of Light. J. Opt. Soc Am. B 2, 1706-1860 (1985) (special issue)
S. Stenholm: Light forces put a handle on the atom: To cool and trap atoms by laser light. Contemp. Phys. 29, 105 (1988)
9.223 V.G. Minogin, V.S. Letokhov: *Laser Light Pressure on Atoms* (Harwood, London 1987)
9.224 W.D. Phillips, H.J. Metcalf: Cooling and trapping of atoms. Sci. Am. 256/3, 36 (1987)
P.L. Gould, P.D. Lett, W.D. Phillips: New measurements with optical molasses, in [Ref.9.26, p.64]
9.225 D.J. Wineland, W.M. Itano, J.C. Bergquist, J.J. Bollinger: Trapped Ions and Laser Cooling, NBS Technical Note 1086 (NBS, Washington, DC 1985)
9.226 D.J. Wineland, W.M. Itano, R.S. VanDyck, Jr.: High-resolution spectroscopy of stored ions, in *Advances in Atomic and Molecular Physics*, Vol.19, ed. by D.R. Bates, B. Bederson (Academic, New York 1983)
9.227 D.J. Wineland, W.M. Itano, J.C. Bergquist, J.J. Bollinger, J.D. Prestige: Spectroscopy of stored ions, in *Atomic Physics 9*, ed. by R.S. Van Dyck, Jr., E.N. Fortson (World Scientific, Singapore 1985) p.3
9.228 H.G. Dehmelt: Proposed 10^{14} $\Delta\nu>\nu$ laser fluorescence spectroscopy on Tl+ mono-ion oscillator II. Bull. Am. Phys. Soc. 20, 60 (1975)
9.229 Th. Sauter, W. Neuhauser, R. Blatt, P.E. Toschek: Observation of quantum jumps. Phys. Rev. Lett. 57, 1696 (1986); See also Phys. Scr. T 22, 129 (1988); and [Ref.9.26, p.121]
9.230 J.E. Bergquist, R.G. Hulet, W. M. Itano, D.J. Wineland: Observation of quantum jumps in a single atom. Phys. Rev. Lett. 57, 1699 (1986); See also Phys. Scripta T22, 79 (1988) and [Ref.9.26, p.117]
9.231 W. Nagourney, J. Sandberg, H. Dehmelt: Shelved optical electron amplifier: Observation of quantum jumps. Phys. Rev. Lett. 56, 2797 (1986); See also [Ref.9.26, p.114]
W. Nagourney: The mono-ion oscillator: An approach to an ideal atomic spectrometer. Comments At. Mol. Phys. 21, 321 (1988)
9.232 J.C. Bergquist, F. Diedrich, W.M. Itano, D.J. Wineland: Hg+ single ion spectroscopy, in [Ref.9.27, p.274]
9.233 V.I. Balykin, V.S. Letokhov, Yu. B. Ovchinnikov, A.I. Sidorov: Focusing of an atomic beam and imaging of atomic sources by means of a laser lens based on resonance-radiation pressure. J. Mod. Opt. 35, 17 (1988)
V.I. Balykin, V.S. Letokhov, V.G. Minogin: Laser control of the motion of neutral atoms and optical atomic traps. Phys. Scr. T 22, 119 (1988)
9.234 K. Cloppenburg, G. Hennig, A. Mihm, H. Wallis, W. Ertmer: Optical elements for manipulating atoms, in [Ref.9.26, p.87]
9.235 S.N. Atutov, S.P. Podjachev, A.M. Shalagin: Diffusion pulling of Na vapor into the light beam. Opt. Commun. 57, 236 (1986)

9.236 Kh. Gel'mukhanov, A.M. Shalagin: Sov. Phys. - JETP **51**, 839 (1980)
9.237 H.G.C. Werij, J.P. Woerdman, J.J.M. Beenakker, I. Kuscer: Demonstration of a semipermeable optical piston. Phys. Rev. Lett. **52**, 2237 (1984)
G. Nienhuis: Theory of light-induced drift and the optical piston. Phys. Rev. A **31**, 1636 (1985)

Chapter 10

10.1 L.J. Radziemski, R.W. Solarz, J.A. Paisner (eds.): *Laser Spectroscopy and its Applications* (Dekker, New York 1987)
10.2 H. Medin, S. Svanberg (eds.): Laser Technology in Chemistry, Special issue. Appl. Phys. **B46**, No.3 (1988)
10.3 W.C. Gardiner, Jr.: The chemistry of flames. Sci. Am. **246**/2, 86 (1982)
W.C. Gardiner, Jr. (ed.): *Combustion Chemistry* (Springer, Berlin, Heidelberg 1984)
J. Walker: The physics and chemistry underlying the infinite charm of a candle flame. Sci. Am. **238**/4, 154 (1978)
10.4 M. Gehring, K. Hoyermann, H. Schacke, J. Wolfrum: Direct studies of some elementary steps for the formation and destruction of nitric oxide in the H-N-O system. 14th Symp. on Combustion (Combustion Institute, Pittsburgh, PA 1973)
10.5 A. G. Gaydon, H. G. Wolfhard: *Flames, their Structure, Radiation and Temperature* (Chapman and Hall, New York 1979)
10.6 J. Wolfrum: Chemical kinetics in combustion systems: The specific effect of energy, collisions, and transport processes. 20th Symp. on Combustion (Combustion Institute, Pittsburgh, PA 1985)
10.7 A. C. Gaydon: *The Spectroscopy of Flames* (Chapman and Hall, New York 1974)
10.8 D.R. Crosley (ed.): *Laser Probes for Combustion Chemistry*, ACS Symp. Ser. Vol.134 (Am. Chem. Soc., Washington 1980)
10.9 A.C. Eckbreth, P.A. Bonczyk, J.F. Verdiek: Combustion Diagnostics by Laser Raman and Fluorescence Techniques, Progr. Energy Comb. Sci. **5**, 253 (1979).
10.10 J.H. Bechtel, C.J. Dasch, R.E. Teets: Combustion research with lasers, in *Laser Applications*, ed. by R.K. Erf, J.F. Ready (Academic, New York 1984)
J.H. Bechtel, A.R. Chraplyvy: Proc. IEEE **70**, 658 (1982)
10.11 T.D. McCay, J.A. Roux (eds.): Combustion diagnostics by nonintrusive methods. Progr. Astronautics and Aeronautics, Vol. **92** (1983)
10.12 A.C. Eckbreth: *Laser Diagnostics for Combustion Temperature and Species* (Abacus Press, Turnbridge Wells 1987)
10.13 K. Iinuma, T. Asanuma, T. Ohsawa, J. Doi (eds.): *Laser Diagnostics and Modelling of Combustion* (Springer, Berlin, Heidelberg 1987)
10.14 M. Aldén, H. Edner, S. Svanberg, T. Högberg: Combustion studies with laser techniques, Göteborg Institute of Physics Reports GIPR-206 (Chalmers University of Technology, Göteborg 1980)
10.15 M. Aldén, H. Edner, G. Holmstedt, T. Högberg, H. Lundberg, S. Svanberg: Relative distribution of radicals and temperature in flat flames, studied by laser-induced fluorescence and BOXCARS spectroscopy. Lund Reports on Atomic Physics LRAP-1 (Lund Institute of Technology, Lund 1981)
10.16 D.R. Crosley, G.P. Smith: Laser-induced fluorescence spectroscopy for combustion diagnostics. Opt. Eng. **22**, 545 (1983)
K. Schofield, M. Steinberg: Quantitative atomic and molecular fluorescence in the study of detailed combustion processes. Opt. Eng. **20**, 501 (1981)

10.17 R. Lucht: Applications of laser-induced fluorescence spectroscopy for combustion and plasma diagnostics, in [Ref.10.1, p.623]
10.18 N.S. Bergano, P.A. Janimaagi, M.M. Salour, J.H. Bechtel: Picosecond laser-spectroscopy measurement of hydroxyl fluorescence lifetime in flames. Opt. Lett. **8**, 443 (1983)
10.19 M. Aldén, H. Edner, P. Grafström, H.M. Hertz, G. Holmstedt, T. Högberg, H. Lundberg, S. Svanberg, S. Wallin, W. Wendt, U. Westblom: Imaging measurements of species concentrations, temperatures and velocities in reactive flows using laser-induced fluorescence, in *Lasers 86*, ed. by K.M. Corcoran, D.M. Sullivan, W.C. Stwalley (STS Press, McLean, VA. 1985) p.209
10.20 M. Aldén, H. Edner, G. Holmstedt, S. Svanberg, T. Högberg: Single-pulse laser-induced OH fluorescence in an atmospheric flame, spatially resolved with a diode array detector. Appl. Opt. **21**, 1236 (1982)
10.21 M.J. Dyer, D.R. Crosley: Two-dimensional imaging of OH laser-induced fluorescence in a flame. Opt. Lett. **7**, 382 (1982)
G. Kychakoff, R.D. Howe, R.K. Hanson, J.C. McDaniel: Quantitative visualization of combustion species in a plane. Appl. Opt. **21**, 3225 (1982)
G. Kychakoff, R.D. Howe, R.K. Hanson: Quantitative flow visualization technique for measurements in combustion gases. Appl. Opt. **23**, 704 (1984)
G. Kychakoff, K. Knapp, R.D. Howe, R.K. Hanson: Flow visualization in combustion gases using nitric oxide fluorescence. AIAA J. **22**, 153 (1984)
G. Kychakoff, R.D. Howe, R.K. Hanson, M.C. Drake, R.W. Pitz, M. Lapp, C.M. Penney: Visualization of turbulent flame fronts with planar laser-induced fluorescence. Science **224**, 382 (1984)
R.K. Hanson: Combustion diagnostics: Planar imaging techniques, in *Proc. 21st Symp. on Combustion*, Munich 1986 (The Combustion Institute Pittsburgh, PA 1986)
B. Hiller, R.K. Hanson: Simultaneous planar measurements of velocity and pressure fields in gas flows using laser-induced fluorescence. Appl. Opt. **27**, 33 (1988)
C. Véret (ed.): *Flow Visualization IV* (Springer, Berlin, Heidelberg 1987)
10.22 M. Aldén, H. Edner, P. Grafström, S. Svanberg: Two-photon excitation of atomic oxygen in a flame. Opt. Commun. **42**, 244 (1982)
M. Aldén, H.M. Hertz, S. Svanberg, S. Wallin: Imaging laser-induced fluorescence of oxygen atoms in a flame. Appl. Opt. **23**, 3255 (1984)
10.23 R.P. Lucht, J.P. Salmon, G.B. King, D.W. Sweeney, N.M. Laurendeau: Two-photon-excited fluorescence measurement of hydrogen atoms in flames. Opt. Lett. **8**, 365 (1983)
10.24 M. Aldén, A.L. Schawlow, S. Svanberg, W. Wendt, P.-L. Zhang: Three-photon excited fluorescence detection of atomic hydrogen in an atmospheric pressure flame. Opt. Lett. **9**, 211 (1984)
10.25 J.E.M. Goldsmith: Two-step saturated fluorescence detection of atomic hydrogen in flames. Opt. Lett. **10**, 116 (1985)
J.E.M. Goldsmith, R.J.M. Anderson: Imaging of atomic hydrogen in flames with two-step saturated fluorescence detection. Opt. Lett. **11**, 67 (1985)
10.26 M. Aldén, S. Wallin, W. Wendt: Applications of two-photon absorption for detection of CO in combustion gases. Appl. Phys. **B33**, 205 (1984)
10.27 J.E.M. Goldsmith: Resonant multiphoton optogalvanic detection of atomic hydrogen in flames. Opt. Lett. **7**, 437 (1982)
J.E.M. Goldsmith: Recent advances in flame diagnostics using fluorescence and ionization techniques, in [Ref.10.28, p.337]
P.J.H. Tjossem, T.A. Cool: Chem. Phys. Lett. **100**, 479 (1983)
10.28 W. Persson, S. Svanberg (eds.): *Laser Spectroscopy VIII*, Springer Ser. Opt. Sci., Vol.55 (Springer, Berlin, Heidelberg 1987)

10.29 K. Tennal, G.J. Salomo, R. Gupta: Minority species concentration measurements in flames by the photoacoustic technique. Appl. Opt. 21, 2133 (1982)
A.C. Tam: Applications of photoacoustic sensing techniques. Revs. Mod. Phys. 58, 381 (1986)
10.30 R.K. Hanson, P.A. Kuntz, C.H. Kruger: High-resolution spectroscopy of combustion gases using a tunable IR diode laser. Appl. Opt. 16, 2045 (1975)
K. Knapp, R.K. Hanson: Spatially resolved tunable diode-laser absorption measurements of CO using optical Stark shifting. Appl. Opt. 22, 1980 (1983)
10.31 Special Issue on Computerized Tomography. Proc. IEEE 71, 291-435 (March 1983)
Special Issue on Industrial Applications of Computed Tomography and NMR Imaging. Appl. Opt. 24, 23 (1985)
10.32 K.E. Bennett, G.W. Faris, R.L. Byer: Experimental optical fan beam tomography. Appl. Opt. 23, 2678 (1984)
10.33 H.M. Hertz, G.W. Faris: Emission tomography of flame radicals. Opt. Lett. 13, 351 (1988)
10.34 H.M. Hertz: Experimental determination of 2-D flame temperature fields by interferometric tomography. Opt. Commun. 54, 131 (1985)
10.35 A. Rose, G.J. Salamo, R. Gupta: Photoacoustic deflection spectroscopy: A new specie-specific method for combustion diagnostics. Appl. Opt. 23, 781 (1984)
H. Sonntag, A.C. Tam: Time-resolved flow-velocity and concentration measurements using a travelling thermal lens. Opt. Lett. 10, 436 (1985)
G.W. Faris, R.L. Byer: Beam-deflection optical tomography. Opt. Lett. 12, 72 (1987)
G.W. Faris, R.L. Byer: Beam-deflection optical tomography of a flame. Opt. Lett. 12, 155 (1987)
10.36 M. Lapp, C.M. Penney: Raman measurements on flames, in *Advances in Infrared and Raman Spectroscopy*, ed. by R.J.H. Clark, R.E. Hester (Heyden, London 1977)
10.37 R.W. Dibble, A.R. Masri, R.W. Bilger: Combust. Flame 67, 189 (1987)
J.J. Valentini: Laser Raman techniques, in [Ref.10.1, p.507]
10.38 M. Aldén, H. Edner, S. Svanberg: Coherent anti-Stokes Raman spectroscopy (CARS) applied in combustion probing. Phys. Scripta 27, 29 (1983)
10.39 D. Klick, K.A. Marko, L. Rimai: Broadband single-shot CARS spectra in a fired internal combustion engine. Appl. Opt. 20, 1178 (1981)
G.C. Alessandretti, P. Violino: Thermometry by CARS in an automobile engine. J. Phys. D 16, 1583 (1983)
10.40 L.A. Rahn, S.S. Johnston, R.L. Farrow, P.L. Mattern: CARS thermometry in an internal combustion engine, in *Temperature*, Vol.5, ed. by J.F. Schooley (AIP, New York 1982)
10.41 M. Aldén, S. Wallin: CARS Experiment in a full-scale ($10 \times 10 m^2$) industrial coal furnace. Appl. Opt. 24, 3434 (1985)
10.42 B. Attal, M. Pealat, J.P. Taran: J. Energy 4, 135 (1980)
10.43 A. C. Eckbreth: CARS thermometry in practical combustors. Combust. Flame 39, 133 (1980)
10.44 A.C. Eckbreth, P.W. Schreiber: Coherent anti-Stokes Raman spectroscopy (CARS): Applications to combustion and gas-phase diagnostics, in *Chemical Applications of Non-Linear Raman Spectroscopy*, ed. by A.B. Harvey (Academic, New York 1981)
R.J. Hall, A.C. Eckbreth: Coherent anti-Stokes Raman spectroscopy (CARS): Application to combustion diagnostics, in *Laser Applications*, ed. by J.F. Ready, R.K. Erf (Academic, New York 1984) Vol.5
10.45 H. Haragushi, B. Smith, S. Weeks, D.J. Johnson, J.D. Wineforder: Measurement of small volume flame temperature by the two-line atomic fluorescence method. Appl. Spectr. 31, 156 (1977)

R.G. Jolik, J.W. Daily: Two-line atomic fluorescence temperature measurements in flames: An experimental study. Appl. Opt. 21, 4158 (1982)

M. Aldén, P. Grafström, H. Lundberg, S. Svanberg: Spatially resolved temperature measurements in a flame using laser-excited two-line atomic fluorescence and diode-array detection. Opt. Lett. 8, 241 (1983)

10.46 J. Pender, L. Hesselink: Phase conjugation in a flame. Opt. Lett. 10, 264 (1985)

P. Ewart, S.V. O'Leary: Detection of OH in a flame by degenerate four-wave mixing. Opt. Lett. 11, 279 (1986)

10.47 R.M. Osgood, S.R.J. Brueck, H.R. Schlossberg (eds.): *Laser Diagnostics and Photochemical Processing for Semiconductor Devices* (North Holland, Amsterdam 1983)

D. Bäuerle (ed.): *Laser Processing and Diagnostics*, Springer Ser. Chem. Phys., Vol.39 (Springer, Berlin, Heidelberg 1984)

D. Bäuerle, K.L. Kompa, L.D. Laudé (eds): *Laser Processing and Diagnostics II* (Physique, Les Ulis 1986)

D. Bäuerle: *Chemical Processing with Lasers*, Springer Ser. Mat. Sci., Vol.1 (Springer, Berlin, Heidelberg 1986)

L.D. Laudé, D. Bäuerle, M. Wautelet (eds): *Interfaces under Laser Irradiation*, NATO ASI Series (Nijholl, Dordrecht 1987)

W.G. Breiland, M.E. Coltrin, P. Ho (eds): *Laser-Based Studies of Chemical Vapor Deposition*, Proc. Soc. Photo-opt. Instrum. Eng. 385, 146 (1983)

10.48 K.L. Kompa, J. Wanner: *Laser Applications in Chemistry* (Plenum, New York 1984)

V.S. Letokhov (ed.): *Laser Analytical Spectrochemistry* (Hilger, Bristol 1986)

T.R. Evans (ed.): *Applications of Lasers to Chemical Problems* (Wiley, New York 1982)

S. Svanberg: Laser spectroscopy applied to energy, environmental and medical research. Phys. Scr. T 23, 281 (1988); Appl. Phys. B 46, 271 (1988)

10.49 E.R. Pike, H.Z. Cummins (eds.): *Photon Correlation and Light Beating Spectroscopy* (Plenum, New York 1974)

10.50 L.E. Drain: *The Laser Doppler Technique* (Wiley, Chichester 1980)

10.51 E. Durst, A. Melling, J.H. Whitelaw: *Principles and Practice of Laser-Doppler Anemometry*, 2nd edn. (Academic, London 1981)

10.52 C.J. Dasch, J.A. Sell: Velocimetry in laminar and turbulent flows using the photothermal deflection effect with a transient grating. Opt. Lett. 11, 603 (1986), and references therein

R. Miles, C. Cohen, J. Connors, P. Howard, S. Huang, E. Markovitz, G. Russel: Velocity measurements by vibrational tagging and fluorescent probing of oxygen. Opt. Lett. 12, 861 (1987)

10.53 B. Hiller, J.C. McDaniel, E.C. Rea, Jr., R.K. Hanson: Laser-induced fluorescence technique for velocity field measurements in subsonic gas flows. Opt. Lett. 8, 474 (1983)

10.54 U. Westblom, S. Svanberg: Imaging measurements of flow velocities using laser-induced fluorescence. Phys. Scripta 31, 402 (1985)

U. Westblom, A. Aldén: Spatially resolved flow velocity measurement using laser-induced fluorescence from a pulsed laser. Opt. Lett. 14, 9 (1989)

10.55 E.J. McCartney: *Absorption and Emission by Atmospheric Gases* (Wiley, New York 1983)

10.56 E.J. McCartney: *Optics of the Atmosphere; Scattering by Molecules and Particles* (Wiley, New York 1976)

10.57 L.S. Rothman et al.: AFGL atmospheric absorption line parameters compilation: 1982 Version. Appl. Opt. 22, 2247 (1983)

L.S. Rothman et al.: The HITRAN database: 1986 Edition. Appl. Opt. 26, 4058 (1987)

10.58 W. Bach, J. Pankrath, W. Kellogg (eds): *Man's Impact on Climate* (Elsevier, Amsterdam 1979)
10.59 R. Revelle: Carbon dioxide and world climate. Sci. Am. 247/2, 33 (1982)
S.H. Schneider: Climate modeling. Sci. Am. 256/5, 72 (1987)
R.A. Houghton, G.W. Woodwell: Global climatic change. Sci. Am. 260/4, 18 (1989)
S.H. Schneider: The changing climate. Sci. Am. 261/3, 38 (1989)
B.J. Mason: The greenhouse effect. Contemp. Phys. 30, 417 (1989)
10.60 T.E. Graedel, D.T. Hawkins, L.D. Claxton: *Atmospheric Chemical Compounds: Sources, Occurrence, Bioassay* (Academic, Orlando 1986)
10.61 R.M. Harrison, R. Perry (eds.): *Handbook of Air Pollution Analysis*, 2nd edn. (Chapman and Hall, London 1986)
10.62 R.P. Wayne: *Chemistry of Atmospheres* (Clarendon, Oxford 1985)
10.63 J.H. Seinfeld: *Atmospheric Chemistry and Physics of Air Pollution* (Wiley, New York 1986)
10.64 D.A. Killinger, A. Mooradian (eds.): *Optical and Laser Remote Sensing*, Springer Ser. Opt. Sci., Vol.39 (Springer, Berlin, Heidelberg 1983)
10.65 R.M. Measures: *Laser Remote Sensing: Fundamentals and Applications* (Wiley, New York 1984)
10.66 E.D. Hinkley (ed.): *Laser Monitoring of the Atmosphere*, Topics Appl. Phys., Vol.14 (Springer, Berlin, Heidelberg 1976)
10.67 V. Zuev, I. Naats: *Inverse Problems of Lidar Sensing of the Atmosphere*, Springer Ser. Opt. Sci., Vol.29 (Springer, Berlin, Heidelberg 1983)
10.68 R.M. Measures: In *Analytical Laser Spectroscopy*, ed. by N. Omenetto (Wiley, New York 1979)
D.K. Killinger, N. Menyuk: Laser remote sensing of the atmosphere. Science 235, 37 (1987)
W.B. Grant: Laser remote sensing techniques, in [Ref.10.1, p.565]
T. Kobayashi: Techniques for laser remote sensing of the environment. Rem. Sens. Rev. 3, 1 (1987)
R.M. Measures (ed.): *Laser Remote Chemical Analysis* (Wiley-Interscience, New York 1988)
E. Zanzottera: Differential absorption lidar techniques in the determination of trace pollutants and physical parameters of the atmosphere. Crit. Rev. Anal. Chem. 21, 279 (1990)
S. Svanberg: Environmental monitoring using optical techniques, in *Applied Laser Spectroscopy*, ed. by M. Inguscio, W. Demtröder (Plenum, New York 1990)
10.69 S. Svanberg: Lasers as probes for air and sea. Contemp. Phys. 21, 541 (1980)
10.70 S. Svanberg: Fundamentals of atmospheric spectroscopy, in *Surveillance of Environmental Pollution and Resources by Electromagnetic Waves*, ed. by T. Lund (Reidel, Dordrecht 1978)
10.71 R.T. Menzies, R.K. Seals, Jr.: Science 197, 1275 (1977)
10.72 E.D. Hinkley: Laser spectroscopic instrumentation and techniques: Long path monitoring by resonance absorption. Opt. Quant. Electr. 8, 155 (1976)
10.73 M.C. Alarcon, H. Ito, H. Inaba: All-optical remote sensing of city gas through CH_4 gas absorption employing a low-loss optical fibre link and an InGaAsP light emitting diode in the near-infrared region. Appl. Phys. B43, 79 (1987)
10.74 M.L. Chanin: Rayleigh and resonance sounding of the stratosphere and mesosphere, in [Ref.10.64, p.192]
C. Granier, G. Megie: Daytime lidar measurement of the mesospheric sodium layer. Planet Space Sci. 30, 169 (1982)
C. Granier, J.P. Jegou, G. Megie: Resonant lidar detection of Ca and Ca^+ in the upper atmosphere. Geophys. Res. Lett. 12, 655 (1985)
K.H. Fricke, U. v. Zahn: Mesopause temperatures derived from probing the

hyperfine structure of the D_2 resonance line of sodium by lidar. J. Atm. Terr. Phys. **47**, 499 (1985)

U. von Zahn, P. von der Gathen, G. Hansen: Forced release of sodium from upper atmosphere dust particles. Geophys. Res. Lett. **14**, 76 (1987)

L.A. Thompson, C.S. Gardner: Laser guidestar experiment at Mauna Kea Obervatory for adaptive imaging in astronomy. Nature **328**, 229 (1987)

10.75 L.J. Radziemski, T.R. Loree, D.A. Cremers, N.M. Hoffman: Time-resolved laser-induced breakdown spectrometry of aerosols. Anal. Chem. **55**, 1246 (1983)

J.A. Millard, R.H. Dalling, L.J. Radziemski: Time-resolved laser-induced breakdown spectrometry for the rapid determination of beryllium in berylliumcopper alloys. Appl. Spectr. **40**, 491 (1986)

D.J. Cremers, L.J. Radziemski: Laser plasmas for chemical analysis, in [Ref.10.1, p.351]

10.76 K. Fredriksson, B. Galle, K. Nyström, S. Svanberg: Mobile lidar system for environmental probing. Appl. Opt. **20**, 4181 (1981)

10.77 K. Fredriksson, I. Lindgren, S. Svanberg, G. Weibull: Measurements of the emission from industrial smoke stacks using laser radar techniques. Göteborg Institute of Physics Reports GIPR-121 (CTH, Göteborg 1976)

10.78 H. Edner, K. Fredriksson, A. Sunesson, S. Svanberg, L. Unéus, W. Wendt: Mobile remote sensing system for atmospheric monitoring. Appl. Opt. **26**, 4330 (1987)

10.79 D.J. Brassington: Measurement of the SO_2 absorption spectrum between 297 and 316 nm using a tunable dye laser. Lab. Note No. RD/L/N184/79 (Central Electricity Res. Labs., Leatherhead 1979)

D.J. Brassington: Sulphur dioxide absorption cross section measurement from 290 nm to 317 nm. Appl. Opt. **20**, 3774 (1981)

10.80 J. Pelon, G. Megie: Ozone monitoring in the troposphere and lower stratosphere: Evaluation and operation of a ground based lidar station. J. Geophys. Res. **87**, 4947 (1982)

G.J. Megie, G. Ancellet, J. Pelon: Lidar measurements of ozone vertical profiles. Appl. Opt. **24**, 3454 (1985)

O. Uchino, M. Tokunaga, M. Maeda, Y. Miyazoe: Differential absorption-lidar measurement of tropospheric ozone with excimer-Raman hybrid laser. Opt. Lett. **8**, 347 (1983)

O. Uchino, M. Maeda, H. Yamamura, M. Hirono: Observation of stratospheric vertical ozone distribution by a XeCl lidar. J. Geophys. Res. **88**, 5273 (1983)

J. Werner, K.W. Rothe, H. Walther: Monitoring of the stratospheric ozone layer by laser radar. Appl. Phys. **B32**, 113 (1983)

10.81 M.P. McCormick: Lidar measurements of Mount St. Helens effluents. Opt. Eng. **21**, 340 (1982)

M.P. Mc Cormick, T.J. Swisser, W.H. Fuller, W.H. Hunt, M.T. Osborn: Airborne and groundbased lidar measurements of the El Chichon stratospheric aerosol from 90° N to 56° S. Geofisica Internacional **23-2**, 187 (1984)

M.R. Rampino, S. Self: The atmospheric effects of El Chichon. Sci. Am. **250**/1, 34 (1984)

E.E. Uthe: Application of surface based and airborne lidar systems for environmental monitoring. J. Air Pollut. Control Assoc. **33**, 1149 (1983)

10.82 R.L. Byer, E.K. Gustafson, R. Trebino (eds.): *Tunable Solid State Lasers for Remote Sensing*, Springer Ser. Opt. Sci., Vol.51 (Springer, Berlin, Heidelberg 1985)

10.83 D.H. Hercules (ed.): *Fluorescence and Phosphorescence Analysis* (Interscience, New York 1966)

P. Pringsheim: *Fluorescence and Phosphorescence* (Interscience, New York 1949)

10.84 J.B. Birks: *Photophysics of Aromatic Molecules* (Wiley, New York 1970)

10.85 I. Berlman: *Handbook of Fluorescence Spectra of Aromatic Molecules*, 2nd edn. (Academic, New York 1971)

10.86 J.R. Lakowicz: *Principles of Fluorescence Spectroscopy* (Plenum, New York 1983)
E.L. Wehry (ed.): *Modern Fluorescence Spectroscopy*, Vols.1 and 2 (Plenum, New York 1976)
10.87 L. Celander, K. Fredriksson, B. Galle, S. Svanberg: Investigation of laser-induced fluorescence with application to remote sensing of environmental parameters. Göteborg Institute of Physics Reports GIPR-149 (CTH, Göteborg 1978)
10.88 S. Svanberg: Environmental diagnostics, in *Trends in Physics*, ed. by M.M. Woolfson (Hilger, Bristol 1978) p.119
10.89 Govindjee, R. Govindjee: The absorption of light in photosynthesis. Sci. Am. 231/6, 68 (1974)
D.C. Youvan, B.L. Marrs: Molecular mechanisms of photosynthesis. Sci. Am. 256/6, 42 (1987)
H.K. Lichtenthaler, U. Rinderle: The role of chlorophyll fluorescence in the detection of stress conditions in plants. CRC Crit. Rev. Anal. Chem. **19**, Suppl.1, S29-85 (1988)
F.E. Hoge, R.N. Swift: Airborne simultaneous spectroscopic detection of laser-induced water Raman backscatter and fluorescence from chlorophyll a and other naturally occuring pigments. Appl. Opt. **20**, 3197 (1981)
F.E. Hoge, R.N. Swift, J.K. Yungel: Active-passive airborne ocean color measurement 2: Applications. Appl. Opt. **25**, 48 (1986)
10.90 R.A. O'Neill, L. Buja-Bijunas, D.M. Rayner: Field performance of a laser fluorosensor for the detection of oil spills. Appl. Opt. **19**, 863 (1980)
G.A. Capelle, L.A. Franks, D.A. Jessup: Aerial testing of a KrF laser-based fluorosensor. Appl. Opt. **22**, 3382 (1983)
10.91 H.H. Kim: Airborne laser bathymetry. Appl. Opt. **16**, 45 (1977)
J. Banic, S. Sizgoric, R. O'Neill: Airborne scanning lidar bathymeter measures water depth. Laser Focus 23/2, 40 (1987)
K. Fredriksson, B. Galle, K. Nyström, S. Svanberg, B. Öström: Underwater laser-radar experiments for bathymetry and fish-school detection. Göteborg Institute of Physics Reports GIPR-162 (CTH, Göteborg 1978)
10.92 S. Montán, S. Svanberg: A system for industrial surface monitoring utilizing laser-induced fluorescence. Appl. Phys. **B38**, 241 (1985)
10.93 S. Montán, S. Svanberg: Industrial applications of laser-induced fluorescence. L.I.A. ICALEO **47**, 153 (1985)
P.S. Andersson, S. Montán, S. Svanberg: Remote sample characterization based on fluorescence monitoring. Appl. Phys. **B44**, 19 (1987)
10.94 E.S. Yeung: In *Adv. Chromatography* 23, Chap.1 (Dekker, New York 1984)
E.S. Yeung: In *Microcolumn Separations: Columns, Instrumentation and Ancillary Techniques*, ed. by M.V. Novotny, D. Ishii (Elsevier, Amsterdam 1985) p.135
E. Gassman, J.E. Kuo, R.N. Zare: Electrokinetic separation of chiral compounds. Science **230**, 813 (1985)
M.C. Roach, P.H. Gozel, R.N. Zare: Determination of methotrexate and its major metabolite, 7-hydroxylmethotrexate, using capillary zone electrophoresis and laser-induced fluorescence detection. J. Chromatography **426**, 129 (1988)
10.95 J.S. McCormack: Remote optical measurements of temperature using fluorescent materials. Electr. Lett. **17**, 630 (1981)
10.96 J.C. Hamilton, R.J. Anderson: In situ Raman spectroscopy of Fe-18Cr-3Mo(100) surface oxidation. Sandia Combustion Research Program Annual Rept. (Sandia, Livermore, CA 1984)
10.97 R.K. Chang, T.E. Furtak: *Surface-Enhanced Raman Scattering* (Plenum, New York 1982)
10.98 M. Moskovits: Surface-enhanced spectroscopy. Rev. Mod. Phys. **57**, 783 (1985)
10.99 Y.R. Shen: Ann. Rev. Mat. Sci. **16**, 69 (1986)

10.100 Y.R. Shen: Applications of optical second-harmonic generation in surface science, in *Chemistry and Structure at Interfaces*, ed. by R.B. Hall, A.B. Ellis (Verlag-Chemie, Weinheim 1986) p.151
10.101 W. Yen, P.M. Selzer (eds): *Laser Spectroscopy of Solids*, 2nd. ed., Topics Appl. Phys., Vol.49 (Springer, Berlin, Heidelberg 1989)
W.M. Yen (ed.): *Laser Spectroscopy II*, Topics Appl. Phys., Vol.65 (Springer, Berlin, Heidelberg 1989)
10.102 F.R. Aussenegg, A. Leitner, M.E. Lippitsch (eds.): *Surface Studies with Lasers*, Springer Ser. Chem. Phys., Vol.33 (Springer, Berlin, Heidelberg 1983)
10.103 K. Kleinermanns. J. Wolfrum: Laser Chemistry - What is Its Current Status?, Angew. Chem. Int'l Ed. Engl. 26, 38 (1987)
10.104 A.M. Ronn: Laser chemistry. Sci. Am. 240/5, 102 (1979)
10.105 V.S. Letokhov: *Nonlinear Laser Chemistry*, Springer Ser. Chem. Phys., Vol.22 (Springer, Berlin, Heidelberg 1983)
10.106 E. Grunvald, D.F. Dever, P.M. Keeher: *Megawatt Infrared Laser Chemistry* (Wiley, New York 1978)
10.107 A. Zewail (ed): *Advances in Laser Chemistry*, Springer Ser. Chem. Phys., Vol.3 (Springer, Berlin, Heidelberg 1978)
10.108 V.S. Letokhov: Laser-induced chemistry - basic nonlinear processes and applications, in [Ref.10.2, p.237]
10.109 R.L. Woodin, A. Kaldor (eds.): *Applications of Lasers to Industrial Chemistry*. SPIE **458** (SPIE, Bellingham, WA 1984)
10.110 J.A. Paisner, R.W. Solarz: Resonance photoionization spectroscopy, in [Ref.10.1, p.175]
J.A. Paisner: Atomic vapor laser isotope separation, in [Ref.10.2, p.253]
10.111 V.S. Letokhov: Laser separation of isotopes. Ann. Rev. Phys. Chem. 28, 133 (1977)
V.S. Letokhov: Laser isotope separation. Nature 277, 605 (1979)
J.L. Lyman: Laser-induced molecular dissociation. Applications in isotope separation and related processes, in [Ref.10.1, p.417]
10.112 H.G. Kuhn: *Atomic Spectra* (Longmans, London 1962)
10.113 N. Bloembergen, E. Yablonovitch: Collisionless multiphoton dissociation of SF_6: A statistical thermodynamic process, in *Laser Spectroscopy III*, ed. by J.L. Hall, J.L. Carlsten (Springer, Berlin, Heidelberg 1977)
10.114 R.V. Ambartzumian, V.S. Letokhov, G.N. Makarov, A.A. Puretsky: Laser separation of nitrogen isotopes. JETP Lett. 17, 63 (1973); JETP Lett. 15, 501 (1972)
10.115 J.-L. Boulnois: Photophysical processes in recent medical laser developments: A review. Lasers in Med. Sci. 1, 47 (1986)
J.-L. Boulnois: Photophysical processes in laser-tissue interactions, in *Laser Applications in Cardiovascular Diseases*, ed. by R. Ginsburg (Futura, New York 1987)
10.116 D. Sliney, M. Wolbarsht: *Safety with Lasers and Other Optical Sources* (Plenum, New York 1980)
ANSI: Laser Standards designed Z 136.1 - 1973 (American National Standards Institute, Wash. 1983).
10.117 L. Goldman (ed.): *The Biomedical Laser: Technology and Clinical Applications* (Springer, Berlin, Heidelberg 1981)
10.118 S. Martellucci, A.N. Chester: *Laser Photobiology and Photomedicine* (Plenum, New York 1985)
10.119 J.A. Parrish, T.F. Deutsch: Laser photomedicine. IEEE J. QE-20, 1386 (1984)
10.120 T.J. Dougherty: In *CRC Critical Reviews in Oncology/Hematology*, ed. by S. Davis (CRC, Boca Raton, FL 1984)
10.121 Y. Hayata, T.J. Dougherty (eds.): *Lasers and Hematoporphyrin Derivative in Cancer* (Ikaku-shoin, Tokyo 1983)

10.122 R. Pratesi, C.A. Sacchi (eds.): *Lasers in Photomedicine and Photobiology*, Springer Ser. Opt. Sci. Vol.22 (Springer, Berlin, Heidelberg 1980)
10.123 A. Andreoni, R. Cubeddu (eds.): *Porphyrins in Tumor Phototherapy* (Plenum, New York 1984)
10.124 Ch.J. Gomer (ed.): Proc. Clayton Foundation Conf. on Photodynamic Therapy (Childrens Hospital, Los Angeles 1987)
10.125 S. Svanberg: Medical diagnostics using laser-induced fluorescence. Phys. Scr. T 17, 469 (1987)
10.126 S. Svanberg: Medical applications of laser spectroscopy. Phys. Scr. T 26, 90 (1989)
10.127 A.E. Profio, D.R. Doiron, O.J. Balchum, G.C. Huth: Fluorescence bronchoscopy for localization of carcinoma in Situ. Med. Phys. 10, 35 (1983)
10.128 H. Kato, D.A. Cortese: Early detection of lung cancer by means of hematoporphyrin derivative fluorescence and laser photoradiation. Clinics in Chest Medicine 6, 237 (1985)
10.129 J.H. Kinsey, D.A. Cortese: Endoscopic system for simultaneous visual examination and electronic detection of fluorescence. Rev. Sci. Instr. 51, 1403 (1980)
10.130 P.S. Andersson, S.E. Karlsson, S. Montán, T. Persson, S. Svanberg, S. Tapper: Fluorescence endoscopy instrumentation for inproved tissue characterization. Med. Phys. 14, 633 (1987)
10.131 S. Andersson-Engels, A. Brun, E. Kjellén, L.G. Salford, L.-G. Strömblad, K. Svanberg, S. Svanberg: Identification of brain tumours in rats using laser-induced fluorescence and haematoporphyrin derivative. Laser Med. Sci. 4, 241 (1989)
10.132 P.S. Andersson, S. Montán, S. Svanberg: Multi-spectral system for medical fluorescence imaging. IEEE J. QE-23, 1798 (1987)
10.133 S. Udenfriend: *Fluorescence Assay in Biology and Medicine*, Vol.I (1962), Vol.II (1969) (Academic, New York)
K.P. Mahler, J.F. Malone: Digital fluoroscopy: A new development in medical imaging. Contemp. Phys. 27, 533 (1986)
G.M. Barenboim, A.N. Domanskii, K.K. Turoverov: *Luminenscence of Biopolymers and Cells* (Plenum, New York 1969)
D.M. Kirschenbaum (ed.): *Atlas of Protein Spectra in the Ultraviolet and Visible Regions*, Vol.2 (IFI/Plenum, New York 1974)
10.134 R.R. Alfano, B.T. Darayash, J. Cordero, P. Tomashefsky, F.W. Longo, M.A. Alfano: Laser induced fluorescence spectroscopy from native cancerous and normal tissue. IEEE J. QE-20, 1507 (1984)
R.R. Alfano, G.C. Tang, A. Pradhan, W. Lam, D.S.J. Choy, E. Opher: Fluorescence spectra from cancerous and normal human breast and lung tissues. IEEE J. QE-23, 1806 (1987)
10.135 Y.M. Ye, Y.L. Yang, Y.F. Li, F.M. Li: Characteristic autofluorescence for cancer diagnosis and the exploration of its origin. Proc. CLEO'85 (Baltimore, MD)
10.136 S. Montán: Diploma paper, Lund Reports on Atomic Physics LRAP-19 (Lund University, Lund 1982)
P.S. Andersson, E. Kjellén, S. Montán, K. Svanberg, S. Svanberg: Autofluorescence of various rodent tissues and human skin tumour samples. Lasers in Med. Sci. 2, 41 (1987)
10.137 R.R Alfano, W. Lam, H.J. Zarrabi, M.A. Alfano, J. Cordero, D.B. Tata, C.E. Swenberg: Human teeth with and without caries studied by laser scattering, fluorescence and absorption spectroscopy. IEEE J. QE-20, 1512 (1984)
10.138 F. Sundström, K. Fredriksson, S. Montán, U. Hafström-Björkman, J. Ström: Laser-induced fluorescence from sound and carious tooth substance: Spectroscopic studies. Swed. Dent. J. 9, 71 (1985)

10.139 C. Kittrell, R.L. Willett, C. de los Santon-Pacheo, N.B. Ratliff, J.R. Kramer, E.G. Malk, M.S. Feld: Diagnosis of fibrous arterial atherosclerosis using fluorescence. Appl. Opt. 24, 2280 (1985)
R.M. Cothren, G.B. Hayes, J.R. Kramer, B. Sachs, C. Kittrell, M.S. Feld: Lasers Life Sci. 1, 1 (1986)
10.140 P.S. Andersson, A. Gustafson, U. Stenram, K. Svanberg, S. Svanberg: Monitoring of human atherosclerotic plaque using laser-induced fluorescence. Lasers Med. Sci. 2, 261 (1987)
S. Andersson-Engels, A. Gustafson, J. Johansson, U. Stenram, K. Svanberg, S. Svanberg: Laser-induced fluorescence used in localizing atherosclerotic lesions. Laser Med. Sci. 4, 171 (1989)
10.141 J.M. Isner, R.H. Clarke: The current status of lasers in the treatment of cardiovascular disease. IEEE J. QE-20, 1406 (1984)
J.M. Isner, P.G. Steg, R.H. Clarke: Current status of cardiovascular laser therapy. IEEE J. QE-23, 1756 (1987)
10.142 M.R. Prince, T.F. Deutsch, M.M. Mathews-Roth, R. Margolis, J.A. Parrish, A.R. Oseroff: Preferential light absorption in atheromas in vitro: Implications for laser angioplasty. J. Clin. Invest. 78, 295 (1986)
10.143 S.R. Meech, C.D. Stubbs, D. Phillips: The application of fluorescence decay measurements in studies of biological systems. IEEE J. QE-20, 1343 (1984)
10.144 M. Yamashita, M. Nomura, S. Kobayashi, T. Sato, K. Aizawa: Picosecond time-resolved fluorescence spectroscopy of hematoporphyrin derivative. IEEE J. QE-20, 1363 (1984)
V.S. Letokhov (ed.): *Laser Picosecond Spectroscopy and Photochemistry of Biomolecules* (Hilger, Bristol 1987)
10.145 L. Stryer: The molecules of visual excitation. Sci. Am. 257/1, 32 (1987)
10.146 D.B. Tata, M. Foresti, J. Cardero, P. Thomachefsky, M.A. Alfano, R.R. Alfano: Fluorescence polarization spectroscopy and time-resolved fluorescence kinetics of native cancerous and normal rat kidney tissues. Biophys. J. 50, 463 (1986)
10.147 S. Andersson-Engels, J. Johansson, S. Svanberg: The use of time-resolved fluorescence for diagnosis of atherosclesotic plaque and malignant tumours. Spectrochim. Acta, in press
10.148 S. Andersson-Engels, J. Johansson, K. Svanberg, S. Svanberg: Fluorescence diagnostics and photochemical treatment of diseased tissue using lasers. Pt.I, Anal. Chem. 61, 1367A (1989); Pt.II, ibid. 62, 19A (1990)
10.149 S. Andersson-Engels, J. Johansson, U. Stenram, K. Svanberg, S. Svanberg: Malignant tumor and atherosclerotic plaque diagnostics using laser-induced fluorescence. IEEE J. QE (October 1990)

Subject Index

Ablation 342
Absorbance 133
Absorption 40
– edge 73
– measurements 242
Accidental coincidence 85
Acid rain 304
Acousto-optic modulator 258, 265
Acousto-optic spectroscopy 251, 292
Active mode-locking 260
Active remote sensing 145
ADC 71, 263
Air pollution 129, 303, 323, 318
– measurements 71, 147, 252, 318
Airy distribution 109, 202
Alexander's dark band 63
Algae fluorescence 331
Alignment 93
Alkali atoms 5
Alkaline-earth atoms 12
Analogue-to-digital converter 71, 263
Angioplasty 342
Anharmonicity 35, 58, 339
Anomalous dispersion 272
Anti-bunching 46
Anti-crossing spectroscopy 177
Anti-reflection coating 121
Anti-Stokes Raman effect 57, 232
Aperture synthesis 191
Apodization 114
Arc lamp 88
Artheroesclerosis 348
Astrophysical spectroscopy 152, 190, 268
Atmospheric absorption 128, 130, 146, 188, 193
Atmospheric chemistry 129
Atmospheric composition 156, 318
Atmospheric dispersion 128

Atmospheric remote sensing 146, 188, 318
Atmospheric scattering 61, 322
Atmospheric turbulence 147
Atomic absorption spectroscopy 135
Atomic alignment 93
Atomic beam
–, collimated 279
– fountain 298
– lamp 89
– magnetic resonance 160, 283
Atomic clock 168, 170, 171, 182
Atomic orientation 169
Auger spectroscopy 83
Auto-correlation 191, 316
Auto-ionizing state 248
Avalanche diode 116

Balmer-Rydberg formula 295
Band head 53
Bathymetry 131
BBO 227
Beam-foil spectroscopy 90, 269
Beam-laser technique 269
Beer-Lambert's law 132, 320
Bending mode 36
Binding energy 73, 83
Birefringence 125, 227, 246
Black-body radiation 46, 94, 100, 192
Blaze angle 105
Bloch equations 181
Blumlein circuit 208
Bohr atomic model 4, 68
Bohr magneton 7
Bolometer 116
Boltzmann distribution 53, 181, 212
Boltzmann's law 42, 134
Born-Oppenheimer approximation 30

Boxcar integrator 263, 314
BOXCARS 233, 314
Bragg cell 258, 260, 316
Bragg relation 69
Branches 52
Breit formula 176, 276
Breit-Rabi diagram 19, 173, 256
Breit-Rabi formula 18, 26
Bremsstrahlung 66
Brewster's angle 64, 125, 201, 217
Broadband dye laser 245, 246, 314
Broadening
–, collisional 87, 188
–, Doppler 86, 285
–, Fourier 295
–, Holtsmark 88
–, homogeneous 46, 86
–, inhomogeneous 46
–, Lorentz 87
–, transit time 278, 298
Building-up principle 12

Caesium clock 168, 171, 286
CARS 232, 312
Cascade decay 92, 238, 254
Cavity 196
– dumping 261
Ccd detector 119
Central field 4, 12
Characteristic wavelength 96
Chemical activation 336
Chemical analysis
–, ESCA 81
–, laser methods 248, 251, 302, 334
–, NMR 182
–, optical methods 131
Chemical applications 68, 80, 134, 141, 182, 248, 315, 335
Chemical shifts 68, 74, 80, 84, 182
Chemical vapour deposition 315
Chlorofluorocarbons 129
Chromatography 332
Clear sky spectrum 99
Coherence 40
– length 198
–, spatial 198
–, temporal 198
Coherent anti-Stokes Raman spectroscopy 144, 232, 312

Coherent detection 319
Coherent excitation 92, 175, 274
Coherent Raman spectroscopy 144, 312
Colliding pulse technique 262
Collimated atomic beam 279
Collimation ratio 279
Collisional broadening 87, 188
Collisional ionization 239, 248
Collisional processes 64, 65
Collisional relaxation 170, 251
Collisional transfer 309
Coloured glass filters 112
Combustion
– chemistry 302
– diagnostics 302
– engine 314
Comet
– band 158
– spectrum 156
Comparator 115
Compression
–, kinematic 284
– of optical pulses 262
Computer-compatible tape 151
Conversion factors 3
Cooling 299
Core polarization 9
Corneal refractive surgery 342
Correlation spectrometer 148
Cross-over resonances 290
Crystal
– field 203
–, organic 72
– spectrometer 69
Crystallization 299
Curve of growth 152
Czerny-Turner spectrometer 105

Dark current 115
Deflection
–, laser-beam 310
–, magnetic 161, 283
–, recoil 282
Degenerate four-wave mixing 230, 315
Delayed-coincidence technique 263, 271
Delves cup 138
Depolarization 256
Detectors 115

Deuterium lamp 94, 141
Diagonalization 27, 166
DIAL 325
Diamagnetic atoms 168
Dielectric layer 119
Difference frequency generation 227
Differential absorption lidar 325
Diffraction
– limited divergence 200
– order 104
Diffusive pulling 300
Diode
– array 78, 118, 241, 307, 314
– -pumped lasers 207
Discharge
–, gas 88
–, hollow cathode 88, 135, 249
–, oven 168
–, radio-frequency 89
–, spark 88
Dispersion 101
– curve 177, 290
Dissociation energy 35, 339
Doas 147
Dobson instrument 129
Doppler
– broadening 87, 154, 198, 279
– burst 315
– -free laser spectroscopy 286
– -limited techniques 241
– shift 86, 155, 193, 269, 316
– width 87
Doubly excited states 266
Doubly refractive materials 125, 227
Dry ice 116
Dye 212, 218, 229
Dynodes 115

Ebert spectrometer 105
Edlén formula 128
Effective charge 81
Einstein coefficients 42
Elastic scattering 56, 60
Electric dipole
– moment 20
– operator 41
Electric fields 20

Electric quadrupole moment 23
Electron
– beam excitation 270
–, equivalent 13, 16
–, non-equivalent 13
– spectrometer 76
– storage ring 96
– synchrotron 95
Electrophoresis 332
Elementary reactions 303
Emission
–, spontaneous 42
–, stimulated 40, 195
Emissivity 94
ENDOR 187
Energy-dispersive spectrometer 70
Enhancement cavity 227
Environmental applications 134, 141, 145, 188, 248, 251, 318, 330
Equivalent width 152
ESCA 75, 99
ESR 186
Etalon 202, 212, 219
EXAFS 74, 99
Excitation methods 237
Exponential decay 258
External fields 16, 25

Fabry-Pérot interferometer 107, 201
Fano resonance 248
FEL 98
Felgett advantage 114
Fermat's principle 107
Fermi contact interaction 23
Fermi Golden Rule 41
Fermi level 74
FID 183
Field ionization 21, 239
Filter 119
–, coloured glass 122
–, interference 119
–, liquid 123
–, neutral density 124
Fine-structure 8, 18, 256, 281
Finesse 110
Flame 137, 303, 305
– absorption spectroscopy 134

– emission spectroscopy 134
– opto-galvanic spectroscopy 249
– temperature 137
Flash-lamp 204, 258
Flop-in,out 164, 283
Fluorescence 47, 238, 280, 292, 329, 343
Forbidden transitions 48, 154, 299
Fortrat parabola 53
Four-wave mixing 230, 232
Fourier analysis 45, 198, 274
Fourier broadening 295, 296
Fourier transform 45, 114, 260, 276
– NMR 183
– spectrometer 112
Franck-Condon principle 54
Free-electron laser 98
Free induction decay 183
Free spectral range 110, 112, 201
Frequency
– doubling 226, 334
– mixing 227
– modulation spectroscopy 293
– stabilization 286
Fresnel rhomb 126
Fresnel's formulae 125
Fusion 224, 268

g factor 25
Gain profile 201
Gas correlation spectrometer 149
Gaussian curve 87
Geiger counter 69, 167
Ghost lines 105
Giant pulse generation 205
Glan-Taylor polarizer 125
Glan-Thompson polarizer 125, 289
Glass
– absorption 127
– dispersion 102
Graphite oven 138
Grating 104
–, echelle 105
– equation 104
–, holographic 105
–, replica 105
Gray wedge 149
Grazing incidence 106

Green flash 64
Greenhouse effect 318
Ground configuration 10, 12

Haloes 64
Hamiltonian 4
Hanle effect 177, 257, 269
Harmonic oscillator 35, 45, 52
Heat-pipe oven 243, 247
Heisenberg uncertainty principle 44, 77, 164, 170, 252, 260
Hematoporphyrin derivative 343
Heterodyne detection 117, 319
HgCdTe detector 116
High-contrast transmission spectroscopy 288
High-frequency deflection method 270
Highly ionized atoms 90, 92
Hole-burning 285
Hollow cathode 88, 135, 249
Hologram 114
Hook method 272
Hot bands 58
HPD 343
Hubble space telescope 154
Hund coupling cases 34
Hydrogen maser 171
Hydrospheric remote sensing 189, 330
Hyperfine anomaly 167
Hyperfine structure 21, 85, 166, 173, 180, 277, 278, 281, 288
–, magnetic 21
–, electric 23

ICP 139
IFEL 98
Image intensifier 118
Imaging fluorescence spectroscopy 119, 307, 317, 348
Impactor 71
InAs detector 116
Incandescence 305
Incoherent scattering 175
Independent particle model 12
Index of refraction 60, 102, 128, 227
Induced transmission 119

Inductively coupled plasma 139
Inelastic scattering 57
Infrared detectors 116
Inner electrons 66
InSb detector 116
Intercombination lines 48
Interference filter 119
Interferogram 112
Interferometer
–, confocal 111
–, Fabry-Pérot 107
–, Fourier 112
–, Mach-Zehnder 272
–, Michelson 112, 239
–, multi-pass 112
–, radio-astronomic 192
Intermediate coupling 15, 20
Intermediate field 18, 26
Intermediate frequency 319
Intermodulated experiments 292
Intra-cavity absorption 243, 285
Invar 110
Inverse free-electron laser 98
Inverse hook method 273
Inversion layer 147
Inverted Lamb dip 286
Inverted population 195
Inverted structure 9, 23, 281
Ion
– accelerator 90
– trap 298
Isotope separation 283, 337
Isotopic shift 27, 337

Jaquinot advantage 114
jj coupling 14

Kayser 2
KDP 227
Kramer-Kronig relations 272

Lamb dip 286
Lamb shift 46, 290, 295
Lambda doubling 34, 193
Land remote sensing 150
Landé
– interval rule 13, 22
– factor 25, 174
– formula 17

LANDSAT satellite 151
Langmuir-Taylor detector 167
Laplace operator 4
Larmor frequency 178, 183, 275
Laser
– amplifier 196, 206, 216, 295
– communication 226, 331
– Doppler velocimetry 315
– -driven fusion 90, 206, 224
– -enhanced ionization 248
– -induced breakdown spectroscopy 323
– -induced chemistry 335
– -induced fluorescence 305, 315, 329, 343
– isotope separation 337
– medicine 341
– modes 200
– remote sensing 318, 330
– safety 343
– spectroscopy 235
Lasers 195
–, alexandrite 221
–, argon-ion 211
–, chemical 224
–, CO 224
–, CO_2 209, 222
–, colour-center 221
–, $Co:MgF_2$ 221
–, copper 209
–, diode 207, 224
–, dye 212
–, emerald 221
–, excimer 208
–, fixed frequency 203
–, four-level 205
–, gold 209
–, He-Cd 210
–, He-Ne 209
–, HF 224
–, HgBr 331
–, ion 210
–, krypton-ion 212
–, Nd:YAG 206
–, nitrogen 208
–, ring 220, 262
–, ruby 203, 221
–, semiconductor 224
–, spin-flip Raman 212
–, TEA 223

399

–, three-level 203
–, tunable 212
–, waveguide 223
–, X-ray 197
LCAO method 30
LDV 315
LEI 248
Level-crossing spectroscopy 174, 252, 268
Level labelling 245
Lidar 322
LIF 305, 315, 329, 343
Lifetime
–, mean 44
– measurements 267
Light-induced drift 300
Light shifts 170
Light sources 85
–, continuum 94
–, laser 195
–, line 86, 235
–, natural 99
Limb absorption 147, 188
Line width
–, collisional 88
–, Doppler 87
– measurements 268
–, natural 44, 86, 172
Liquid filter 122
Liquid nitrogen 116
Lithium niobate 229
Littrow mount 102, 105, 214, 223
Local oscillator 319
Lock-in amplifier 252, 254, 287
Long-path absorption 320
Lorentz broadening 87
Lorentz condition 39
Lorentzian curve 46, 88, 177, 179, 181, 276, 290
LS coupling 12
Lyot filter 217, 260

Mach-Zehnder interferometer 272
Magic angle 184, 276
Magnetic moment 7, 16
–, effective 164, 283
Magnetic resonance 159
–, atomic beam 160
–, electron-nuclear 187

–, electron spin 186
– imaging 184
–, nuclear 182
–, optically detected 188
Magnetometer 171, 186
Many-electron systems 9
Maser
–, ammonia 196
–, cosmic 192
–, hydrogen 171
–, rutile 191
Maxwell distribution 86
Medical diagnostics 66, 71, 132, 184, 341
Mercury lamp 95
Metastable state 85
Meteorological applications 189, 318, 327
Metre definition 286
Michelson interferometer 112, 239
Microchannel plate 78, 118
Microdensitometer 88, 115, 245
Microwave
– radiometry 188
– remote sensing 188
– spectroscopy 188
Mie scattering 60, 323
Mirages 64
Mirrors 121
Mode
– hopping 202
– locking 259
– separation 201
Molecular bands 55
Molecular structure 29
Moment of inertia 32, 52
Moon spectrum 99
Morse potential 34
Moseley's law 68
Multi-channel analyser 70, 72, 78, 118, 254, 264
Multi-channel quantum defect theory 15
Multi-element analysis 71, 139
Multi-pass cell 142
Multi-photon absorption 237
Multi-photon dissociation 339
Multiple scattering 71, 253, 272
Multiplicity 12, 31
Multipole field 43
Multi-spectral scanning 150

Multi-step excitation 237
Muonic atoms 28

Natural radiation sources 99
Natural radiation width 44, 72, 86, 172, 268
Nephelometry 61
Nernst glower 94, 141
Neutral density filter 124
NEXAFS 75
NMR 182
Nuclear magneton 22
Nuclear reactor 337
Nuclear spin 21, 164, 166
Non-crossing rule 19, 177
Nonlinear laser spectroscopy 286
Nonlinear optical phenomena 226
Nonlinear third-order processes 230, 312

O branch 59
"Old" atom detection 276
Oil
– film measurements 333
– spill detection 189, 330
Optical components 119
Optical density 124, 133
Optical detectors 115
Optical double resonance 172, 252, 268
Optical heterodyne detection 319
Optical klystron 98
Optical materials 127
Optical multi-channel analyser 118
Optical parametric oscillator 230
Optical piston 300
Optical pumping 168
Optical remote sensing 145, 312
Optical spectrometers 101
Optical spectroscopy 85
Optical transient measurements 262
Opto-acoustic spectroscopy 251, 292, 309
Opto-galvanic spectroscopy 248, 292, 309
Oscillator strength 268
Overlapping orders 105, 111
Overtone 52
Ozone 129, 320

π radiation 50, 169, 172
P branch 52, 223
Paramagnetic substances 187, 299
Parametric amplifier 230
Parametric oscillator 212, 230
Parity 47
– violation 290
Particle
– extinction 322
– optics 61, 323
– pollution 71, 303
Paschen-Back effect 18, 26, 172
Passive mode-locking 261
Passive remote sensing 145, 188
Pauli principle 12, 30
PbS detector 116
PDT 346
Periodic table 132
Perturbation theory 8, 19, 37
Phase
– conjugation 230
– -locked loop 241
– matching 227, 233
– -shift method 271
Phosphorescence 47
Photocathode 115
Photoconductivity 116
Photodynamic therapy 346
Photoelectric effect 68, 115
Photoelectron spectroscopy 75
Photographic plate 115
Photoionization 210, 239, 282, 338
Photomultiplier tube 69, 115
Picosecond laser spectroscopy 266, 307, 349
Piezo-electric detection 252
Pile-up 264
PIXE 71
Planck radiation law 42, 94
Planetary atmospheres 156
Plasma 268
–, laser-produced 90
Plasmon 84
PLL 241
Pockels effect 205, 258
Point group 36
Polarizability constant 20, 56, 226, 282
Polarization
–, emitted light 49, 169

– spectroscopy 289
Polarizers 125
Polarizing beamsplitter 126
POLINEX 292
Polyatomic molecules 35
Polyvinyl film polarizers 127
Population inversion 195
Powder spectroscopy 142, 252
ppb 71, 132
ppm 71, 132
ppt 132
Precession 7, 17, 159, 178, 275, 276
Pressure
– broadening 88, 188
– scanning 110, 217
Prism
–, Amici 103
–, corner-cube 104
–, dove 104
–, Glan-Taylor 126
–, Glan-Thompson 126, 289
–, Pellin-Broca 104
–, penta 104
–, right angle 103
–, spectrometer 101
–, straight view 104
Pulsar 194
Pump-probe technique 266
Purification 336
Push-broom sensor 151
Pyroelectricity 116
Pyrolysis 336

Q branch 56
Q switching 205
QED 5, 42, 46, 295
Quadrupole moment 23
Quantum-beat spectroscopy 93, 274
Quantum defect 6
Quantum efficiency 115, 117
Quantum electrodynamics 5, 42, 46, 295
Quantum jumps 300
Quantum number
–, azimuthal 4
–, effective 6
–, magnetic 17
–, principal 4

–, rotational 33
–, vibrational 35
Quartz
– absorption 127
– dispersion 102
Quasar 155
Quenching 64, 186, 305, 310

R branch 52, 223
Radar 189
Radiation processes 37
Radiationless transitions 213, 251
Radical
–, atmospheric 147
–, flame 305
–, interstellar 193
Radio astronomy 190
Radioactive isotopes 85, 167, 170, 283, 285, 341
Radio-frequency
– lamp 89, 253
– spectroscopy 159
Rainbow 62
– angle 63
Raman scattering 56
Raman spectra 58
Raman spectroscopy 142, 310, 334
Raman, stimulated scattering 231, 340
Raman, surface-enhanced spectroscopy 334
Ramsey fringe technique
– optical domain 296, 298
– radio-frequency domain 165, 168
Rayleigh scattering 56
Recoil 282
Reduced mass 27, 33
Refocusing 90
Refractory elements 168, 268
Relaxation
–, collisional 170
–, longitudinal 182
–, transversal 182
Remote sensing
–, active 145
–, atmospheric 146, 318
–, hydrospheric 150, 330
–, land 150

–, microwave 188
–, optical 145, 318
–, passive 145
REMPI 249
Resolving power 101, 104, 110, 114
Resonance 40
 – absorption 47
 – ionization mass spectroscopy 251
 – ionization spectroscopy 249
 – methods 159
 – multi-photon ionization 249
 – Raman scattering 144
 – scattering 37
Resonator 199
 – configurations 200
 – modes 199
Retardation 108
Retarder plate 126
Retroreflector 104, 239, 320
RIMS 251
Ring laser 220, 262
RIS 249
Rotating frame 160
Rotating polarizer 254
Rotational energy 32
Rotational Raman spectra 59
Rotational transitions 51
Rotator
–, elastic 33, 51
–, rigid 33, 51
Rowland circle 72, 77, 107
Rydberg atoms 20, 238, 282, 294
Rydberg constant 4, 290

σ radiation 50, 169, 172
S branch 59
SAR 190
Satellite 145, 151, 154, 190
Saturable absorber 261
Saturation 236
 – interference spectroscopy 290
 – spectroscopy 285
Schawlow-Townes limit 203
Schrödinger equation 4, 37
Scintillation counter 69
Sea water
 – absorption 131
 – elements 72
 – fluorescence 330

Second order Doppler effect 278, 298
Selection rules 43
–, atoms 48, 166, 294
–, molecules 52
Self-consistent field 15, 81
Self-phase modulation 262
Sensor 145, 150
Sextupole magnet 168, 283
Seya-Namioka spectrometer 106
Shaded to the red, violet 53
Shake-up lines 78
Shelving 299
Short optical pulses 258
Side-bands 293
Single-atom detection 251, 299
Single-mode operation 202, 220
SLAR 189, 330
Solvation energy 83
Soot formation 303, 310
Spark 88
Spectral resolution instruments 101
–, Fabry-Pérot interferometer 107
–, Fourier transform spectrometer 112
–, grating spectrometer 104
–, prism spectrometer 101
Spectrophone 251
Spectrophotometry 141
Spin echo 184
Spin-orbit interaction 8, 12
SPOT satellite 151
Spot size 201
Sputtering 88
Squeezed states 46
Standard addition method 136
Standing wave 201
Star
 – classification 152
 – spectra 153
Stark effect 20, 27, 282
Stern-Gerlach's experiment 161
Stimulated emission 40
Stimulated Raman scattering 231
Straggling 92
Stratosphere 318
Stray light 105, 107, 252, 272
Streak camera 266
Stokes beam 232
Stokes Raman effect 57, 232
Stokes shifting 55

Stretching mode 36
Submarine communication 131, 331
Sum-frequency generation 227
Sun spectrum 99, 129
Supernova 155
Super-radiant laser 208
Surface spectroscopy 83, 84, 331, 334
Susceptibility 226, 312
Swan bands 305
Symmetry properties 35, 46
Synchronous pumping 260
Synchrotron radiation 74, 95, 258
Synthetic aperture radar 190

TAC 263
Tagging 316
TEA 223
TEM 200
Temperature measurements 58, 151, 189, 310, 312, 315, 334
Term 12
Thermal population 169, 182
Thermal velocity 161, 279
Thermionic detection 239, 294
Thermistor 116
Thermocouple 116
Thermodynamic equilibrium 42
Theta pinch 90
Time-resolved spectroscopy 258
Time standard 168, 171
Time-to-amplitude converter 263
Tissue absorption 342
Tokamak 90
Tomography 184, 310
–, computer-aided 186
–, NMR 184
Transient digitizer 262, 323
Transit time broadening 164, 278, 298
Transition probability 36, 267
– determination 88, 272
Trapped ions and atoms 298
Tumour
– detection 343
– treatment 345
Tungsten lamp 94, 141
Turbidimetry 61
Two-line fluorescence 315

Two-photon absorption 237, 246
Two-photon spectroscopy 246, 293, 309
Two-photon transitions 39, 237

Ultra-sensitive laser spectroscopy 251, 293
Undulator 98

Vacuum
– spectroscopy 128, 230
– state 46
Van der Waals molecules 29
Vapour pressure 89, 162
Variance 44
Vector model 16, 25
Velocity
– -changing collisions 292
– of light 286
– measurements 315
–, thermal 279
Vibrational energy 34
Vibrational-rotational Raman spectra 59
Vibrational-rotational spectra 52
Vibrational transitions 51
Vidicon 78, 119, 309
Virtual levels 57, 237, 310
Visibility 61, 131
VLBI 192
VUV 128, 230

Water
– absorption 131
– remote sensing 150
Waveguide laser 223
Wavelength
– meter 239
– setting 239
White cell 142, 242
White excitation 253
Wiggler 98
Wolf-Rayet star 154

XANES 75
Xenon lamp 95

XPS 75
X-ray
- absorption spectroscopy 73
-, characteristic 66
- continuum 66, 90
- emission spectroscopy 68
- fluorescence 68
- laser 197
- lithography 90
- microscopy 90
- monochromatization 77
- photoionization 210
- spectrometers 69
- spectroscopy 66
- transitions 49

Young double-slit experiment 63, 175, 198, 275
Yttrium-aluminum garnet 206

Zeeman
- effect 17, 25, 155, 256
- quantum beats 275
Zeldovich mechanism 303
Zodiacal light 99

Laser Spectroscopy

W. Persson, S. Svanberg (Eds.)
Laser Spectroscopy VIII
Eighth International Conference,
Åre, Sweden, 1987

1987. XX, 474 pp. 315 figs. (Springer Series in Optical Sciences, Vol. 55) Hardcover DM 99,– ISBN 3-540-18437-6

T.W. Hänsch, Y.R. Shen (Eds.)
Laser Spectroscopy VII
Seventh International Conference,
Hawaii, 1985

1985. XV, 419 pp. 298 figs. (Springer Series in Optical Sciences, Vol. 49) Hardcover DM 96,– ISBN 3-540-15894-4

H.P. Weber, W. Lüthy (Eds.)
Laser Spectroscopy VI
Sixth International Conference,
Interlaken, Switzerland, 1983

1983. XVII, 442 pp. 258 figs. (Springer Series in Optical Sciences, Vol. 40) Hardcover DM 106,–
ISBN 3-540-12957-X

A.R.W. McKellar, T. Oka, B.P. Stoicheff (Eds.)
Laser Spectroscopy V
Fifth International Conference,
Jasper Park Lodge, Alberta,
Canada, 1981

1981. XI, 495 pp. 319 figs. (Springer Series in Optical Sciences, Vol. 30) Hardcover DM 77,– ISBN 3-540-10914-5

H. Walther, K.W. Rothe (Eds.)
Laser Spectroscopy IV
Fourth International Conference,
Rottach-Egern, FRG, 1979

1979. XIII, 652 pp. 411 figs. 19 tabs. (Springer Series in Optical Sciences, Vol. 21) Hardcover DM 86,–
ISBN 3-540-09766-X

J.L. Hall, J.L. Carlsten (Eds.)
Laser Spectroscopy III
Third International Conference,
Jackson Lake Lodge, Wyoming,
1977

1977. XI, 468 pp. 296 figs. (Springer Series in Optical Sciences, Vol. 7) Hardcover DM 69,– ISBN 3-540-08543-2

Ultrafast Phenomena

E. Klose, B. Wilhelmi, (Eds.)

Ultrafast Phenomena in Spectroscopy
Proceedings of the Sixth International Symposium, Neubrandenburg, German Democratic Republic, August 23-27, 1989

1990. XII, 326 pp. 245 figs. 10 tabs. (Springer Proceedings in Physics, Vol. 49)
Hardcover DM 118,– ISBN 3-540-52781-8

C.B. Harris, E.P. Ippen, G.A. Mourou, A.H. Zeweil (Eds.)

Ultrafast Phenomena VII
Proceedings of the 7th International Conference, Monterey, CA, May 14-17, 1990

1990. XXII, 554 pp. 435 figs. 14 tabs. (Springer Series in Chemical Physics, Vol. 53)
Hardcover DM 119,– ISBN 3-540-53049-5

T. Yajima, K. Yoshihara, C.B. Harris, S. Shionoya (Eds.)

Ultrafast Phenomena VI
Proceedings of the 6th International Conference, Kyoto, Japan, July 12-15, 1988

1988. XXII, 616 pp. 487 figs. (Springer Series in Chemical Physics, Volume 48)
Hardcover DM 128,– ISBN 3-540-50469-9

G.R. Fleming, A.E. Siegman (Eds.)

Ultrafast Phenomena V
Proceedings of the Fifth OSA Topical Meeting, Snowmass, Colorado, June 16-19, 1986

1986. XIX, 551 pp. 427 figs. (Springer Series in Chemical Physics, Volume 46)
Hardcover DM 99,– ISBN 3-540-17077-4

D.H. Auston, K.B. Eisenthal (Eds.)

Ultrafast Phenomena IV
Proceedings of the Fourth International Conference, Monterey, California, June 11-15, 1984

1984. XVI, 509 pp. 370 figs. (Springer Series in Chemical Physics, Vol. 38)
Hardcover DM 106,– ISBN 3-540-13834-X

K.B. Eisenthal, R.M. Hochstrasser, W. Kaiser, A. Laubereau (Eds.)

Picosecond Phenomena III
Proceedings of the Third International Conference on Picosecond Phenomena, Garmisch-Partenkirchen, Federal Republic of Germany, June 16-18, 1982

1982. XIII, 401 pp. 288 figs. (Springer Series in Chemical Physics, Vol. 23)
Hardcover DM 72,– ISBN 3-540-11912-4

R. Hochstrasser, W. Kaiser, C.V. Shank (Eds.)

Picosecond Phenomena II
Proceedings of the Second International Conference on Picosecond Phenomena, Cape Cod, Massachusetts, USA, June 18 – 20, 1980

1980. XII, 382 pp. 252 figs. 17 tabs. (Springer Series in Chemical Physics, Vol. 14)
Hardcover DM 81,– ISBN 3-540-10403-8